Common Sense

Common Sense

Intelligence as Presented on Popular Television

EDITED BY LISA HOLDERMAN

LEXINGTON BOOKS

A division of
ROWMAN & LITTLEFIELD PUBLISHERS, INC.
Lanham • Boulder • New York • Toronto • Plymouth, UK

LEXINGTON BOOKS

A division of Rowman & Littlefield Publishers, Inc.
A wholly owned subsidiary of The Rowman & Littlefield Publishing Group, Inc.
4501 Forbes Boulevard, Suite 200
Lanham, MD 20706

Estover Road
Plymouth PL6 7PY
United Kingdom

British Library Cataloguing in Publication Information Available

Library of Congress Cataloging-in-Publication Data

Holderman, Lisa, 1967–
Common sense : intelligence as presented on popular television / edited by Lisa Holderman.
p. cm.
Includes bibliographical references and index.
ISBN-13: 978-0-7391-1521-3 (cloth : alk. paper)
ISBN-10: 0-7391-1521-9 (cloth : alk. paper)
ISBN-13: 978-0-7391-1522-0 (pbk. : alk. paper)
ISBN-10: 0-7391-1522-7 (pbk. : alk. paper)
1. Intellect on television. I. Title.
PN1992.8.I66H65 2008
791.45'653—dc22 2007045498

Printed in the United States of America

∞™ The paper used in this publication meets the minimum requirements of American National Standard for Information Sciences—Permanence of Paper for Printed Library Materials, ANSI/NISO Z39.48–1992.

For John, Natalie, and Sean

Contents

Introduction

Intelligence on Television

Lisa Holderman

Given prevalent popular opinion of most television programming, one might raise an eyebrow at a book title that connects the terms "intelligence" and "television." Popular TV monikers such as the "boob tube" and "vast wasteland" offer up, among others, images of dim-witted personalities existing within mind-numbing programming. However, representations of intelligence in many forms abound on popular television programming. This volume explore just this—by examining different TV genres and in using various methods and theoretical perspectives, this book offers insight into important cultural beliefs regarding intelligence (the power of which may be farther reaching than one might imagine) and examines the complexity of popular television images as they both verify and, more often, vilify intelligence. The proliferation of certain images and the relative absence of others in fictional, reality, and fact-based television teach mass audiences, among other things, what and whom they should value and what they should expect from their lives. Given this, how intelligence is demonstrated, portrayed, and evaluated in the public sphere is especially significant.

Representations of intelligence and influences of such representations are complex—they cannot easily be distilled into dichotomous categories of good/bad, influential/meaningless, positive/negative, powerful/powerless, etc. In the first place, "intelligence" on television programs exists in many forms. While well-known television "nerds" and "geeks" may immediately come to mind, intelligence is also situated in a host of characters, including teachers, experts, professors, scientists, students, and other characters and personalities who think critically and/or possess knowledge in its many forms. This spectrum of representation alone indicates that images of smartness are multifaceted and are not easily reduced to one particular stereotype. In the second place, constructions of intelligence intersect with constructions of other social statuses, such as social class, race, gender, age, and sexual orientation, among others, so that mediated stories about intelligence and the characters within these stories are not simply one-dimensional. The intersection of intelligence with other social statuses creates differing issues of power in the constructions created and the impact of those constructions. Finally, the impact or function of stories about, and images of, intelligence in popular media is complicated, as the various chapters in this book reveal.

The first section lays the foundation for the anthology and provides both theory and data to give the reader a sense of the issues at hand, the existing scholarship in this area, and the significance of this line of scholarship. In Chapter One, Sari Thomas and I offer an essential introduction to this issue and provide the necessary grounding upon which each of the subsequent chapters, despite theoretical and methodological differences, build. We contend that cultural representations of and stories about intelligence (myths, really) are *central* to our understanding of *all* cultural storytelling and mythology and, in using a functional-constructivist perspective, explains the function of this intelligence mythology within contemporary culture in mediated *and* non-mediated contexts.

In Chapter Two, Susan Kahlenberg adeptly differentiates between different types of intelligence shown on popular television programming: academic intelligence ("book smarts"), practical intelligence ("street smarts"), and technical intelligence ("techno smarts"). Kahlenberg constructs a profile of the media-constructed intellectual through a systematic analysis of the portrayals of these types of intelligence on prime-time, fictional television programming, which adds an important piece of data-based scholarship to this area of study.

The second section of the book focuses on the ways in which representations of intelligence intersect with representations of social class, gender, and youth culture. In Chapter Three, Mary T. Conway explores the intersection of social class and intellectuality in her close reading of the complex and often seemingly contradictory narratives in *The Simpsons*. Specifically, Conway skillfully demonstrates how the philosophy of Pragmatism is manifested in *The Simpsons* and the ways this philosophy allows the program to ridicule both intellectuality and stupidity, to defend practical knowledge, to condemn capitalists, while promoting myths of meritocracy and the American Dream.

Kylo-Patrick R. Hart continues this section of the anthology in Chapter Four by focusing on the complex merging of gender and intelligence. His careful analysis of the character Brenda Chenowith on HBO's highly-praised series *Six Feet Under* explores the ways in which a powerful and intellectually brilliant female character eventually becomes re-constructed as non-threatening and is removed as any serious danger to the existing patriarchal social order.

In Chapter Five, Holly Randell-Moon examines popular beliefs about intelligence as they are represented in the highly-successful series *Buffy the Vampire Slayer*. Randell-Moon argues that there exists an important relationship between subjectivity and knowledge and that these concepts play a critical role in making representations of intelligence culturally relevant. Her brilliant examination of the "nerd" character, Willow, reveals the ways the series represents intellectuality, particularly in connection to gender and youth, and the ways that intelligence is, ultimately, related to power.

Following up on issues of youth culture, Amy Richards Franzini's analysis in Chapter Six explores the connection of academics, intelligence, and interpersonal success within the representations of teen characters on three television dramas. Franzini poses and answers the question, "Is School Cool?" as it is constructed on television (specifically on *Beverly Hills, 90210*, *Dawson's Creek*, and *The O.C.)* and skillfully analyzes her overall findings that academics are mostly ignored on teen television programming and that most teen characters are portrayed as exceedingly average in terms of intelligence.

The third section of the book concentrates on intelligence within science fiction, representations of scientists in popular programming, and the impact of science-based fictional programming on audiences' beliefs about science. In Chapter Seven, Cynthia Walker and Amy Sturgis explore the "cerebral hero" in (mostly science fiction) television drama. These authors demonstrate the complexity of studying intelligent characters by exploring the ways specific smart characters are constructed and/or received by audiences as sexually attractive. Their analysis of popular television characters such as Illya Kuryakin of *The Man From U.N.C.L.E.*, Mr. Spock of *Star Trek,* and Fox Mulder of the *X-Files,* among many others, illustrates the ways in which smart characters are constructed/received as "sexy," and explores the shortcomings associated with these smart characters as well.

In Chapter Eight, Christine Mains explores the military/science debate within popular science fiction programming, with a specific focus on the series *Stargate SG-1*. Using rhetorical narratology, Mains critically analyzes the ways in which television science fiction generates narrative tension by portraying scientific knowledge as secondary to military needs and examines the ways that scientist characters struggle with their scientific pursuits under the pressures of their military positions.

In Chapter Nine, Gary Pettey and Cheryl Campanella Bracken extend the section on representations of scientists by using experimental method to assess the priming effects of science-focused entertainment programming. Specifically, these authors explore the notion of the "*CSI* Effect," which has garnered much media attention as it suggests a profound impact of the television program *CSI: Crime Scene Investigation* on audience beliefs about and interest in forensic science. Pettey and Bracken discuss important theoretical implications of their overall findings that exposure to an episode of *CSI* led to more satisfaction with science and with government expenditures on scientific pursuits.

The final section of this books focuses on issues and portrayals of intelligence on "reality-based" television. In Chapter Ten, I examine the popular portrayal of intellectual expertise on top-rated popular U.S. television talk shows and interpret data that reveal "intellectual experts" are treated more negatively than "non-intellectual" experts and that experts on these shows essentially are treated no differently than non-expert guests. I suggest the existence of a "leveling phenomenon" on popular television, which contributes to social-order maintenance by weakening the status of intelligence.

In Chapter Eleven, Marilyn Ellzey and Alison Miller explore portrayals and storylines on reality television programs to reveal the ways they reflect the ideology of intelligence within American culture. Through a content analysis of three reality programs, *The Apprentice*, *Beauty and the Geek*, and *The Scholar*, these authors provide excellent data on, among other important variables, the physical attributes and personality traits of intelligent (and non-intelligent) people on these shows and explore the implications of these findings.

Finally, in Chapter Twelve, Joan Connors completes this section nicely with her analysis of representations of intelligence on The Learning Channel's transformation reality program, *Faking It*. Through a content analysis of 18 episodes, Connors evaluates the notion of intelligence as it is shown through the participants' preparation for their metamorphoses and considers how the program differentiates between (and disproportionately values) "social" and "intellectual" competences.

Altogether the chapters of this book piece together a clearer picture of the ideology of intelligence as it is represented on popular television. Given that thus far only a small number of scholars have analyzed the portrayal of intelligence in popular media, this anthology helps significantly to develop this important line of scholarship. Taking into consideration that intelligence and knowledge are core ingredients in issues of power, it is crucial to understand how television, arguably the most potent and popular storyteller in history, constructs these concepts for mass audiences.

Part I

Surveying Smarts on the Television Landscape

Chapter One

The Social Construction of Modern Intelligence

Sari Thomas
and
Lisa Holderman

This chapter provides an overview of our culture's modern and popular creation of "intelligence." We begin with an introduction to the chapter's philosophical underpinnings and then review processes and mechanisms by which intelligence is *constructed* as well as the *function* of this social construction.

Preliminary Clarifications and Caveats

No thesis should be written *as if* it were the only possible interpretation of the subject. This first introduction addresses methodological and theoretical matters *without* specific regard to "intelligence." The preface covers much ground, but stops at each subject very briefly—only to lessen the need to litter subsequent analyses with piecemeal definitions and qualifications that, otherwise, would be needed by anyone unfamiliar with the functional-constructivist perspective employed.

Television and Other Life

This volume is generally dedicated to *television's* representation of intelligence, but this overview chapter considers all means through which the characterization of intelligence is formed. No undue regard is given to system, medium, genre or format, and attention is also directed to *non*-media contexts. This breadth doesn't reflect arbitrary eclecticism, but rather, an attempt at methodological rigor. If research is comparative, one *must* distinguish among information sources. However, concern for the "social construction" of any phenomenon cannot logically be (pre-)confined to a single source—particularly if one's fundamental philosophy postulates that social structure unites and integrates all systems and institutions. Indeed, the discovery of coherence and significant pattern among and across various contexts is one way of affirming the rectitude of one's premise.

Although this chapter is introductory, theoretical and discursive, its arguments were stimulated by canonical theory and research, and sustained by numerous data sets and resources collected over the past three decades, during which time there has, unfortunately, remained a relative dearth of comparable analyses.[1] Despite the aforementioned "sampling" breadth, over half of the research informing this text addresses media representation—and the lion's share of that work concentrated on television stories. Given this, and the general orientation of this volume, it remains that TV is unquestioningly the preeminent single (or "case") resource from which to derive *general* cultural themes. As a dissemination system, television still reaches the largest, most widespread population, and becomes something of an ideological common denominator. As discussed in some detail later, people adamantly disavow both their intimacy with TV as well as its tacit "instructional" function, but such denials don't prevent it from being Americans' most visited, waking-life destination, or a cultural hub resulting precisely from its tacit ubiquity and the presumed ideological "meaninglessness."

Philosophical Issues

While this chapter analyzes a popular *concept*, this doesn't mean that our analytic perspective is equally familiar or popular. Indeed, to the extent that it is uncommon, it demands preliminary explication. To this end, we supply this introduction to clarify terms and concept used.

Ontology and Social Construction

Ontology, as used here, indicates analysis of how reality is imagined. "Social construction" implies that we assume, a "constructivist" perspective—but, as we shall explain, a *functionalist*-constructivism. Constructivism (which includes competing varieties) is normally not an idea to be dispatched in a brief paragraph. Fortunately, for our purposes, what's critical is that we treat all concepts and conduct as having *only* the existence derived from complex social-legitimation processes. This contrasts with a more popular belief that there is a reality "behind" existence independent of what and how people know the universe. But, rejecting this idea of some constant, universal truth, we make particular use of the idea of "reification." In everyday life, it's very common for people casually to remark: "Oh, it's only natural . . ." or "It's just human nature . . . ." People usually offer such observations when they—and everyone they know—repeatedly and exclusively respond a certain way in certain contexts. "Naturalization" is the *in*accurate assumption that something is a product of "human nature." Thus, "naturalization" means believing *acculturated* behavior is genetic, as if it comes with membership in the species. However, for anything *to be* human nature or innate, it must be unambiguously universal, i.e., it must operate in *all* cultures. Indeed, some sociobiologists have argued that universality is just a threshold criterion; it is a pre-condition of human nature, but not proof of it.

Both naturalization *and* reification indicate a *failure* to acknowledge the role of *social construction*. While naturalization, specifically, misreads *learned* conduct as "human nature," reification treats a socially-constructed phenomenon *as if* it were *not* an artifact or consequence of human formation. To "reify" an event is to assign it "real" or "essential" status independent *of* social construction. These are, of course, similar, but not identical phenomena. While constructivists might be more likely to take note of these errors, they are not terms exclusive to constructivism.

Structure, Order, and Evolution

As noted above, only *part* of this chapter's perspective is *constructivist*. Our version of structural-functionalism constitutes the other part. Some textbooks describe constructivism and functionalism as mutually exclusive, but neither constructivism nor functionalism is monoistic. The variants to which we subscribe are eminently capable of integration. On the heels of our introduction to social construction or reproduction, this elementary summary is meant mainly to take us through to the very end of the chapter where we finalize our assessments of function. To start, we address the concepts of society and structure. Although these are commonplace terms, they are used distinctively in structural theory, in general, and structural-functionalism, specifically. It might be helpful to consider—at least with respect to the concept of society—the familiar meaning derived from our culture's "default" philosophy: *rationalism*. Everyone operates on the basis of an invisibly internalized overarching philosophy, and for most people in the U.S., rationalism is that philosophy. It is "default" because it is that which people normally and automatically learn. Rationalism is so automatic that its basic premises are naturalized and, thus, rarely acknowledged, much less articulated because they seem so involuntarily routine. As we will see with respect to "society," rationalism and structural philosophies are not just different, but diametrically opposite.

The fundamental premise of rationalism is that an individual's subjective thought is the *self-generating* source of ideas (as in, "I think, therefore . . . ") Directly connected to this premise is the fact that, to a rationalist, "society" is a metaphor; it represents "everybody taken together" or "considered at once." Mathematics provides a different parallel: we could say that rationalism relies on a plane/Euclidean calculation of society—where the "whole" (society) is equal to the sum of its parts (the sum of all individual members). Although most people can't identify rationalism, they will, nevertheless, identify with this "definition" of society. If we extend this simple premise, then, any social phenomenon analyzed from a rationalist perspective is represented as the sum total of individual postures toward the analyzed phenomenon. Postmodern rationalism, for instance, explains the interpretation of a text by asserting that "there are as many texts as there are readers." Taking this one step further, then, the social *function* of any phenomenon is, from the rationalist viewpoint, identical to all the individual, subjective functions *totaled*. While this sort of "additive" or "cumulative" logic is familiar to most people, it is not employed by structural theorists. Plane/Euclidean premises are *in*compatible with structural analyses, which, instead, must rely on other geometric principles.[2] With a differential geometry, one might find that in a three-dimensional sphere, parallel lines meet, triangles have more than 180° and, most pertinently, a whole is *not* equal to the sum of its parts, but rather most vitally, to the sum of its parts *plus their interrelationships*. We know that an atom includes protons, neutrons, and orbiting electrons, etc.; yet the behavior of an atom is determined *not* by the sum of their subatomic parts, but by all the parts *as they are particularly arranged.* Is this simply a conceptual flourish with no "practical" consequence? Well, for one instance, atomic organization can critically affect an element's stability.

From a structural perspective, "society" is no mere figure of speech, but rather, that level of association among (in this case) people at which all the information/organization necessary for its survival and reproduction is self-contained. If we had to find an appropriate *structural* metaphor for society, then, it would be that of a self-regulating and self-reproducing *enclosure* encompassing all citizens along with their usual customs, roles, rules, and values. As we will explain, the regulatory influences on this enclosure are such that the customs, rules, etc., typically remain sufficiently stable over time so as to prevail *before and after* the lifespans of the citizens who live and embody them. This is one very important way in which the societal whole is greater than the sum of its individual constituents. From a structural perspective, then, people *are* frequently envisioned as transient inhabitants of reasonably-established social roles, and thus, they are seen more as products of society than vice versa. In other words, although this self-organizing and self-reproducing society, of course, *involves* individual actors, it does so at a level transcending their intentions and, often even, their cognizance.[3] In later discussion, we distinguish more precisely between gradual and stable evolution, on the one hand, and the mechanical notion of preservation, on the other. When we refer to "sufficient stability," we figuratively address how one's body typically maintains both its anatomic and metabolic "composure" in a normal tendency to survive. Yet, no matter how our bodies tend to preserve and even restore, nothing averts the gradual changes it experiences; nor do bodies uniformly escape the severe (and sometimes intentional) mayhem to which they might be exposed. Thus, references to stability and maintenance are characterizations of the normal and of the aggregate.

These issues of stability, endurance, and social evolution are frequently expressed in terms of the apparatus of "social structure." Social structure can be said *to be* those additional interrelationships that must be integrated with (the sum of the) individuals in order to constitute society; social structure can be understood as the framework organizing and interconnecting society's individual constituents. However, what is emphasized here is that those "additional relationships" or that organizing "framework" must simultaneously be interpreted as those factors—or that formation—which provides the stability and endurance. In other words, it is those things—social structure—that persists and, ultimately, transcends the lives of individuals; it is, obviously, in this sense that social structure is characterized as an apparatus of social reproduction. In textbooks, social structure might otherwise be conceived as that enduring matrix configuring interrelationships among social institutions, thereby legitimating, inculcating, organizing and governing social concepts and conduct.[4] Sometimes the term "social order" is used seemingly in place of social structure. However, for our intents and purposes, we need to distinguish the terms.

Most all extended definitions of social structure will include references to role configurations, interrelationships, complexes of norms and values, and so forth. What may further modify most of these terms—such as configuration, interrelationship, complex—is the notion of *differentiation*. It is unilaterally understood that societies, in theory and practice, distinguish among roles, strata, and statuses; indeed, this differentiation occupies a significant portion of sociological and anthropological research. How a culture organizes all types of possible divisions (based on such variables as sex, age, occupation, etc.) is certainly part of its governing matrix; however, at least here, social order is reserved for that differentiation regarded as *hierarchical* and which, more specifically, is generally counted as social-class differentiation. There exists an historically long and substantial literature offering competing analytic accounts of the precise correlates of social class, e.g., income, occupation, education, influence. While, in our analysis of intelligence we often consider social-class hierarchy, that consideration is largely confined to a primitive-but-familiar, binary differentiation, distinguishing the empowered from the (relatively) disempowered. To the extent that further elaboration is relevant, we attend to that in the immediate text. What is most salient is that the term "social order," then, is used to refer to social structure's *social-class hierarchy*.

Social order, then, may be interpreted as an ordinal adaptation or version of social structure; however, the two concepts represent more than identical phenomena viewed from different perspectives. Social theories importantly frame *both concepts simultaneously*. Figuratively, one might even suggest that all grand, social-structural theory attempts to *reconcile* social structure *with* social order. Specifically, a logical contradiction exists between the ideas of (1) an evolving, enduring, and relatively stable social structure with (2) the concentration of resources and influence in a very narrow portion of the population. While all social theory may not have historically arisen from overt analysis of class struggle, this structure/order paradox has long been a provocative and core inconsistency that has, ultimately, evoked three major, theoretical "responses." One response, common to Cartesian foundations, has been the denial of the structural premises, to wit, imagining instead a society in constant transformation wherein individuals are continually engaged in contradiction and conflict. A second response doesn't challenge cultural stability, but involves what might be called "last-resort rationalism," i.e., acknowledging an historical class struggle that, eventually accumulates and ultimately stimulates a new level of class polarization propelling major revision of the offending social order. This second perspective's orientation becomes rationalist when the only means by which the contradiction may be resolved is through autonomous individual motivation and activities. The first and second interpretations have in common both an overt sanguinity about the immediate or shorter-term delivery of egalitarianism as well as a presumption of structurally-independent, individual agency. Thus, it is in this conceptual arena dominated that structural functionalism provides a third interpretation of the structure/order paradox—an interpretation in which the approach's central, analytic framework is embedded.

One may contest the first type of response—the "structural denial"—on two interrelated grounds. The first (and lesser) argument suggests that scholarly characterizations do not necessarily mirror life as normally experienced. Correspondingly, immediate reports of normal experience are not necessarily compatible with methodical records-keeping. In either case, *to compare* (or contest) assessments of change (or stability), it is essential that "measurements" be taken of the same space with identically-calibrated instruments. As we discuss later, cultural temperocentrism often influences people to aggrandize and attribute historical significance to events in their lifetimes; yet, rationalism-driven accounts considers contemporaneous, subjective assessments as appropriately integrative with other records. Canonical, functionalist characterizations of culture change (or, certainly, those on which we base our assessments) are derived from the longer view of comparative, methods-based records. More importantly, few analytic approaches deny the persistence (from, *at least,* the Industrial Revolution) of the resilience of the basic class-differentiation/struggle on which almost all our discussions of social change are hinged. Thus, one might conclude that arguments about the (in)comparability of different change-inventorying-techniques speak to the empirical inferences drawn by structural-functionalists. However, we would contend that to argue against structural-functionalism because of a seeming paradigmatic emphasis on continuity creates a *straw* epistemological debate. Social-structural methods in general don't, fundamentally, preclude or even covertly mitigate against, the discovery of social transformation, conflict or contradiction. As we discuss again later, social-maintenance processes arose and evolved *because* social change is endemic to behavior, society, and life. It is illogical to see stability as even a sociobiological or ecological *objective* much less (and perversely) as a political goal of theorists. Rather, the "objective" of living systems is simply to live—that reasonable constancy and dependability serves survival better than flux and volatility is an (empirically-based) inference, not a motivation. However, just as living systems have a natural tendency toward self-preservation, their lives also demands evolution; the wholly homeostatic society is simply unachievable. To varying degrees and in different ways, all societies face exigencies and must regularly confront structurally-dysfunctional forces that, if left unattended, might precipitate its decline or disintegration. To reconcile these two ecologically-adaptive "drives"—toward both preservation and evolution—relative and aggregate stability is a research finding and, consequently, an assumption with respect to adaptability. Nevertheless, to study social stability is to study social change, to study social evolution is to study social maintenance, and to subject any phenomenon to functional analysis is to study its role in any or all of these processes.

Structure and Function

Of course, we need to be less sweeping in articulating the methodological level of functionalism.[5] In application, then, "function" is *always* shorthand for *social* function; the perspective is *invariably* social. This means that no (correct) answer is derivable from inventories of individual preferences or subjective reactions.[6] This also means, then, that the question of function always translates to "how does the analyzed phenomenon *serve cultural reproduction*?" "How does the given phenomenon contribute to the maintenance or erosion of the social structure?" To the extent we're interested in cultural reproduction and the social *order,* we might rephrase the question as: How does hierarchical class-division (entailing the subjugation of many to create wealth for a very few) sustain? Functionalist analysis responds to such questions by addressing how any and all social institutions and processes foster the maintenance (or decomposition) of those structure/order conditions. To the extent that, as William Graham Sumner once said, "the first task of life is to live," we typically find more continuation and reinforcement than deterioration, and that should no more be confused as a research *objective* or political endorsement than the historical observation of the failure of a *coup d'etat* should be mistaken for celebration.

The methodological orientation may be applied to any methodically-observed phenomenon so its patterns of conduct may be derived. For example, one could ask: "how does the *social construction of intelligence* contribute to the maintenance or erosion of the social order?" Employing a style not unlike Foucault's social "archaeology," it is possible to search heuristically for various social contexts or mechanisms in or by which the larger phenomenon is constructed and, then, to collect *all relevant patterns for functional analysis.* For, again, the alternate mathematics of space must be applied, and it is essential to examine not just a set of patterns, but also, how those patterns are integrated into the social-structural matrix.

Stories and Myth

Although we don't exclusively attend to specific media, genres, and so forth, *stories* conveyed across these and other categories have been examined. We use the term "story" in the generic, colloquial sense—as a narrative or account of ideas and/or events—but we do not use "*mythology*" as it is commonly invoked. Colloquially, "myth" indicates an *untrue,* but popular, belief—a widespread lie. However, here (and in most social theory), myth refers to a belief (1) that is not only widely held, but is part of a larger, coherent organization of common beliefs (mythology), and (2) the authenticity of which is unquestioningly *presumed* so that it is ontologically *beyond* truth (or falsity). We could say, then, that myths are *reified* cultural premises or knowledge. The notion of "coherence" refers here to "*mythological* uniformity"—an abiding ideological pattern within and among cultural premises. In other words, the culture's central (and, thus, important) ideas will manifest mutual ideological reinforcement, independent of form and context. Mutually reinforcing does not mean identical, though; coherency may be achieved in a number of ways, including exact repetition, physical resemblance, metaphoric similarity, compatibility, and complementariness. We will see such variations when examining how intelligence is characterized in different types of stories, but by way of introduction, we will offer more general examples.

If we consider our cultural myths about, say, wealth, there is, *among* them, disparate topics made similar with respect to downplaying the effectiveness of money, e.g., "money doesn't buy happiness," *or* "money is the root of all evil," *or* "all you need is love," or "can't buy me love," etc. And, these variations may be manifested in any context where money is characterized, e.g., music lyrics, movies, books, folklore and so on. Moreover, there is a coherence among all myths dealing with those social assets limited to a very small portion of the population such as wealth, beauty, genius, etc. All myths of this type "reveal" non-affirming, off-putting, and negative assessment of the given asset, i.e., what we have seen with respect to wealth, also occurs with respect to "skin deep" beauty and geeky genius.

Regarding mythological uniformity, popular stories and narratives are exemplary with respect to observing how patterns are structurally integrated. Indeed, we routinely use stories and narration to illustrate various issues in this introduction precisely because storytelling: (1) is a mechanism to which we turn in several contexts to analyze the reproduction of intelligence, (2) is an important mechanism by which cultural values are routinely and widely disseminated, (3) involves processes in which most people engage and about which they are, therefore, familiar, (4) has a familiar, default (rationalist) orientation with which functional-constructivism may be simply compared, (5) involves systemically coherent patterns that are relatively easy to demonstrate, and (6) as a result of this coherence, consists of compelling patterns that typically reinforce traditional ideas and conduct, and, therefore, that contribute to social-order maintenance.

As intimated above, the default, rationalist perspective on *how* people deal with narratives is probably pretty predictable. Stories—whether they're delivered via TV, movies, novels, press, etc.—are seen as events with which people intentionally choose to interact with the objective of being entertained, distracted, or informed. In fact, this same perspective provides the organizational basis for the media industries, themselves. So, part of its familiarity is that the perspective encompasses a lot of cultural activity. Still, in the face of this all-inclusive understanding, stories may be, alternatively, viewed as endlessly propagated, repeated, circulated, and publicized vehicles which invisibly diffuse cultural values and worldview.

Secondary analysis of data from a number of disciplines (e.g., anthropology, sociology, linguistics, etc.) shows that cultures normally manifest this systemic coherence in terms of mythological uniformity. Ironically, the imprint of cultural uniformity is *so* strong that even ardent rationalists (e.g., postmodernists) have, arguably, been forced to reinvent social structure through concepts like "interpretive community." Also, as noted, the ideological elements are simultaneously reflected in all forms of artifacts and conduct—narrative and *non*-narrative. By strict logic, these findings don't speak directly to the inherent-ness of storytelling, but the extension and coherence of worldview might be thought, at the very least, to work against any lingering plausibility that such uniformity results from serendipitous subjectivity. In fact, each culture's ideological coherence might be thought to recommend *enlarging* even our initial proposition: rather than merely supporting the inherent nature of storytelling, one might hypothesize that *storytelling—as an ideological mechanism—is a natural outgrowth of cultural development.*

Learning Theory and the Irony of Popular Rationalism

We have discussed how regular consumers as well as non-structural scholars minimize the social *function* of stories. They suggest, for example, that cultural values *are* consciously *inserted* by writers, producers, etc., and that, on the other end, individual consumers exercise personal control over their consumption, interpretation, and use of any information embedded in stories. People quite vehemently and resentfully reject the idea of their ideological indoctrination—especially through entertainment. What's apparently offensive is the implication of cognitive *susceptibility*—a vulnerability that most adults equate with simple-minded, childish, unsophisticated gullibility. Moreover, that susceptibility significantly encroaches on the comfort of rationalistic self-determination. Thus, it's not learning, but *unintentional* learning that so violates people's beliefs in their own autonomy. From all this comes a special definition of learning, to wit, the acquisition of "new" information—understanding something formerly unknown. Children are allowed to be active learners because they are, presumably, unfinished, and *have* so much more of the unknown *ahead* of them. Indeed, this explains why research on media effects is almost invariably formulated in terms of *children*. Also, the design of popular storytelling (apart, perhaps, from news and "educational TV") has no superficial resemblance to anything that Americans have learned to identify as instructional. Thus, there is a whole complex of reasons that makes adult-socialization-via-popular-narratives unattractive.

Of course, this popular characterization of learning is not empirically supportable and it's contradicted by *learning theory* with respect to two assumptions: (1) the assumption that learning is confined to new or unknown information, and/or (2) imagining that learning abates in adulthood. While learning *includes* apprehending novel information, this is neither the most common nor necessarily the most important part of the process. As any introduction to behavioral science will teach, reinforcement is the common and vital component of all forms of learning. Reinforcement, in its broadest sense, refers to the social and/or neurophysiological corroboration of sensory- and/or cognitively-*identified* relationships. Ironically, while the significance of reinforcement in the learning process cannot be over-emphasized, it is a virtually forgotten element in our default rationalism. While people generally accept the principle of reinforcement as an abstract truth, they don't typically see it operating in their own everyday lives. When *it is* identified, it is usually in connection with an isolated and atypical experience, e.g., realizing that one can't use her high-school learned foreign language because, in the proceeding, twenty years, there's been no practice or reinforcement. However, people rarely, if ever, identify their *routine exposure* to values and ideas as new or reinforced learning.

It is important here also to distinguish embedded values from intentional "social engineering." Again, it must be recalled that rationalist philosophy individuates *everyone*—senders and receivers; both sides of the process are seen as autonomous.[7] Functionalist analysis certainly does not *exclude* intentional ideological cultivation, but conversely, it also makes every attempt to record the less visible, systemic forces that govern both production and interpretation. The *consciousness* of cultural processes (or lack thereof) applies to owners and workers, activists and the apolitical, the creators and their audiences. All behavior is socialized and, therefore, overwhelmingly likely to reflect some aspect of the systemic relationships constituting social structure. Popular beliefs about thinking outside "boxes"—like 1960s rhetoric about being "outside the system"—must be understood as referencing a figurative, if not imaginary, location. We use such expressions merely to indicate aberrant conduct, but as the compelling research on schizophrenia has shown, all of human communication is learned (or socially laminated) inside *some* social system. Thus, when considering various *creators*—producers, writers, critics, and scholars—they, too, must be seen as *products* of the stories they are professionally charged to "create" or analyze. In this sense, functional-constructivism may regard "independent social engineers" only as a reified construct.

That it is an accepted and understood role means that people *may* self-consciously *try to be* such engineers; interestingly, though, when they make such attempts, the identification of their product must be adjusted. Specifically, we distinguish the overt, explicit and intentional moralizing that constitutes "*didacticism*"—the practice of being *morally instructive*—from the unintended and patterned reflection of worldview (or the embedding of cultural mythology). *All stories reflect worldview*; didacticism is optional. The popular TV show *Seinfeld* prided itself on being relatively "low" on didactic content; however, none of this necessarily affected the show's embedded mythology. Didacticism is hard *not* to notice. Few can watch an episode of, say, *The Brady Bunch* and, from that, *not* deduce its intended moral. Yet, although the tacit moralizing is hardly the stuff of which *TV Guide* descriptions are made, it will appear in any given episode; but of course it's unlikely that producers and writers even recognize, let alone contemplate, its construction and production. In the decade it was produced, for instance, it would have been preposterous for an episode about older brother, Greg, to feature Marcia in a non-plot-functional background scene, casually calling a boy for a date. It would be preposterous because all the "invisible" enacted decisions would, invariably, represent culturally-appropriate ideas, styles, and values. Except for scholarly analysts, few people contemplate the narrative flotsam and jetsam that, nonetheless, constitute an overwhelming proportion of disseminated information. Finally, there are several reasons that it is the didacticism we notice: we see it because that is how we expect "messages" to be delivered—that is, because it's usually in the form of an especially pronounced message. Formally, didacticism is invariably presented through a narrative device we're trained to recognize—plot or sub-plot. A story's "background" myths and values, though, needn't have much narrative connection. Also, didacticism often emerges from "topical" issues to which our attention is already heightened. We don't notice the ever-present *reinforcement* of already-learned values. Still, what is dramatic background, central casting, set design, etc. may nonetheless serve as *ideological* "ambience." In this sense, didacticism may be seen as a style of writing, whereas the values embedded in storytelling evolve quite naturally from the routine (and typically unconscious) choices involved in mundane labor. Throughout the remainder of the chapter, when we refer to ideological values, mythology, etc., we rarely address didactics (and, then, never without so specifying).

Not only is it important to distinguish between didactics and implicit ideology, we need also to integrate into that distinction our earlier arguments about rationalist conceptions of interpretation. Specifically, characterizing popular stories as "entertainment," denying adult learning, failing to recognize reinforcement, emphasizing consumers' autonomous control over material, and confining "politics" to *deliberate* opinions intentionally inserted by creators, may be said, collectively, to represent a (psychological) veneer obscuring the function of storytelling and/or keeping worldview and mythology *off* the radar of social "perception." Ironically, such obscuring and blocking, can be said *to enhance* the ideological/acculturative function of stories and myths. In other words, the influence of implicit ideological patterns probably becomes more compelling when 1) what is *always* ideological is, instead, promulgated as benign relaxation and escape and when (2) all narrative material is characterized as texts that anyone may independently manipulate to fit his or her own needs. If (conservative) resistance to social change *were* someone's political objective, this rationalist construction of adult learning as conscious, autonomous, and intentional would be that theory which would best facilitate those prescriptive ends. The belief that scholarly theory can empower people may be one of the most disempowering beliefs if it's popularized. Obscuring implicit ideology and denying its power as reinforcement obviously negates the value of its study; together, it provides a recipe for the unproblematic distribution of social-maintenance ideology.

The Social Function of Reality

There are, of course, junctures when people—most commonly scholars and/or activists—do analyze popular images in terms of the non-didactic, hidden ideological agenda. Probably the most common analytic strategy involves comparing the popular, (typically, fictional) images with "real life" and, then, concluding that popular stories involve "over-representation," "under-representation," or "misrepresentation." This strategy is certainly not obscure; most people are eminently familiar with *the idea* of this sort of comparison. And, while it's possible that the accuracy of a count might be challenged, virtually no ado is made with respect to this comparison's epistemological implications. Specifically, two critical questions don't get asked: (1) What *is* "real life" in the context of these studies?, and (2) *Why* is it appropriate to compare whatever this "real life" is with these other (typically, popular, fictional) images? When articulated in the abstract, these questions may seem profound or complex, but it is fairly basic logic that is consistently eluded. Although the logic is not profound, it is critical to the connection between social construction and function.

Obviously, "real life" must be packaged as a routine variable to serve as a comparative standard. As social constructivists, we could hardly reject that limitation since, after all, all events are ideas with assigned dimensions However, we *may,* appropriately, inquire if what is packaged as "real life" is suitable for the comparison that's implied by the research. At its simplest, when scholars compare stories to "real life," they compare two sets of data, one of which everyone seems automatically to equate with "real life." Among those trained to question such equations, this might be restated as: one set of data is legitimated as approximating "real life" more than that to which it's compared. From either statement, we may conclude that there exists for assessing recognized narrative forms, an unwritten but socially-legitimated, hierarchy of "realism"—there are presumptions about how different types of storytelling reflect "reality." We could offer a list of specific examples of narrative forms that most people could rank order according to this unwritten hierarchy. Unranked, such a list might consider: U.S. census reports, documentary films (à la Michael Moore), industrial films, CNN news, major network-news, local TV news, *The New York Times*, *New York Daily News*, *New York Post*, *The National Enquirer*, *Time*, *The*

Nation, *The Oprah Winfrey Show*, *Sixty Minutes*, and *The New England Journal of Medicine*. No matter what order is assigned, though, we'd probably agree that, at one point or another, every entry on this list has had its validity and/or reliability challenged. In all, then, we can recognize that the ordering is (1) negotiable, and (2) all sources are representations or approximations. These facts, in turn, also make it clear that, as it is discussed in this context, "real life" is constructed from authored, organized, and edited chronicles. Or, more pointedly, as used in such comparisons, "real life" represents media forums that are, *just like those media forums with which they're compared*, subject to gatekeeping and agenda-setting. To illustrate these important inferences, we could consider, for one example, how recent campus violence has revived public discussion of universities monitoring and controlling their "crime blotters." This discussion involves how some universities, to insure that the most reassuring image of campus safety is projected to concerned families, "adjust" the statistics over which their private police forces have control. Similarly, reports of declining, urban, crime, have, sometimes, been attributed to changes in how authorities account for incidents. Still, we know that security logs, police blotters, and, certainly, FBI reports are assessed as among the most realistic accounts and are regularly used, for example, in comparing content analyses of television violence.

That the "real life" discussed in so much research is, technically, a media account is something that might give us some pause. People generally don't think of real life as a mediated artifact; indeed, the two are colloquially contrasted as opposite, e.g., "I'm not talking about what happens on TV, I'm talking about *real* life." However, the as-yet un-discussed, second, epistemological question comes with an answer that is not meant to give pause, but to halt discourse. That second question, again, is: *why* do we have this invisible presumption that—whatever this "real life" is—it should serve as a standard against which other cultural accounts should be compared? There is a profoundly important and accepted presumption embedded in this comparison that needs very much to be excavated.

We should begin this excavation with what seems to be a logical inference, to wit, if we hold anything "up to" real life . . . if we measure anything's variance from or with real life . . . we must be assuming that the thing compared *is supposed* either to differ from or replicate real life. If this were not the case, there would be no reason for comparison. From here, then, we know that in the case of media comparisons, we expect real life to be *accurately reflected*. We know this from any or all of the instances when this comparative technique has been employed. For example, through the 1970s and early 1980s, feminists often spoke and wrote about the under-representation of women on television. This discourse clearly indicated that, since there were many more women in real life than as represented on TV programs, the number of women appearing on TV should be proportionately increased. While we suspect that, for many, this logic is eminently (although covertly) sensible, the next steps tend to muddy up the prevailing logic. One direction in which this analysis may proceed is crudely related to function. Specifically, when we consider such expectations of accurate media reflection, to what should we attribute the motivation? In other words, although countless studies and arguments have made similar demands after comparing one set of stories to "real life," why are they concerned about such accurate correspondence? Before rattling off a routine answer, let's consider a follow-up to the above example.

At some point in the 1980s, there was what might best be described as a glut of corporate-bitch characters on TV; both day and evening programming was flush with manipulative, money-, power-, and sex-hungry women climbing corporate ladders in miniskirts and padded shoulders. At this time, it was also, ironically, reported that the proportion of executive TV roles for female actors *exceeded* those available in the "real" corporate work force. A major feminist figure, who had much experience decrying the under-representation of women in the media, was asked (probably with tongue-in-cheek) if she would now be concerned with *reducing* the number of female-executive roles on TV so that the real-life parity for which she so vigorously campaigned could be achieved. Not known for her sense of humor, she did not take kindly to the query. Despite the tone of the question or the reaction to it, there is something embedded in this anecdote that speaks to the motivation behind comparisons with "real life." One feature of the example given, though, is clear: comparisons to real life are used to achieve something desired, *but dismissed* if the comparative data don't serve these desired ends. In other words, despite the rhetoric, verisimilitude, in and of itself, is not the quest. Thus, epistemologically, it's clear that the motivation behind real-life comparisons *is not* some (unexplained) belief in the presumed similarity of life in and out of mediation. This doesn't mean that interest groups should not be concerned about media representation, but rather, that such concern need not drag *reality* into it. Perhaps the appropriate argument in this context is that stories should only feature politically-proper representations. Such an argument, at least, logically and straightforwardly represents the objectives of scholars and activists inappropriately using real-life as a comparative standard. Of course, this position would have first to determine all facets of rectitude; however, government and industry's extant intrusion into story creation with respect to the representation of violence has, arguably, already established a model for using rectitude as the standard criterion for narrative assessment. Nonetheless, this revised model would not serve functional-constructivism insofar as neither real life nor rectitude is relevant research standards.

Indeed, in this context, functional-constructivism has no research objectives pertaining to narrative creations. The scholarly devil's advocate might urge structural scholars, nonetheless, to embrace the abuse of realism insofar as represents (perhaps unintentionally) another set of voices confirming that people *do learn* from entertainment media, i.e., why else would anyone worry about over-, under-, or mis-representations? However, such imagined "gains" are far and away offset by a far more problematic issue: even if the rhetoric of over-, under-, and mis-representation does not represent scholars' or activists' earnest concern about verisimilitude between the media and everyday life, that general attachment—the presumed rectitude of real-life as a logically-appropriate criterion against which to assess popular stories—doggedly survives. It is the unquestioned persistence of that bond that importantly obscures both the function of narratives and how reality itself is reified. While, of course, obfuscation of these connections and ideas contributes to cultural reproduction by helping *to keep* a hidden curriculum unobserved, it does not contribute to worthy scholarship. In contrast, functional-constructivism offers a complete theoretical alternative to understanding the function of popular narratives and their relationship to that which is commonly codified as real life.

How Stories Broker Reality

Consumers, presumably, *live* in the real world—the one with which media stories are (thus far, inexplicably) compared. They, *we*, are *surrounded* by the very existence that is supposed to be replicated. And there is the proverbial rub, to wit, this persistent connection promotes the idea that stories should symbolically reproduce this world that's already eminently available, and in fact, inescapable. What makes this promotion even more alarming is that despite its constant, non-mediated availability, the culture has already assigned distinct media genres (such as news and public affairs) that are already transmitted through several media, in several different programs, networks, platforms, etc., twenty-four hours a day, seven days a week. This genre (news, etc.) is dedicated to replicating at least those aspects of the real world deemed the most serviceable knowledge. Indeed, this genre is commonly regarded as the noblest in our culture's pantheon of popular media genres. It and educational programming are the only two genres characterized as cultivating "intelligence" (as opposed to vegetation). And to the extent that nobility may be daunting, the genre *also* produces different, "softer" variations of both content and style. Ironically, other than real life itself, news and public affairs information is probably the hardest knowledge *to avoid*. Few online carries don't open to headline news, and, even if one assiduously attends to cable-fiction, news is the only information for which regular broadcast programming is interrupted. Moreover, other than network IDs and promotions, news is the only information you'll find crawling along the bottom of a screen tuned to something else. Our argument, then, is that the ubiquity of all forms of knowledge socially-deemed to be "real" is what makes us further question the logic of any claim that the function of all *other* stories is *also* to replicate "reality." Given all the "reality" available and inescapable, what cultural function could be served by employing *all* storytelling to that same end? Or put another way, why would it logically be the objective of stories to provide information to which we have access without stories? Or, even if one argues that life-immersion is not the same as condensed storytelling, why is it the objective of stories to provide information about things to which we have dedicated a whole, very active and prestigious storytelling genre? Indeed, given the canons of professional journalism, news is more formally dedicated to replicating reality than any loose rules of correspondence employed by reality-comparison researchers. What also doesn't make sense is why such pervasive and expensive storytelling mechanisms are tapped to record faithfully what is already and constantly in our midst? For that matter, why even *have* stories and non-reality genres? We suggest that "reality" certainly does *not* adequately respond to the function of this symbolic world in which most Americans spend most of their lives. Perhaps the horse is already dead, but in case not, let us offer a somewhat subtler puzzle concerning the rectitude of this assumption or default belief. If replicating real life is stories' function then why is it so frequently *not* achieved in "entertainment" stories? At least our functionalist premises would suggest that anything so vital and needed would have to reasonably succeed to serve social structure. Were this, indeed, the case, we wouldn't have all the scholars and critics bemoaning the over-, under-, or mis-representation of so many things.

We suggest, instead, that nothing here has gone awry; it is not our culture that's "gone wrong," but rather the persistent presumption of stories appropriate connection to real life. We would offer the following instead: it is *not* the cultural function of *any* stories (including news) to reproduce the world from which they emerge, i.e., regardless of textbook rhetoric saying otherwise. We can find no source for such rules of correspondence because it's *not* a rule. It is so often violated because the objective of storytelling is something else—something that is *not* so often violated. A functional-constructivist interpretation speaks not only about the function of stories, but to the function of our popular rhetoric about stories.

"Reality" is all around us. However, the stories *about* reality (news) are *stories—not reality*. For all intents and purposes, we do not (as functionalists) distinguish news from reality TV or, for that matter, reality TV from fiction. In other words, the function of stories—of created, narrative packages—is the same, but there *are* functional reasons for their segregation. In terms of ideology and values, research repeatedly shows that news tells similar stories to fiction accounts. Sometime there is a division of labor, i.e., the culture may rely more on a given genre for certain kinds of information. For instance, election and voting (the democratic process) is critical ideological content in our culture. Clearly, fiction typically does not provide up-to-the-minute counts of elections, but the basic values presented in a program like *The West Wing*, for instance, are indistinguishable from those set in the agendas of newspapers. We have concentrated elsewhere on news, but our point is that the function of *all* popular media stories is *not* to replicate "reality." *To the contrary, the function of our stories is to prevent mundane familiarity with real-life patterns!* Moreover, we produce stories to supplement and, even to supplant those views or ideas we would obtain from the surrounding "real life" or from relatively "untreated" data from which police or census reports are constructed. In a limited sense, it is correct to say that stories *are* there to distract and amuse, but *not* for the sheer hedonism of entertainment or relaxation (as popular rhetoric implies). That rhetoric, like the stories, themselves, serves to distract us from whatever truths the systematic analysis of everyday conduct, systems, and institutions *would* yield—had we the easy access, motivation and ability to interpret all of it. The objective of stories (in any genre) is to construct for the culture an alternate universe—a universe that may appear superficially similar, but where verisimilitude is absolutely not required, and is often, in fact, wholly inappropriate. One introductory example should suffice.

Census data tell us that poor astronomically outnumber rich. Communication scholars have told us that the reverse is true for media representations (in any genre); TV and movies, it is reported, "over-represent" wealth and "under-represent" relative poverty. It has been implied, and even argued outright, that instead, the popular media should reflect the workers who constitute its viewing audience. But let us consider what stories would look like *if* their creation were held to, say, census-report standards. In that scenario, the face of both news and "entertainment" would be one in which wealth is incredibly scarce. Stories would have to show that the chance of significant upward mobility is hugely improbable. Stories would also need to begin dramatizing characters who are made content by money and the things it buys—and with that, the rich people would not be any *less* happy than the poor. They might even be much happier. Harmony and unity within poor families would not be any stronger or more prevalent than shown for the wealthy. In short, the stories would be unlike any to which we've been regularly exposed. Not only would they be very unfamiliar, but they'd be pretty bleak. Of course, this hypothetical picture is very different from what *is* represented. Our popular TV, film, books, news, and music, etc., systematically construct a world that is very different from "real-life." Instead, we see a world in which there are far fewer poor people, and where even the poor—everyone and anyone—can "make it." The stories to which we're familiar are ones where money rarely buys anyone happiness and where the poor are far happier than the rich in every way: the poor are even more like to be saved (holy redemption). The Bible tells us so.

We refer to the absence of correspondence between "real-life," on the one hand, and commonly told stories, on the other, as "social bookkeeping." Social bookkeeping is metaphorical reference to what we'll call the *public* "ledgers"—the records in the front office that may and often differ from those hidden in the vault. As we will discuss with respect to "intellectual quarantine," few people have the combination to the vault, and fewer people seem to have the desire to know its contents. While some may find our description duplicative of Marx's notion of *false consciousness*, we purposely have not borrowed that term because of the distinctions between false consciousness and what we describe here. As noted, we rarely interpret social ledgers as emerging from the intentions of one class to exploit another; although that effect may be achieved, it is the intentionality we reject. Moreover, also as noted, we never characterize even the vault-ledgers as un-authored "reality." Rather, we suggest that it is typical in a stable society that all classes know and accept essentially the same accounts. While it *is* true that mythology is primarily oriented to the disempowered, there are other reasons for the unilateral saturation of cultural beliefs. For one, any social system will operate more smoothly if *all* of its constituents subscribe to the same or similar realities—if only to avert "knowledge leaks."

As we will show (using intelligence as the illustration), the function of stories is absolutely *not to duplicate* what's already here, *but rather, to "adjust" people to it* (i.e., what's here)—to rationalize the world that people must live in and respect if the order is to be duplicated. From this perspective, it is misleading for anyone (including ourselves) to suggest that these stories *mis*represent "real life" because misrepresentation implies that people don't understand that the media are there to provide entertainment, relaxation and escape. To the contrary, people embrace the entertainment rhetoric which allows them to ignore the ideological implications of what's consumed. What these stories—*and* the popular rhetoric developed around them—in fact do is contribute to social balance and stability. Through reconstruction, our popular stories make the world in which we work palatable, thereby allowing everyone (particularly those who work most for the least reward) to continue to serve the social order and, thus, to contribute to social stability. We connect many of these issues specifically to intelligence in this chapter's final section.

Imagining Intelligence

At this point, we have a common theoretical basis from which to analyze our subject matter. Inasmuch as our approach relies, in part, on the constructivism just described, the fundamental premises on which material has been collected were heuristically derived. For this reason, our tendency is to show the work—the derivation of equations and the evolution of definitions. After all, this chapter is about how intelligence has come to be imagined in everyday life and, thus, to work from any strict deductive, *a priori* premises would wholly defeat half of our philosophical orientation. With this said, it will probably be helpful to note at the outset a couple of our heuristically-derived guiding principles and boundaries.

Our investigation started with only the word—"intelligence"—which, itself, has a long and reasonably stable history in the English language. From the word, we established both temporal and lexical referents. What we're calling "modern" intelligence refers to a conceptual evolution in progress since, roughly, the height of the (second) Industrial Revolution, and developing most rapidly since the second half of the twentieth century—the period in which our analyses are, therefore, concentrated. Although our study is not historical, the general rationale for these broad, temporal markers is explained in the course of the relevant topics. Moreover, as indicated, intelligence—*the idea* that we came to reveal—

was found to have a time-honored referent, to wit: "traditional intellectuality and erudition" involving the kind of knowledge and/or learnedness typically associated with the values (even more than the precise content) of classical instruction. For lexical simplicity, often we refer to this time-honored understanding simply as "traditional intelligence." Surely, it is not our objective to ignore those other "features" that most people would now, probably, argue additionally or alternatively define intelligence. Indeed, such challenges are the point! In other words, we attempt to reveal the modern construction of intelligence by examining what has "become" of the traditional referent. Thus, how "traditional intellectuality and erudition" has been used, understood, and replaced, is what we use to *reveal* the social construction of modern intelligence. Of course, all the foregoing introduction should similarly explain that our ultimate objective in revealing this particular conceptual construction is to understand its relationship to major social-structural elements, and, of course, its function *vis-à-vis* the continuity of the social order. Studying intelligence is hardly arbitrary; as we hope to show, the field of referents with which that term has, historically, been associated, makes it one of the most important constructs—if not *the* most important construct—in the popular lexicon

Narrative and Proverbial Representations

In this section we mainly examine intelligence as represented in popular stories and aphorisms. Because there is not a single criterion by which all representations can be divided, we open this section with a summarized sense of the overall narrative pattern. This is followed by subdivisions, each illustrating a specific representational feature or issue. In all, our examination here crosses media genres and history in what we suggest is, nonetheless, a coherent distribution. Whenever relevant, this coherence is used to demonstrate the mythological uniformity of intelligence.

The Anti-Intellectual Tradition

Particularly since the second half of the twentieth century, our culture—and, thus, its mythology and prominent stories—is basically anti-intellectual. In other words, traditional intellectuality has, increasingly, become devalued and discouraged—at least as such values and encouragement are provided by popular aphorisms and narratives. It's important to note that, in asserting this, we specifically do not conflate or confuse knowledge with *education*. There is no question, for instance, that our parents and mentors, not to mention a plethora of public-service announcements and situation-comedy didactics, *extol* the benefits of education; "Don't drop out!," every MTV generation is urged (probably only a little less than it's encouraged to vote). These pro-education sentiments are readily acknowledged; *however*, it should also be acknowledged that such sentiments most typically (if tacitly) reference the consequences of academic *credentialing* (and, possibly, of networking, perseverance and/or delayed gratification); what is not usually being praised are the consequences (hypothetical) positive consequences of traditional intellectual enrichment. Indeed, such consequences are not even, typically, *intimated,* in relation to credentialing. Indeed, some popular comedies are virtually *predicated* on distinguishing the administrative process, on the one hand, from intellectual enrichment, on the other, e.g., *Back to School* or *Billy Madison*. Indeed, if one examines any movie in which college graduation is a central, narrative element—from *The Graduate* through *Reality Bites*—the ironic truth of this distinction is often one of the film's "messages." But there are many other contexts in which to detect this distinction and the more fundamental anti-intellectualism of which it's a part. We now look at some of these other instances, beginning with a context that doesn't necessarily involve narrative or mediation.

Folklore: Aphorisms about Knowledge

Intelligence is an ancient concept always generating reflection. Expressions like "ignorance is bliss," "what you don't know won't hurt you," "too smart for your own good," or the British "too clever by half," have been familiar to people *without* benefit of media. Principles of this sort pepper most Western languages, and such aphorisms were circulating *before* the advent of what now constitute mass media. Of course, the media were, ultimately, shouldered with the responsibility of promulgating a lot of "folk" wisdom. However, that wisdom must be seen as pre-dating major, mass-communication technology.

Teachers

That teachers may be associated with intellectuality and erudition is obvious. (In fact, to the extent that they are *not* becomes, as we shall see, becomes an important independent variable.) George Gerbner, in a rare comprehensive analysis of media-represented education, poignantly observes that romance and sex are compelling symbolic indices of vitality; in stories, they generally reinforce that to which they're attached. Thus, Gerbner argues, how a character fares in "love" speaks to his or her overall credibility. This is both eminently reasonable and is a coherent pattern in popular narrative.

In demographic analyses of prime-time TV fiction, anywhere from half to three-quarters of *all* adult characters are married. However, when married characters are further analyzed *by profession,* educators are significantly *un*partnered compared to adults in other professions. As discussed in the introduction, we have little interest in the degree to which media statistics mirror documentary information; however, lest that be considered the rationale for TV's marital patterns, it happens that the census statistics show educators to be married significantly *more* than average. Thus, the media pattern here is *so* contrary that it seems especially poignant.

School- or teacher-set dramas and comedies have appeared on TV since its early days—from *Our Miss Brooks* and *Mr. Peepers*, through *Room 222* and *Mr. Novak*, onto *Saved by the Bell*, *Head of the Class*, or *Fame*, to later series like *Boston Public*. Additionally, TV shows from *Leave It to Beaver* through *The Wonder Years*, *The Simpsons*, or *Third Rock from the Sun*, may not be principally set in school, but they, nonetheless, feature educators in continuing roles. In *all* the examples mentioned (just like in dozens of less-prominent programs featuring educators) one is hard-pressed to locate the married teacher—and shows like *Room 222*, *Fame*, or *Boston Public* (like lesser-known series like *Teachers* or *The Bronx Zoo*) included *many* roles for education professionals, often without one successful marriage among them. This is not a *televisual* artifact either, because movies, also, prominently portray teachers as an unmarried lot—from classics like *Good Morning, Miss Dove* through *To Sir, with Love*, as well as later entries from *Porky's* through *Good Will Hunting* or *Krippendorf's Tribe*. Occasionally a teaching-character *is* married, but such romantic "luck" is intriguingly correlated with the subject taught; the partnered teacher is far less likely to have strong, *academic* affiliation. Accession on the marriageability index is positively linked to disdain for the academic element. Characters cast as physical-education teachers or sports coaches have the best chances of making it to the altar. Through his long-running series, *Coach*, Hayden Fox, takes a wife and, not long after his marriage, *leaves* academe altogether for the major leagues. Gerbner observed a more general, but similar pattern: *leaving* academe places characters in the way of the passion, romance and success that cannot be found while on campus. This escape route is classic, dating back at least as far as a mousy William Powell doffing his academic robes from the Missouri Conservatory of Music in exchange for a hep suit befitting his exodus to Tin Pan Alley. More recently, the story involves fumbling, elbow-patched Dan Acroyd donning flamboyant suits and name—*Dr. Detroit*—in his life-awakening vocational shift from professor to sex procurer. Probably most famously, many generations recall *The Nutty Professor* (Lewis *or* Murphy) whose sexuality emerges upon becoming his own version of Mr. Hyde—which should be remembered as a parallel image dating back to Victoria's reign. Other media teachers sometimes marry, but more subtle than their attachment to sports, might be an intellectually-compromised, professional charge. A well-known example here is TV's Mr. Kotter and his assignment to the barely-teachable "sweathogs." In any case, Kotter's happy marriage remains a media anomaly. More typically, stories of commitment to intellectual pedagogy demand lives devoted to the task—whether facing merely innocent or callow youth (like Miss Dove) or, when humbled by students' emotional and/or economic deprivation (like educators in *Boston Public*).

Beyond such sublimation-in-dedication is yet another, significant story permutation: the ignominy of sexual degradation particularly reserved for "fallen" teachers—felled by the inability to cope with intellectually-incompatible sexual forces. This is seen early in film history, with for instance, *The Blue Angel*'s Professor Rath, and later with Miss Jean Brodie. More recent intellectual's sexual discomfiture is dramatized in films like *Educating Rita* or *Wonder Boys*. Being libidinally and/or romantically challenged can be portrayed more graphically in movies, but less gut-wrenching portrayals *are* enacted for the smaller screen, and even for pre-adolescents, e.g., when *Saved by the Bell*'s handsome and urbane anthropologist, Professor Lasky, gets dumped by the sexy ingénue/student returning to her un-intellectual, boyish high-school flame.

Potentially-partnered or not, independent of the subject taught, one way or another, media teachers are typically detached from sex and romance. It can be because of sublimated devotion to students (so common to pre-1970s films), placing scholarship ahead of personal life (as the professor in *Altered States* who sacrifices his marriage and physical well-being to his research), the comic interpretation of head-in-the-books obliviousness to sex (like *The Nutty Professor*, or *The Absent-Minded Professor*—who repeatedly forgets his own weddings—or the well-known Professor residing on *Gilligan's Island*). *Women* teachers, seem to have their own stereotype resulting not so much from their devotion or oblivion, but from sheer lack of suitors, as often seen in the archetypical schoolmarm. This gender variation is common and meaningful. Whereas learnedness disorients men, it tends to render women less attractive.

The Academy

Beyond the representation of teachers, stories about school-life tend either to detach intellectuality from the educational process or cast it in a negative light. In the first instance, if you consider dramatic contexts like politics, law, crime, or business, you will find them less likely to serve *merely* as backdrop; typically, those scripts will overtly address the setting and, thus, the drama will *be about* politics, business, etc. A school setting, though, is comparatively less likely to address knowledge and learning. The theatrical docudrama, *Stand and Deliver*, for instance, recounts the tribulations of a teacher who attracts inner-city students to the study of calculus. Unlike many stories set in the academy, this one *does* consider academic material. But, typically, when the virtue of academic knowledge is addressed, the value is largely equated with very practical applications (rather than, say, the derived benefits of analytic reasoning). At first blush, one might think this level of observation holds media stories to too high a standard; how reasonable *is it* to expect education's methodological or critical "secondary function" to be dramatized? While this level of scrutiny is, admittedly, subtle, it sustains as a significant variable when compared across dramatized subjects. Take, for instance, a television series, like *The West Wing*—dealing with the critical underside of government and politics to much critical acclaim. The idea of successfully dramatizing education with complementary narrative attention to pedagogical details and concerns would seem unimaginable—

esoteric, bizarrely idealistic, and suitable only for an early-morning subcommittee discussion televised on C-Span. In media stories, knowledge is recognized without mockery only when it's applied to mundane ends. In the more recent TV series, *Numb3rs*, two brothers—an older, more rugged FBI agent, and a younger, delicate, mathematics professor/FBI consultant—together, solve crimes. The older brother supplies physical power and, often, practical expertise. The professor talks about the extended value of mathematics, logic, reasoning, etc., and he is shown constructing elaborate theorems that, typically, work deductively to isolate the culprit. Thus, his brilliance is useful when it comes to capturing the bad guys. However, his generic possession of logic and reason are not shown to be meaningful; his extraordinary crime-solving brilliance doesn't bleed into *other* areas of life. Advanced math, the audience may observe—like physical (forensic) anthropology (in the onslaught of post-mortem crime series)—is implicitly limited to its intelligible effects. And, even with these obvious limitations, *Stand and Deliver*, *Numb3rs*, and similar stories are still exceptional; more often than not, academic knowledge is portrayed only *as* incomprehensible, obscure, and patently useless beyond the ivory tower. In this sense, *Third Rock from the Sun*'s physics professor is alien in more ways than one.

Although *intellectual* content and skills are often ignored, people are often shown to learn things in academic settings; of course, that learning may be wholly detached from the academic instruction. In *Stand and Deliver*, for example, the central accomplishment is, arguably, not about calculus or even knowledge, but rather, the teacher's ability to motivate children for whom society-at-large has not held out much hope. This "motivation of the unlikely" or the "unexpected success" is a common theme in educational stories e.g., *Dangerous Minds, Freedom Writers,* or even *Welcome Back, Kotter.* In most stories of this ilk, students acquire self-possession and confidence by conquering the unfamiliar or unexpected. Of course, these positive outcomes could have been derived from non-intellectual activity, too—and, in fact, many, if not most "against-the-odds" school sagas *don't* dramatize academic issues, e.g., *Rudy*, *Bring it On*, *Save the Last Dance*, *Wildcats*. In most cases, redemption is delivered through sports (with, perhaps, a particular nod toward football), and if not sports, music and dance. One might be apt to argue that sports and musical performances are inherently more dramatic that intellectual development; however, is that judgment a result of naturalized constructions? Does watching Matt Damon pretend to solve mysterious equations as janitor Will Hunting lack drama because one has no clue what the symbols represent? In any case, the endings to these sports- and music-dominated school stories are often moving and noble, but intellectual enrichment is almost never the heroic force. Indeed, one variation of these stories involves "reverse enrichment," where *teachers* are the ones transformed—from intellectually-lofty, hard-edged educators to openly-sensitive, parental-surrogates, who, in another common "twist," are sometimes even *intellectually* "unblocked" from this reverse-emotional enrichment, e.g., in *Finding Forrester* or *Wonder Boys*. Sometimes the subject of a reverse effect is not the educator, but a different representative of the intellectual elite—the snobbish, over-achieving, pedantic student who receives an "emotional" education, e.g., Selma Blair's character in *Legally Blonde*. (Indeed, the whole premise of *Legally Blonde* is the benevolent intellectual "castration" of Harvard's intellectual image. This deconstruction of cold intellect is, of course, "good" in the context of these stories, and, might even be seen as one variation of the "leaving-academe" theme. In this case, the departure is spiritual and the destination is "humanity."

Of course, there are other contexts in which the academy becomes a stage for non-intellectual narrative. It is a setting for limitless love (e.g., *Love Story*), for coming-of-age (e.g., *Mona Lisa Smile*), for nostalgic refusal-to-come-of-age (e.g., *Animal House*), and often for class struggle (e.g., *Class*, *School Ties*, or *Pretty in Pink*). Interestingly, it is not uncommon to show education as causing *within*-class struggle, e.g., *Educating Rita*, *Good Will Hunting*. A common story involves an uneducated, working-class character pursuing or pushed into higher education where the outcomes are limited. One possibility is "reverse education"—the intellectuals are more deeply transformed by the working-class character than *vice versa*. This includes not only those stories in which scholars are thawed or humanized, but where warm-blooded, but wounded or flawed intellectuals are revitalized by the outsider-student (Robin Williams has a career track enacting this one transformation). By far the most common outcome, though—often in combination with these other correlates—is the working-class character's epiphany that all the world's erudition is not as valuable as that wisdom taught by the proverbial college of hard knocks from which, of course, she or he has already graduated *summa cum laude* (e.g., as seen in movies from *Tammy, Tell Me True* through *Back to School* and beyond). This is the intelligence-mythology's version of a more generic myth: "What we imagine as important or desirable is illusory." If the story is serious or tragic, it ends with sadder-but-wiser intellectual alienation ("You can't go home, again"); the more upbeat ending is the protagonist's opportunity to discover "there's no place like home."

Ironically, while Dorothy's original "there's no place like home" is not, itself, education-linked, *The Wizard of Oz* makes a clear and grand statement on the illusory value of intellectual achievement. When handed diploma parchment (but not the education it typically symbolizes) Scarecrow's lifelong desire for a brain is not only satisfied, but he becomes instantly fluent in intellectual gibberish. More than half-a-century later, portrayals of education still wryly deliver this message. In *Peggy Sue Got Married,* for instance, the title character fantastically relives her youth with adult sensibilities and uses this hindsight (to the teacher's dismay) to inform her classmates that there is no earthly use for algebra in adult life. Of course, such reflection is also rendered less sarcastically, e.g., lyrics to more than one Paul Simon song, or in lyrics to more classic popular songs such as "Wonderful World" (which, interestingly, perhaps, was covered by Art Garfunkel).

Finally, if the value of being intellectually developed weren't sufficiently dubious, one need only consider those stories where intellectual development seems to precipitate insanity. Whether mocking portrayals of academic geniuses drove to distraction by their intelligence—such as the many villains in the old TV show *The Wild, Wild West* (and reincarnated for the more recent film version), or very serious dramatizations of academic-genius-caused-schizophrenia (e.g., John Nash's *(A) Beautiful Mind*, or father and daughter mathematicians in the Broadway and Hollywood productions of *Proof*). The image of extreme scholarly brilliance transformed into "madness" seems, by our mythology, not only plausible, but possibly, a naturalized correlation.

Intellectuals

Throughout all popular media and genres—*in or out* of the academy—intellectuals are typically life's losers. Although research shows that rich people are invariably portrayed as significantly *less* congenial or happy than others, or that criminals and "bad guys" ultimately suffer negative consequences, there seems to be no other identifiable group (particular one that is overwhelmingly lawful and non-violent) depicted as more discontented and deluded than intellectuals. When "brains," beauty, sexiness, stylishness, gracefulness, happiness, and general life-success are constituent narrative elements, the story's equation will, invariably, have intelligence on one side, and all remaining variables on the other. While the Gerbnerian principle that "intellectuality ≠romance and sex" certainly operates symbolically, we might adjust that principle by positioning intellectuality as the fundamental variable—the trait against which *all* the other desirable elements are individually, and, in combination, balanced.

Films like *Revenge of the Nerds* suggest how studiousness translates to problems with appearance, grooming and style, and, of course, it's not a pretty picture. In school stories, the student producing "borrowable" homework or who is smart enough to tutor is also likely to be unattractive. Very often, students in need of academic assistance are the ones sufficiently endowed to trade on their physical assets in obtaining academic help. Invariably, the sexually-unprepared and socially-unsophisticated, studious student is ill-equipped to resist supplying the assistance demanded. Indeed, unless and until a story ends with underdog-victory, the brainy students can't elicit sexual attention unless it's purposely manipulated to exploit their intelligence. The geeks and nerds rarely have the composure or morals to wield their intelligence for sexual (or other) gratification; it's usually the slick player who mediates deals in selling term-papers or exam answers (e.g., *Fast Times at Ridgemont High*). Meanwhile, the smart kids' intellectual exploitation is garnered through predation or pranks. They might be seduced (Urkel in *Family Matters* or Screech in *Saved by the Bell*), bullied (e.g., Albert Brooks' as a child in *Broadcast News,* or Anthony Michael Hall in *The Breakfast Club*), or thoughtlessly made subject of a wager or deal and *then* seduced (e.g., *Never Been Kissed*, *She's All That*, or *Ten Things I Hate about You*). Interestingly, when less-than-pretty child actors (naturally playing brainy kids) mature into good-looking teens, television series often change the *actors' character* (as in *Family Matters* or *The Wonder Years*) ostensibly rather than allow a character, who now has other options, to continue on as the resident brain. Ironically, *Family Matters* actually created *a new* character, Stefan, an alter-ego relative for the über-nerd, Urkel, to accommodate the actor's maturing looks. Either the brainy child could not be allowed to grow into a handsome young man and/or a handsome young man could not be (only) cast as a brainy teen.

There are certainly deviations from this pattern, some of which are more permutation than counter-example. For instance, beautiful, popular, good-natured girls can also be academically successful, as in *Clueless* or *Legally Blonde*. Their academic success is consistent with generic over-achievement—their good grades are generally indicative of compulsive exactitude or practiced manipulation rather than intellectuality. Or, if *all* students are smart (as in *Head of the Class*—premised on a program for the intellectually gifted), an internal hierarchy will likely arise. From a series of heuristic surveys undertaken over two years in several undergraduate classes and involving approximately three hundred students, informants rank-ordered for intelligence a list of many TV characters including the *Head of the Class* cast. Three of the show's characters were ranked among the smartest—the only three who were "character" actors—cast members distinguished by their noticeable lack of good looks. Similarly, the character, Screech, was significantly identified as both the smartest *and* the homeliest *Saved by the Bell* character. These surveys only confirmed identical inferences drawn from critical observation and content analysis. These inferences apply to fiction-TV casting, in general. Kevin's geeky next-door-neighbor, Paul, was also the smart kid on *The Wonder Years* block, as was the large and awkward Margaret on *Leave it to Beaver*, and the less-than-beautiful Zelda on *Dobie Gillis*, and so on. This observation also turns out to be consistent with the (situs) pattern involving the narrative centrality of traditional academic or intellectual issues; the talents of characters exhibiting *non*-academic skills at school (music, dance, etc.) like those on *Fame* (the movie or TV show), for example, are not inversely proportional to beauty, grace, popularity, etc.

Many of these seeming narrative formulae or rules are so old that it's hard to imagine seasoned writers and producers being *un*aware of, at least, the most enduring equations. Yet, we should reemphasize the non-rationalist nature of our analysis, i.e., the irrelevance of creative self-consciousness to what we describe. While writers and producers may, in fact, recognize this or that pattern, our point is that such recognition is not critical. Extending journalism theory's "professional routinization" premise—that workers receive the tacit socialization necessary to function appropriately in a given labor subculture—we assume that people creating our popular narratives, like virtually all reasonably successful workpersons, have internalized a vast, tacit understanding of job specifications.

In addition to learning seating patterns for network meetings, or the kind of scene that's preferred for the opening "teaser" (all information *not* typically described in textbooks), how and when cultural myths are to be embedded in narratives is also part of the professional routine.

Thus far, our inferences have been synthesized from many content analyses, ultimately crossing all genres, time periods, and other standard, distinguishing variables. As suggested, it was critical to our investigation to determine whether or not the hypothesized mythological uniformity could be demonstrated by virtue of ideological coherence across this variety of narrative forms. While this variety is obviously required to check for ideological coherence, it is also true that many hypotheses about that ideology have been instigated by in-depth scrutiny of a specific case. To reflect how *case* analysis is correspondingly useful in terms of revealing specific effects and nuances that might later be included in an analytic instrument, we offer some highlights of one specific and exemplary case. We chose, in this instance, what might be considered the über-example, as it were—a long-running TV series that might be characterized as archetypical of anti-intellectualism in popular, narrative: *Cheers*.

Any principal character in *Cheers* demonstrates the relevant patterns. If we begin with the series' formally-educated characters, we consider two psychiatrists, Lilith and Frasier, as well as assorted recurring guests (e.g., literature professor, Dr. Sumner Sloan). This group of credentialed intellectuals uniformly embodies the sexual repression, social awkwardness, inability to connect interpersonally, pretense, and pomposity that popular narratives commonly associates with serious, academically-derived intelligence. But, *Chee*rs is surprisingly nuanced in its "range" of expressing anti-intellectualism. The cast includes major *un*-credentialed intellectuals, too—postman Cliff and perennial-student-cum-barmaid, Diane. These two characters demonstrate different aspects of intellectual pretense, but remain similar in their constant attempts at dubious pedanticism. In all, both the credentialed *and* self-appointed intellectuals are failed and/or discontented characters (to the extent comedy allows), *and* their malaise is dramatized in stark relief to two other characters in particular: *Cheers*' intellectually-*challenged,* residents, bartenders Coach and Woody. These two barmen are, conversely, the shows' happiest and sweetest spirits. Alone among *Cheer*s' characters, Coach enjoyed a long and *normal* marriage. Woody also gets to love and marry—a beautiful young woman who, not by his design, is the daughter of a wealthy, corporate magnate. Interestingly, Woody's romantic trajectory has yet another counterpoint—with the bar manager, Rebecca. Rebecca's tireless attempts at upward mobility—including predation of Woody's father-in-law—creates an effective contrast with unassuming Woody's effortless life. Rebecca only finds love when she abandons her clever get-rich schemes and marries a plumber. The group's central force is the decidedly *un*-intellectual (and often anti-intellectual) ex-ball-player and bar owner, Sam, who is handsome, graceful, and sexually-empowered. Lastly, the (pointedly) uneducated, senior barmaid, Carla, exemplifies, in her every appearance, how education and intellect are immaterial to wit or presence. With (seemingly) only a reform-school background, Carla relentlessly holds her own with *any Cheers*' character. *Cheers*' most successful spin-off, the very popular *Frasier,* changes locale, but the anti-intellectualism is retained and, arguably, enhanced by the addition of Frasier's younger brother, Niles—a pompous and pretentious fellow-psychiatrist who manages to out-effete his older brother. Through most of the series, the Crane boys haven't much romantic success, and haven't "inherited" the affability or magnetism of their maimed-yet-*manly* ex-cop father. Ironically, a loopy, occultishly-spiritual housekeeper/physical therapist serves, as the grounding influence for the head-in-the-intellectual-clouds, scientifically-trained brothers; she is also, for many years, the object of Niles' unrequited love. Unsurprisingly, even when un-partnered, Niles is incapable of articulating his feelings (much less acting on them). Paralleling the younger brother on *Numb3rs*, neither Niles' nor Frasier's prep school and Harvard-trained skills, much less their professional psychiatric expertise, ever seem to be of much practical use to them. In fact, as they are repeatedly told by their father, colleagues, and friends, most of the brothers' problems seem to be exacerbated, if not caused, by "over intellectualizing."

While intellectual males (such as the "Crane boys") are plagued by various inadequacies and neuroses, including romance and sex "issues," a slightly different spin is given to the plight of female media-intellectuals—very possibly because of complications related to gender-linked expectations regarding romantic passivity. An interesting case in point would be the classic, erotic sexualization of the Victorian-styled, female *librarian*—a profession obviously linked to knowledge and literacy. *As the librarian*, she is unsmiling, indeed, uninviting. Her dark hair is pulled tightly into a neat bun. She wears a starchy-white blouse with a fully-buttoned, high-neckline that is doubly-secured with a black ribbon. Heavy, black-rimmed, reading glasses dominate her face. Her transformation from a woman-of-letters to a sexual vixen (another variation on the flight-from-intellectuality theme) is accomplished not by fully undressing her, but by literally discarding first the literal and symbolic accoutrements of bookish propriety: the hair and prim collar is let loose followed by the ultimate liberation—the removal of her glasses, thereby disconnecting her from the whole environment of written erudition (and from scrutinizing her seducer?). It is not clear whether or not this image launched Dorothy Parker's famous "Men seldom make passes at girls who wear glasses" or the Marilyn Monroe character's preference for stumbling around blind-as-a-bat in *How to Marry a Millionaire* (rather than disclose her need for spectacles), but this is certainly a staple image that has spawned a number of variations. Elements of this image are even worked into stand-up comedy routines—such as that of Dennis Miller, which graphically relates Miller's fantasy of sexual mastery over *Madame Curie*! Although such fantasy may seem to attribute sexiness to female intellectuality, Miller's routine, as well as all other renditions of anecdote, equate eroticism with the *female's* intellectual degradation—as if, quite literally, the transformer could "sex" her brains out. The less graphic and more romanticized variation has a man interrupting a woman's intense, intellectual tirade by kissing her full on the mouth, thereby preventing her speech. This particular maneuver is actually enacted in many stories featuring an intellectually-assertive woman.

Ironically, perhaps, probably the most famous, recent abandonment of intellectual accoutrements involves a male. In his 1940s tweedy, elbow-patched and vested suit with delicate wire-rimmed glasses, we might have little clue that lurking beneath this very proper, professorial exterior is the beautifully rugged, distressed-leather and khaki-wearing, bullwhip-wielding adventurer of Indiana Jones. To transform into a heroic, sexual icon, though, not only must he relinquish his scholarly appearance, but literally depart the sterile academy for exotic points in the "real" world. Of course, unlike the librarian, Professor Jones doesn't transform for anyone's sexual stimulation . . . nor does he lose composure upon abandoning his intellectual self. Indeed, even *without* becoming Indy, tweedy Professor Jones can still set coeds' hearts aflutter. Unfortunately for female counterparts, heat isn't typically generated until intellectual trappings are shed. As Gerbner observed more than three decades ago, anti-intellectualism hits women *particularly* hard.

There are many other stories, probably having their roots in medieval European folktales, featuring the transformations of smart, plain women into sexual sirens. These are not, though, like *Pretty Woman*, incarnations of the Cinderella myth. Rather, in these narratives, serious women, *neither temptresses nor prince-seekers,* are transformed—sometimes against their will and almost always against their better judgment—only to enjoy ultimately (and, sometimes, guiltily) their (re)vamping. This sort of transformation is dramatized in a number of Sandra Bullock roles—from the homely scientist-turned-siren in *Love Potion #9,* to the tough and unadorned FBI-agent-cum-beauty queen in both *Miss Congeniality* roles. Similarly, a teenage version appears in Rachel Leigh Cook's role in *She's All That.* True to gender pattern, these transformations are typically proposed and urged by males. Sometimes, unbeknownst to the strong female, her complicity is urged so that the petitioner may win a wager, e.g., *Ten Things I Hate about You.*

We have, thus far, concentrated on intelligence mythology as it is constructed and reinforced through representations in stories. However, there are certainly additional means by which anti-intellectual sentiment is developed beyond characterizations and portrayals of teachers, education, and intellectuals. We next explore how the love and/or pursuit of knowledge might be "exorcised" from most people's routine existence.

Intellectual Quarantine

This section considers the exorcism of knowledge and erudition from mainstream life, both literally and metaphorically, explicitly and implicity.

The Social Construction of Expertise

Twentieth-century, mass-society scholars, like Lowenthal or Marcuse, lamented what they saw as a regrettable outgrowth of industrialization: the development of inferior product for quick distribution and disposal in a mass economy. Various social critics, such as Boorstin, MacDonald, or Warhol, devised different metaphors to discuss mass industrializiation's *human* product, to wit, that *people* could be widely and instantly celebrated at the click of a media recorder. Arguably, nowhere can the development of inferior 'strains' for mass consumption be more poignantly exemplified than in connection to celebrated intellectuals—media-certified (if not -created) and -distributed "experts."

The popular idea of experts' function seems pretty clear: increasingly, people require more knowledge than they can individually possess so they must substitute someone else's knowledge for that they haven't had the opportunity to develop personally. By the twenty-first century, we have innumerable forums with a huge variety of gatekeepers dealing with an astronomical array of highly-differentiated specialties—each potentially requiring different preparation, knowledge, and experience. The media's role in this process is obviously enormous; we even use mediated interaction to communicate about personal and/or physical conditions, e.g., *Web M.D*. It is important to remember that not only are we personally lacking necessary expertise, but we are probably also lacking the knowledge to determine who actually has the information and abilities we lack. *For* whom should the media gates be opened? *By* whom should our experts be identified and prioritized?

On occasion, the world *is* introduced to profoundly accomplished and respected experts, e.g., Einstein or Salk. But public exposure to the Einsteins or Salks is rare; it has even been suggested that the public is not exposed to legitimately *qualified* experts (other than those they may privately hire), let alone geniuses. Of experts who require academic credentials, those who also achieve popular prominence are, reportedly, less likely to have received affirmation via peer assessment. In other words, not only is (intra-)professional success not requisite for popular expertise, it may, in practice, be incompatible. Around twenty years after Einstein's death, the term "pop physicist" seems to have been coined in connection with Carl Sagan. The very idea of a popular *physicist* seems almost oxymornic. Sagan's ideas about the powers of the universe, though, were popularized in in his PBS series, *Cosmos*, in addition to some highly successful mass-market, trade books. While it is probably essential to distinguish between disgruntled professional envy (especially from the genteel-poor scholars who observe a "colleague" earning a serious salary *as a professional scholar*), attention has also to be paid to claims of unsatisfactory scholarly credentials or ability. Questions about Sagan's ability and status in physics, for example, lead to other considerations. Was Sagan's material so oriented to mass consumption that it did little more than imbue viewers with a false sense of erudition? Do programs like *Cosmos* (or now, *whole cable networks* devoted to intellectual pursuits) merely create (*Cheers*') Cliff Clavens, i.e., do consumers recognize that the popularized treatment is partial and simplified? Does watching the simpler version lead to more serious investigation? Is an "expert's" ability to gain national prominence mostly a matter of his or her audience-friendly (or camera-friendly) style—or any other quality *not* related to *bona fide* expertise? And, perhaps most importantly, how would the average consumer know how to begin making such judgments?

It has been long observed that, to the extent that any subject matter deals with *mundane* concerns, it is *un*likely to be subjected to serious, qualified analysis in public Whereas, to the extent that a subject is esoteric—"quarks," "black holes" or planetary motion, for instance—serious expertise *can* be popularized. Thus, the more seriously expert the knowledge, the less likely it is both to impinge on everyday life, or be popularly disseminated; seriously respected expertise is "saved" for subjects over which people have no control. Since there is no central expert-ranking system, we performed a little exploration about the most consulted self-help/relational gurus on the Internet—as determined by appearing in more than one reference in the first five pages of a returned search, *and* being commonly publicized as "Dr.—" (usually, familiarly, with only the first name), because we, specifically, wanted to consider individuals for whom academic credentials were also a marketing element. We, then, collected biographical information (educational specifics and professional publications/presentations) for the six people who met these two criteria. As indicated by their familiar address, all six held doctorate degrees; however, three of these doctorates were neither Ph.D.'s nor M.D.'s in psychology, psychiatry, or a cognate behavioral, social- or neuro-science discipline. Of the three remaining doctorates, two were from institutions that putatively are only accessbile through the mail and which do not demand prepatory academic credentials. Thus, only one of the six held a psychology Ph.D. from a research university. None of the biographies of all six indicate *any* peer-reviewed activity . . . ever. In fact, ironically, the only listings that featured any professional organization were several references to NATAS—a conference of television executives who hold a fair introducing and promoting shows for syndication, and a mention of a state board of review in conjunction with one expert's alleged professional-license suspension. Clearly, we don't offer these "findings" as remotely rigorous, but rather, as exploratory results assisting the formulation of formal research question. We know, for instance, that we'd want to extend these question to other areas of popular expertise, e.g., do the robed "stars" adjudicating small-claims arbitration programs—or the very opinionated moderators of criminal-law talk shows hold degrees from respected law schools, a strong litigation record, or a good reputation or evaluation from their non-celebrity peers? There certainly is enough indication to suggest that the hypothesized inverse relationship between popularity and profesionally-recognized expertise is a very possible condition.

Despite the fact that popular gurus are ubiqutous—publishing bestsellers and starring in both radio and TV shows—there seems to be little public concern about their professional credentials or expertise—at least beyond the occasional comedy skit in late-night satire. While certain prominent self-help experts have recently been the subject of litigation for *causing* and exaccerbating clients' problems, it has been reported that these reports have not significantly affected overall indices of popularity. Along the same lines, although most states require therapists to be licensed-to-practice, dispensing advice via mass media is unregulated. Similarly, one enounters numerous consumer-warnings about expertise in some areas—like cosmetic surgery, just as consumers are urged to investigate a physicians' credentials, experience, and legal histories before, say, selecting a liposuctionist or breast implanter. Yet, consumer alerts aren't much issued for popular experts attending to other types of problems (some of which may involve life-altering issues). And, arguably, consumers typically know very little about professional standards. In fact, in our routine surveys of students, we were surprised to learn that most of the 300 surveyed believed that (1) mass-media experts need the "right" credentials to offer help, and (2) those experts receiving the greatest media celebration are probably the cream of the profession. The only seemingly paradoxical research findings in this whole context pertains to the use of experts on syndicated talk shows. As anyone who has ever seen such programming can attest, it is common to introduce a relevant expert—usually with announced credentials—after one or more non-celebrity guests have revealed their "problems" to the host and the audience. Ostensibly, the expert is, then, called in to offer professional insights to help resolve the problems. Studies have shown that experts introduced under these circumstances, are not given more time, attention, or—to the extent it can be measured—more credibility than anyone else who might comment on the issues. Whether or not these data are paradoxical requires examining the full set of patterns and their interconnections. Assuming the accuracy of all the above hypotheses and findings, one thing is clear: we need to offer a functional-constructivist analysis of the function of popular experts. Our statement at the outset—that most people likely explain public experts as a means to supplement people's necessary knowledge in an increasingly fragmented world—now seems frightfully limited and naïve. A new analysis would take two things into account: (1) the various circumstances and conditions we've just indicated, and (2) the patterns about intelligence construction obtained from our analysis of popular stories and folklore. Specifically, is there a logical means by which the patterns relevant to public experts meet (are coherent with) those related to popular narratives?

Instead of seeing experts merely as an additional source of information, given what we know about how they are selected, publicized, and treated we might recommend that we understand the symbolic role of public experts. The tendency to rely on second-rate informants (and, keeping "brilliance" away from the public scrutiny) might be seen as a means of adjusting popular standards and expectations about intelligence. Given who people do and do not get to see and hear, it seems fair to suggest that, by and large, consumers are confronted with remarkably ordinary intelligence, at best. The very stylings that media executives find appealing—like being "accessible" or "undaunting"—are also characteristics that make people more like "us." Experts may be shown to be different from us, *in fiction*—but when they are brilliant, they become pompous and obtuse, not unlike Gregory House, the fictional epidemiologist of *House* fame. However, House is still exceptional. More typically, fictional experts may assume superiority, but we learn that they are harmless, sad poseurs, e.g., the Crane brothers. In real-life genres, typically those types are not presented. And, when intimidating experts are given TV shows, for instance, it isn't their brilliance that makes them unapproachable; rather it is plain aggressiveness or stridency that seems menacing, e.g., Drs. Phil or Laura with respect to psychology or Judge Judy or Nancy Grace in law. In this context, then, it seems far more predictable than paradoxical to find that experts are treated like everyone else on syndicated talk shows. In general, in this world where brilliance is concealed and unknown, and where lesser intellectual talents are consistently substituted *in their place*, consumers are consistently provided with standards that, in aggregate, are lower than what is technically available or possible. For most everyone, the only reality pertaining to experts and geniuses is that constructed by the media. Thus, the portrayal of experts contributes to what is equivalent to truth about "great minds," "great thinking," and "great ideas." What is this reality? It is a world where brilliance is concealed and unknown, and where *lesser* intellectual talents are consistently substituted *in their place*. Consumers are consistently provided with depressed intellectual standards. In short, the world as most people know it, then, is not nearly as smart as it could be.

There are many other ways in which intelligence and related phenomena are distanced or segregated from the public-at-large. We will now look at several other instances of this public "quarantine," and, later consider how this "under-representation" of intellectuality (irony intended) coheres with numerous other mechanism involved in the reproduction of intelligence.

The Ivory Tower and the Architecture of Intellectual Isolation

We have already discussed the *representation* of the academy and intellectuals, but here we examine how academic *institutions* and personnel are, in fact, positioned within the general culture. Obviously how an industry is socially treated cannot help but reveal its general cultural value. Interestingly, scholars, members of "think tanks," and others of that ilk are generally pictured as vocationing on a separate stage of existence . . . or in a separate structure known as the "ivory tower." At first blush—just the poetry of that epithet—may make academe seem an idyllic retreat, secluded in pristine clouds, and divorced from the pedestrian trivia of the "real world." However, the closer one looks at "ivory tower," the less heavenly its ammenities seem.

One aspect of academic life that is hardly known beyond the tower walls concerns the economics of academic writing. Scholarly publishing is typically isolated within the overall publishing industry; it is produced by separate divisions, if not separate corporations. Regular (trade) bookstores don't offer most academic books for sale. And, even the more prestigious newspapers are very selective in choosing scholarship for review. It is important to remember here that, for those teaching in research universities and good colleges, publication is no minor concern; it is requisite for professional survival. Yet, scholars are not treated like other writers. Generally, they don't, for example, have literary agents brokering their work. Their percentage of the "take" is not remotely equivalent to advances or royalites given to other writers. Indeed, there aren't other, employed, professional writers who, ultimately, recoup so *little* from their written work. Academic *article*-writing—the mainstay of scholarly publication—largely goes *wholly unpaid*, and often, the writer doesn't even retain copyright. Interestingly, these are facts that seem mainly confined to academics and those that sell their work. Moreover while publishing powerfully affects professional mobility, only *teaching* is typically contracted to be paid at all; thus, other than small honoraria, the prolific scholar is usually not reimbursed more than the one who rarely, if ever, publishes.[8] And, as any publishing readers know, serious scholarly writing, alone, can consume a full work week in and of itself.

Publication does not constitute the academy's only non-mainstreamed socioeconomic pattern. While academic salaries typically fall within good, middle-income, this assessment does not include standard accounting adjustments. Most notably, academic instruction is one of the, if not *the lowest* paid profession in relation to entry-level requirements. Additionally, there are very limited locations, and *only one* season per year, at or during which instructors are marketable. Similar to auditioning for glamorous occupations, the vetting process for these limited positions is typically lengthy and very competitive. Once obtained, an *instructional* position typically includes but two possibilities for advancement during the entire course of employment—and even such advancements are mainly titular involving minimal salary adjustment. Of course, this is not considering the awarding of tenure—which was not conceived as a reward, but as a means guaranteeing academic freedom. What may be ironic in this context is that, while the mainstream workforce is moving toward regulation protecting workers from non-meritocratic termination, the academy is moving toward increasing non-tenured positions (if not the elimination of tenure entirely). This generally peculiar institutionalization of a workforce—particularly one having very demanding job requirements for relatively low reward—is often characterized as resulting in "genteel poverty." Genteel or not, this *relative* undervaluation is symbolically powerful in capitalist society. From a critical-materialism perspective, it would be *the most compelling* sign of our culture's anti-intellectualism.

This relatively weak economic valuation of academics and their work product very much parallels other economic "attitudes" toward education. Recent public debates over "school-tax vouchers" reflect similar assessments. In particular, that citizens feel entitled to tax rebates for *privately* educating their children similarly devalues the academy in the most essential of ways. The pro-voucher logic—that taxation is exclusively for defraying one's personal use of socialized ventures—may be perverse in its own right; however, it connects to a larger issue. If, in our culture, there really were a broad shift in political philosophy whereby citizens no longer embrace at all the notion of a "common good," this vouchering argument should be heard not only from *other citizen* groups, but also, with respect to other public *systems*. What should constituents with *no* children receive, for instance? Why should they pay any school taxes? Or, should non-traveling citizens *without* automobiles have rebated a portion of their taxes normally directed to highway-maintenance? We certainly are not endorsing such movements. Rather, what is peculiarly alarming about *not* witnessing a shift toward campaigns against *all* government-socialization is that, as specifically and unrelentingly aimed at school vouchers, we observe a movement that does not represent a generic principle, but rather, a clear and active bias toward and disregard of education, specifically. In the context of the many, many other mechanisms we reveal, we believe this functions as another anti-intellectual indicator.

Science and Research in Public

True or false, there is probably no general subject *imagined* more demanding of education, erudition, and analytic ability than science. Science is commonly used as part of a metaphor to characterize intelligence, e.g., "It isn't rocket science" (or brain surgery). In this section we consider science and its research in three contexts. First, we look at how journalism is designed "for" science. Second, we explore how scientific- or scholarly-research *controversy* is presented in the news. Third, we examine popular science *fiction*. Thus, we examine the treatment of intellectuality *through* science, by analyzing what are probably the three most popular means by which science is mainstreamed.

The Institution of Journalism and the Representation of Science

Even though social science, and particularly media studies, has not much considered the general representation of "intelligence," there is substantial scholarship about *news* coverage of science and research. Two topics from that literature are presently informative: (1) journalism's *contextualization* of science and/or research among other reportable topics, and (2) journalism's *treatment* of science and research *in* reports. Journalism's contribution here—and in the next section on rhetorical modeling—is especially poignant because of the special, "noble" status assigned to news (among media genres); of all popular media formats, news is the one imagined to best represent and effect intelligence. Thus, here, we are looking at a double-feature: what the "smartest" media form has to say *about* the smartest human endeavors.

Contextualization of Science and Research: Studies repeatedly conclude that, among news topics, science receives comparatively little attention. This condition may, initially, be seen as rooted in journalism's institutionalization. A major, metropolitan daily will typically have many specialized editors and reporters. Whole divisions of news are based on geopolitics—local, regional, Washington, and foreign desks. The Washington desk, alone—in addition to its own editor(s)—will likely include specialized reporters with White House, Senate, Supreme Court, Pentagon, and a host of other "beats." The reporter covering the Supreme Court probably is familiar with appellate processes and is on a first-name basis with several law clerks and administrators. On the city desk, there will be journalists familiar with certain police detectives, members of the town council and/or people in the mayor's office, and so on. Additionally, there are specialized editors and reporters for sports, fashion, food, "society," house and garden, and other newspaper sections, not to mention, economics/market analysts, plus numerous specialty columnists. Specialization in these contexts makes good sense; not *every*one—even with a masters-of-journalism degree—would know, for example, how to describe trends from the runways of Milan or Paris. Similarly, football cannot be ably reported by someone without reasonable knowledge of the game. While all this specialized expertise makes great sense, it is curious that many prestige papers are lacking general-science editors (distinct from recent technology specialists), let alone specially-trained *science reporters*. Among broadcast staff, often the closest one gets to "scientific" personnel is an on-air reporter who delivers *medical* news or weather. On some occasions, these staffers are physicians and meteorologists, respectively; but even they are not available in many markets.

There is little question that, depending on the news area, reportorial skill requires some preparation beyond general newswriting. Some background knowledge appears easier to acquire than other types. For instance, sports enthusiasts probably don't need to take *courses* on football or basketball to know about "downs," "fouls," or penalties, for instance. Probably, a deep interest in sports pre-dates the career choice of sportswriters. Consumers tacitly expect their sports news delivered by someone who is doing more than reading or writing scores, but if they have the same expectations with respect to *science* news, research suggests that their expectations will not be met. Even the minimal scientific background needed to make reasonable inquiries about newsbreaking research is unlikely to be casually acquired through childhood—like sports knowledge might be. Moreover, a survey of journalism departments shows that courses in science reporting are not nearly as common as other specialty courses offered; nor are alternative preparations typically recommended, e.g., a course in research methods. No doubt there is a greater market for news of sports and politics than of science; still the systemic lack of science-journalism preparation probably accounts for why science news has commonly been characterized as little more than iteration of press-conferences or press-releases. Subsequently, it seems reasonable to infer that news consumers, comparatively, are less critically informed about science than most other news topics. This pattern is altered somewhat when specifically considering reports of science controversy, but we will examine that coverage shortly. In the meantime, we would summarize the situation with two interrelated observations. First, stories about science and academic research are those news subjects with the closest relationship to traditional intellectuality and erudition. Second, although preparation for science news seems routinely inferior, this condition is not simply a glitch in the overall scheme of journalism training, but rather, part of a coherent cultural pattern involving the general degradation and/or avoidance of intellectuality and erudition.

Imagining Science: Scholarship has also repeatedly indicated that science is not favorably characterized *in* the media. News stories are said consistently to dramatize the *highly equivocal* nature of the scientific enterprise, i.e., science as being undependable for definitive answers. Scholars who practice science may be very familiar with those caveats required for proper form—which means maintaining the openness and, thus, provisionality of science. Among scientists, this "pose" is understood to guard against the hubris of treating science as religion. Thus, scientific reports and, indeed, its measurements are calibrated so as never to indicate absolute certainty, e.g., statistical significance is always articulated as *approaching* zero. However, normal consumers, understandably, are not familiar with such arcane practices—even as they impinge on media stories. Consider the following two examples. Many TV talk-shows have regularly-featured segments in which a child's paternity is dramatically uncovered before a studio audience. Often the putative father, as well as members of the audience, remains unfazed *after* genetic testing indicates the father's paternity to a certainty. Their nonchalance isn't cold-heartedness—it represents their readings of the results. For instance, when the test proves paternity with 99.4 percent certainty, someone, on camera, might comment that "*even the science*" shows there's still a little doubt. Percentage *is* something people understand; however, that an identification "less than" 100 percent is *not,* technically, *less indicative* of paternity isn't commonly taught—either specifically or among the general principles of probability. Millions of viewers, along with the father, might have turned off their sets still believing that something as critical as paternity is a matter that science cannot establish with certainty. (Although recent news related to model Anna Nicole Smith may have ameliorated that misimpression.) A similar issue arose, for instance, during the still infamous O.J. Simpson double-homicide trial. Scientific testimony identified Simpson's blood with an assigned probability of something like 57 million-to-one—as astronomical certainty insofar as it excludes any earthling *but* Simpson. Following the "not guilty" verdict, jurors were queried as to how they made sense of this damning blood evidence. Among other things, members of the jury pool, as well as numerous persons-on-the-street pointed out that there is still *the chance* of "that one," so it's not "beyond a doubt." While the enormous coverage attended on this trial even included demonstrations of how genetic alleles are matched, this one little, but profound probability lesson was not the focus of attention. Jurors and viewers believed that there was a small chance that it wasn't Simpson's blood. Moreover, we don't know whether or not the prosecutors and journalists, whose responsibility it would have been to clarify this finding, really understand the "language" of probability themselves. In any case, given these threshold handicaps, it's not hard to understand why science may wind up seeming suspicious and equivocal in the press.

Statistics-bound information is a constant in public life, and the general ignorance of just statistical idiom may be blamed, in part, on public education; however, the institution of journalism, also and arguably, contributes, specifically, to the muddling of science, and by extension to general anti-intellectualism. This possibility returns us to matters of professional preparation and expectation and to two ways in which science is differentially constructed for public exposure. First, as noted, research suggests that, in comparison to other areas of reportage, journalists are not commensurately prepared for science-writing. Thus, it makes sense that the public's mystification with science is not ameliorated by information reported in the press. Second, science does not rank high on the news agenda; the relatively little attention directed to science goes to those stories with the potential for creating mundane controversy and division. This limitation speaks *specifically* to science even though a general bias toward "hot" topics applies unilaterally—to any subject. In other words, despite a generic bias toward controversy, there are many subjects that receive routine news coverage, e.g., government, judiciary, sports, fashion, economics, etc. That science and research are *not* on this list of journalistic staples, then, means that the public is *especially* likely to be confronted with science stories *only* as they are controversial. This regular contextualization-in-controversy also, obviously, contributes to characterizing science as equivocal. However, in addition to mystifying and ambiguating science, this controversy-linkage also allows for anti-intellectualism to be spread by news in a wholly different and subtle way.

Rhetorical Modeling in Media Stories

It is fair to say that educators have been, for some time now, preoccupied with the pedagogy of "critical thinking." While that term may be variously defined, one constant observation culled from related educational criticism is that—whatever skills are embodied in critical thinking—they have become increasingly absent among American college students. In an attempt to integrate, prioritize, and generalize this educational criticism, we use "critical thinking," here and throughout the chapter, to reference a broad skill (or collection of skills) that provides the facility to observe, interpret, and reason analytically and coherently. Pedagogically speaking, training *in* critical thinking is analytically separable from instruction supplying finite data for subsequent iteration; critical training provides the intellectual resources to negotiate conceptual problems. The extant educational criticism laments that students seem less and less capable of learning lessons beyond overt content—of lacking the capacity to go beyond what Bertrand Russell called "simple learning." In this sense, then, Peggy Sue and Russell may be integrated: while it is true that many *will* find little earthly use for algebra, this doesn't mean that the skills learned in applying a quadratic equation are not also applicable to more mundane endeavors. Indeed, one of the lessons emerging from recent advances in neuroscience is that learning *one* kind of lesson can neurophysiologically facilitate *other* types of learning experiences. In the case of intellectual pursuits, then, the critical by-products are not *the facts*, but *how* to handle facts, in general. For the present, then, we are concerned with how critical thinking importantly, relates to media coverage of science and research.

Critical thinking is relevant to science in the news especially because this coverage, as indicated, preponderantly addresses controversy. In turn, stories-of-science-controversy are rife with opportunities to exemplify analytic reasoning insofar as they may report conflicting perspectives, which, in turn, invariably involves systematic, inductive reasoning (implicit in the research methods.) News about scientific controversy, then, can provide a relatively rare window through which to observe specifically the relationship between *what* is known and *how* it is known. Through social learning or modeling, journalism may, then, teach critical, analytic technique. Especially because most members of our society consider prestige news, along with certain networks (PBS, Discovery, etc.), comparatively, to have depth and gravitas, the intellectual conduct exhibited *in* these venues might be especially compelling. The question—as it relates to the construction of intelligence—is *how is* critical thinking rhetorically-modeled in these news reports? While our generalizations are based on decades of science coverage, we, specifically, discuss here three, relatively high-profile cases to illustrate the generalizable patterns of this rhetorical modeling.

The first example is interesting because it involves scholarly research imported *after*-the-fact. Specifically, we consider the novel-turned-theatrical film, *The Da Vinci Code*. Reports about *The Da Vinci Code* typically opened by describing the controversy triggered by the book and/or film—a controversy with the Christian (and, more often than not, Roman Catholic) church, i.e., that the fiction characterizes Jesus' life differently from how it has been taught by the churches through the centuries—differences particularly addressing marital history. Reports extending beyond "news-in-brief" typically reported the fictional telling which would be followed by the central contention—typically from a priest (or representative of the church) who often presents a faith-based argument, such as: "It would be *self-evident* to anyone who has developed a personal relationship with Christ that the history as represented in *The Da Vinci Code* is impossible." Once the controversy is thus established, a scholarly and decidedly secular opinion would commonly be introduced into the fray. For instance, a professor or Ph.D.-author might offer an observation concerning how the New Testament is as an anthology, edited down from dozens, if not hundreds of early gospel accounts. From this, the scholar might point out only that the source of the alleged DaVinci code was first chronicled in certain of this excluded testimony. In other cases, archivists might point out certain physical evidence from a secret society that sustained the unorthodox history in obscure journals that have been maintained in some aged, provincial church. Ultimately, attention turns from the secular historians, archivists or archaeologists, to the contesting religious positions, with occasional insertions of scenes from the film or passages from the book and/or person-on-the-street opinion—often those of filmgoers. Depending on the length of the journalistic treatment, this alternating cycle of positions might be repeated—sometimes arranged according to each major point of contention. One common pattern—at least among secular news institutions—is the relatively equal division of space/time to each "side" in the controversy. Interestingly, what also co-occurred in these matching/alternating presentations was the *absence* of discussion (much less value judgment) about the relative merit of how any position was derived. In other words, although the arguments differentially relied on empirical evidence for substantiation, this aspect of the controversy was almost never addressed. This absence would typically persist even if one of the cited parties (typically a scholar) overtly introduced methods as a means of resolving contention. Thus, it would not be a radical inference to suggest that this methodological abstention is by journalistic design. In other words, there appears to be a rule demanding that stories adhere narrowly to the very overt content of debates. From the perspective of rhetorical modeling, then, we might further infer two, interrelated points. First, various epistemological orientations, to wit, faith, systematic empiricism, anecdotal empiricism, and docu-fiction, are equally legitimated in mainstream journalism—at least as determined by allotted attention. Second, not unlike the laments of educators concerning the failure to instruct *critically*, journalism, too, does not model an analytic approach that extends beyond the overt content of the reported issues.

As our initial use of *The Da Vinci Code* exemplifies, how methodological considerations are included in the presentation of public controversy is obviously important in considering how critical thinking is modeled for the popular audience. However, in considering all reportage of science/research controversies, there are additional rhetorical practices that might be said passively to denigrate traditional evidentiary methods of substantiation. In fact, some of these practices might not just be seen as denigrating traditional evidence, but instead and/or in addition, as cultivating *undesirable* rhetorical form. We examine these remaining practices with respect to two more reported controversies: (1) *The Bell Curve*, and (2) "holocaust revisionism." Unlike *The Da Vinci Code*, these stories originally involve *scholarship* that was, atypically, brought to public attention. Again, while the scholarship may be extraordinarily publicized, the kind of journalistic treatment afforded this research is, for all intents and purposes, ordinary.

Both *The Bell Curve* and holocaust revisionism are two bodies of late-twentieth century scholarship that received, comparatively, much media coverage. *The Bell Curve* demographically analyzed American students' standardized intelligence scores and postulated that the racial/ethnic patterns obtained were tied to genetic rather than sociopolitical conditions. Before publication of *The Bell Curve*, a French history professor published scholarly articles challenging the gas-chamber exterminations in Nazi concentration camps—a revision of history most recently given even wider publicity by actor Mel Gibson and his family. The French scholarship questioned whether or not European Jews *had* been systematically slaughtered by Hitler's forces. A few other, European articles from this period either *denied* or significantly *minimized* the count of Jewish dead and/or the precise number of atrocities perpetrated by the Nazis. We refer to this scholarship, *en masse*, as "holocaust revisionism" so as to incorporate all work that questions the genocidal history of the Third Reich. Of course, neither of these bodies of research became news stories as a matter of routine coverage; in both cases, the coverage hinged on research characterized as virulently racist. *The Bell Curve* was criticized as scientific rationalization of racial inequality and the revisionists were sometimes decried as neo-Nazis attempting to rehabilitate the Third Reich. Beyond such epithets, *The Bell Curve* authors, Herrnstein and Murray, were often alleged to have been recipients of earlier support from politically-conservative foundations so as to validate their racist intentions. Similarly, holocaust revisionist, Faurrisson, was often depicted as having a family history of alleged Nazi sympathizing and collaboration. What was much rarer to encounter in any popular report was any attention paid to contradictory data or methodological problems—arguments that could be found in more obscure, scholarly reactions. For instance, there were suggestions that problems with Herrnstein and Murray's regression analysis provided inaccurate IQ patterns. Correspondingly, other historians had observed that the Germans had always been meticulous bookkeepers and that *records* from the camps were there to confirm the genocide. This last point doesn't so much offer contradictory evidence, but merely suggests that the whole question is falsifiable. Other indirect contradictions were offered—such as the recorded testimony from war-crimes trials in which Nazi *defendants*, themselves, repeatedly acknowledged (and even provided) accounts that were used to construct the original, *un*revised history. Despite these methodological annotations, the popular-news coverage of *The Bell Curve* and holocaust revisionism concentrated, instead, on the more obviously contentious *characterizations* of the authors and their putative racism.[9]

In considering all the patterns relating to these controversies—*The Bell Curve*, holocaust revisionism, and *The Da Vinci Code*—we can designate four important patterns of rhetorical modeling: First, as noted, scholarly research typically only *becomes* news when it can be made controversial; this conditional presentation contributes to the portrayal of science and scholarship as invariably ambiguous and contentious. Second, when information about the controversy *is* reported, attention is virtually never paid to the sources' sources, as it were; in other words, evidence-based contentions are typically treated identically to those arguments not derived from conventional evidence. And here, we should consider that, with *breaking*-news, for instance, reporters are not only trained to secure demonstrable evidence before reporting information, but to confirm any new ideas or "leads" suggested to them. Science/research controversies can be looked at in this context—as possibilities that the reporter is free to explore for his story. That the procedures don't, nonetheless, apply or transfer is, arguably, telling. Third, the rhetoric of covering scholarly-research controversy, more typically than not, concentrates on *ad hominem* characterizations. Thus, not only are evidentiary issues generally evaded, but the reports concentrate on elements that really stand apart from the fundamental evidence. However, by virtue of this "side-tracked" concentration, the rhetorical modeling arguably conflates *ad hominem* reactions with fundamental elements of controversy. As a result, we should not be surprised if contests of ideas in everyday life are, increasingly, met with *ad hominem* reactions. Fourth, another means by which popular reports of research controversies avert critical issues is by quick-focusing attention on a highly-charged social issue, such as racism; i.e., topics that are so culturally sensitive, they overshadow or overtake other possible issues. Perhaps "hot-issue detouring" may be seen as a variant of *ad hominem* rhetoric; however, the hot issue can be reported without or, in addition to demonizing particular individuals. Together with *ad hominem* rhetoric, pushing the topical hot-buttons demonstrates how *to elude* the substantive issues. These four patterns of rhetorical modeling suggest that those news stories with the *greatest* propensity to teach or increase intellectually-critical sensitivity are, in fact, analytically *unaggressive* and *unchallenging*—thereby not providing logical or intellectually-sound models of inquiry. While consumers *may* learn isolated facts from science news, they are unlikely to be offered any of the ancillary intellectual resources that travel with these facts.

Science Fiction: Frankenstein's the Doctor

As noted from the start, in our culture, science is that subject and/or profession that is most intimately with intelligence. Nevertheless, journalism, as analyzed above, tends to disseminate relatively little science, and even that little is, arguably, not presented with much intellectual strength. *However,* one place where we *do* find science commonly presented—and even analyzed—is popular *science fiction*. Most every science fiction narrative addresses the value of both science and intelligence. Given Hollywood's seeming love of model-building and special effects, it might seem that its science fiction would demonstrate a profound love for and interest in science and technology. However, systematic examination has repeatedly shown that *not* to be the case. Rather than offering strings of isolated examples, in this instance, let's consider an entire narrative, instead.

> Ancient, peaceful, and physically small botanists descend on earth to scout new vegetation for their famine-plagued planet. Trying to be unobtrusive, they land in the dead of a clear, warm night in a lush, dark, uninhabited park. Upon de-saucering, they scatter, each equipped with specimen bags to collect flora for transplantation to their atmosphere. Their pilot remains aboard as lookout. While the silent visitors neatly dig for specimens, the pilot sounds the alarm, upon discovering their detection by predatory humans. In eager pursuit of the spaceship, the humans are traveling in a monster truck especially equipped with a fiercely lit cab as well as alien-detection devices. Hearing the alarm, the fearful botanists scurry back to their ship. The one furthest afield is amid extracting an interesting green and decides to take the few extra seconds to bag it. Unfortunately, after completing the task, his little legs don't return him quickly enough; the ship must jettison so as to flee the pursuing hunters, and abandons (at least temporarily) the old botanist on earth.
>
> Stranded and hiding in a residential alley, the creature is discovered by a young boy with whom he bonds and by whom he is protectively hidden in a bedroom closet. Over the next couple of days, the alien, the boy and the boy's two, young siblings forge a friendship, but the botanist increasingly requires his own atmosphere. Unfortunately, as the botanist's health deteriorates, the hunters' devices alert them to the alien's precise whereabouts. Back, once again, in the monster truck, they drive to the once safe-house and immediately proceed to find the alien and take him prisoner. We recognize, as they force their way through the house that they are not merely hunters, but government-sponsored hunters cloaked in *science* and the presumption that their intellectual curiosity about extra-terrestrial life should take priority over the emotional bonds forged between random children and the alien. However, as the botanist's health rapidly declines, the head scientist, moved and *humanized* by the children's love for the creature, metaphorically sheds his white lab coat and, against logic and scientific curiosity, allows the little alien to rendezvous with a spacecraft sent to transport him safely home.

The facts of this wildly-popular story may be emphasized here a little differently, but any changes were only to focus more clearly on the narrative as it, specifically relates to science and technology. Those very familiar with the story may claim that critical facts were not sufficiently emphasized—most notably that E.T. was dying, and that detaining him on earth would have cost his life. At least, this is the challenge most represented whenever this narrative is orally dramatized for an audience. Peculiarly, the hardest point to make to these challengers is that E.T.'s health *is not* an "outside" variable controlled by factors (or even a scientific knowledge or logic) outside the author's domain. It was emphasized, relentlessly, that, in fact, a common narrative strategy in scifi stories is the creation of *just* such an intervening variable (like imminent death) around which dramatic conflict is built. *That conflict* allows science or scientific curiosity to assume the role of antagonist. The coincidence of E.T.'s critical deterioration with the scientists' arrival cannot be regarded as some incontrovertible given; indeed, many other possible scenarios might have been written. For instance, since they *are* scientists, perhaps they might have tried to employ their professional skills—together with ET's knowledgeable assistance—to experiment with chemicals, atmospheric changes, or any number of possibilities at their disposal that might improve (or, at least, extend) E.T.'s health. Instead, the story allows for a science sufficiently sophisticated to develop technology that can track non-earthly protoplasm, but *not* one that promotes collaboration with a peaceful, smarter and communicative being to effect the visitor's survival. Instead of having the scientists engage in pretty much *any* constructive activity, the script confines the them to a path of a single-minded, unsentimental, search-and-(virtual)-destroy mission, the objective of which is never clarified.

One cannot expect regular viewers ever to consider any of these issues independently. Most Americans are familiar with a very long-standing image of a cold and/or dehumanizing science. The traditional academic distinction between sciences and "humanities" is one that tacitly confronts all college students without explanation—even though the origin of this division was not to distinguish science from our civilized and benevolent interests. This science metaphor is paralleled in many contexts, maybe none so poignantly as that which addressed intelligence directly, i.e., the common distinction between "thinking" and "feeling." Despite the fact that neurophysiological evidence doesn't at all support a Cartesian model of mind versus emotion, people still, into the twenty-first century, understand the humane and compassionate realm of emotion to be not only separate from, but opposite of the cold and calculating intellectual domain. Traditional psychiatrists—who are rationalists—will instruct patients to "Tell me what you feel, not what you think." Fusing this issue of segregation with popular aphorism, we talk about "going with our hearts" (as opposed to our brain); and, our hearts, we know, err on the side of kindness and good intentions. Science fiction stories, are probably one of the most common forms in which this general equation is repeatedly taught—especially in connection to *de*humanization through science. Perhaps, because of the suspension-of-disbelief cultivated in science-fiction, people are led to overlook—if not, actually embrace—(otherwise) *bad* behavior, if, in the process, that behavior also opposes scientific dehumanization. Consider a different use of another, briefer, narrative summary—again, revised only for emphasis, *not* content:

> In a future society, a pretty, balletic young man is bursting with *joie de vive* that is, unfortunately, expressing itself through decidedly anti-social tendencies. By

> day he diligently toils in a factory to maintain his flat, beloved, exotic pet, and his music collection; however, by night, he becomes a savage, heinous murdering rapist/thief who takes sadistic pleasure in his violent forages. Ultimately, he is caught by authorities who perform what is, then, routine therapy—behavior modification with electroshock and chemical controls. With the help of this relatively inexpensive, easy and quick treatment, and continued medication, the young man is able to resume his old job and his old apartment in a short time. Also, miraculously, he is never able to resume his moonlighting as a vicious thug because the therapy renders him nauseous-to-the-extreme at even the imagined *hint* of violence; its actual perpetration would be wholly crippling. Side effects, there are a few: he occasionally experiences fatigue and less-than-perfect coordination—as manifested while eating pasta and accidentally spilling red sauce down his shirt. Also, his passion for his boa constrictor and his Beethoven is somewhat diminished. Why, then, do many readers/viewers not see this inexpensive, relatively peaceful, and productive alternative to capital punishment or life imprisonment as a wonderful resolution for both society-at-large and the murdering-rapist thug? Why is there lingering sentiment for his former exuberance?

In fact, the moral tone of *A Clockwork Orange* is thoroughly explicable. People have sentiment for this young man for the very same reason that we have chosen, for decades, *not* to utilize these same rehabilitative techniques despite the fact that they're not futuristic and, in fact, are relatively old. While popular science fiction seems generally to have a patina of enthusiastic futurism, its stories are always subject to contemporary motivation. And, despite metallic clothes, that motivation involves two interrelated myths. As portrayed in stories from *Brave New World* through *E.T.* and beyond, (1) science has the simple capacity to be evil and dehumanizing. Even when it isn't intentionally put to nefarious ends, it has an inherent dangerousness that, typically, overrides its risk. This dangerousness is almost unilaterally manifested in (2) the hidden perils of technology. Story after story warns that technology may *look* intriguing, but is, nonetheless, extremely dangerous and, most importantly, *uncontainable*—no matter how good the intentions of its inventors. Dr. Frankenstein learned this, and so did the astronauts, who once loved HAL. Thus, this general fearfulness and distrust of science, scientists and technology, obviously coheres both with the various other contexts featuring negative images of science and, more generally, other anti-intellectual variants.

Signs of Intelligent Life: The Conquest of Intelligence

As noted, we use "traditional intellectuality and erudition" as a default definition for intelligence throughout this chapter. This default usage serves a methodological and theoretical purpose, to wit, it is easier to work from one standard so that constant qualification isn't requisite, and, most importantly, as should be progressively clearer, we are concerned with values and worldview as they specifically attach to traditional intellectuality and classical instruction. In this section, we examine major, alternate referents attached to "intelligence" as well as two related matters. First, we consider those common indices marking the *absence or obstruction* of intellectual conduct. Second, we consider the recent history of the alternate signs of "intelligence" with respect to ambiguation"—a term borrowed from mathematics and linguistics that may pretty straightforwardly refer to either the active neglect of clarity or, conversely, the development of opacity and ambivalence. We try to show that, especially since the second half of the twentieth century, the notion of "intelligence" has been progressively ambiguated, and that progression appears to be socially strategic. "Divide and conquer" is an age-old strategy believed to establish and maintain power. Whenever an already-established and relatively *un*ambiguous term is borrowed and coupled with a revised referent, it certainly isn't because the human brain is too unsophisticated to have incorporated, instead, a whole new word for the newer meaning. Indeed, working with multiple qualifications of the same sign may be *more* taxing than simply expanding one's vocabulary. Ultimately, we examine the recent history of intelligence's multiple referentiality in terms of the function of its conceptual division and ideological co-option.

The Seeming Paradox of Anti-Intellectualism

Judy Holiday's classic dumb blonde in *Born Yesterday* is highly concerned with being judged as not smart. So too, is the silent-movie diva, Lina Lamont, in *Singing in the Rain* when she indignantly shrills: "Whatta they think I yam? *Dumb* or somethin'?" More recently, TV's Judge Judy asks suspect litigants if she has "stupid" written across her forehead. Just as there are common anti-intellectual aphorisms (like "ignorance is bliss"), there are aphoristic assertions of intellectual *competence*. Although old-fashioned, a person feeling underestimated may protest that she or he "didn't *just* fall off the turnip truck!" Over a two-year period, 290 available undergraduates rank-ordered a list of terms (including "good-looking," "ugly," "sexy," "unsexy," "smart," "dumb," "honest," "dishonest," "nice," "unpleasant," "bitch," and "bastard") according to how they'd *most and least* prefer to be characterized both inter- and intrapersonally. Despite variations of gender as well as inner- vs. other-orientations, being called "dumb" was reportedly as the generically *least*-preferred. We would hypothesize that intelligence is probably that attribute about which people get most defensive. Comparing Internet lists of aphorisms about social "assets" or personal characteristics, most "defensive" quotes appear for intelligence. It generally seems that one of the last things anyone wishes to project or believe is that she or he is dumb or stupid. This may seem too obvious to document; however, in this chapter it's critical with respect to contradicting our central argument: How *can* it be argued that we're an *anti*-intellectual society when people seriously *care* whether or not others believe they've recently ridden with turnips? Throughout this section, we shall cumulatively attempt to demonstrate that that this "paradox" is illusory; People can value being smart while being anti-intellectual because of intelligence's multiple referentiality.

Intelligence's Third-Person Effect

> Eighty-eight percent of the population agrees with the following statement: "I am smarter than the average person."
>
> —*Politically Incorrect* Audience Poll

> Three percent of the population believes that it ranks less-than-average in intelligence.
>
> —The East-West Institute for Self Understanding

Although there's a dearth of empirical data on the social construction of intelligence, all available resources indicate that the above quotes reasonably represent popular sentiment; a widespread "third-person effect" operates in our culture with respect to intelligence. Most people don't report feeling intellectually inferior; in fact, most everyone reports imagining him or herself smarter than an inferior aggregate. Is this just false bravado masking intellectual insecurity? Except for the Clifford Clavens or Frasier Cranes, most people readily acknowledge areas absent *erudition*, but most records indicate that, when it comes to basic intellectual resources, people say they're at least as good as the next guy. The conceptual framing of intelligence is key to whether this common belief is accurate or delusional. As we shall see, the variants of intelligence permit *both* outcomes.

The First Boundary: Aptitude and its Reification

Fundamentally, an individual's intelligence is thought to be bounded by something commonly called "aptitude." Standardized instruments—such as the IQ or SAT measure aptitude in, arguably, the only context it can be measured: predictively. These tests are constructed to indicate how the test-taker will succeed in another or other contexts. Aptitude is often considered synonymous with competence, and both terms are typically *contrasted* with performance, however, *operationally*, both are performance. Thus far, there is no other controlled means of measuring aptitude but tests; All tests measure performance. While this may be logical, this logic doesn't seem to influence, substantially, how people talk and think about abilities, competence, and so forth. In fact, although we could find no systematic data to support this hypothesis, it *seems* that people, generally and often, talk about their abilities *as distinct from* performance. In everyday parlance, aptitude seems to be a reification of some metaphysical state of preparedness. One probably doesn't have to strain to recall countless, mundane examples of how people talk about aptitude or competence—theirs or yours—as something that involves personal agreement more than "objective" appraisal. We will later provide more concrete illustrations of such presumptions, but first, we consider the function of aptitude's reification. That people either ignore or reject aptitude as a concrete performance, that they commonly sever aptitude's empirical bonds, serves as the basic or threshold stage in the "ambiguation" of intelligence. It begins the process of obscuring intelligence's defining boundaries. More concretely, the ambiguation of intelligence blurs distinctions between what's good, high, or strong versus what's weak, low, or inferior.

The reification of aptitude particularly commenced during second half of the twentieth century—amidst the growth of the "counter-cultural do-your-own-thing" ethos (as well as eduinflationary changes discussed later). Most simply, the *de*tachment of aptitude from performance became a means to validate selected individuals and also to rationalize substandard performance. This conceptual revision contended, in part, that a person may, indeed, possess intellectual ability *despite* bad intellectual performances. And, by severing aptitude from performance, opinion is the method of assessment—as demonstrated by those assessments characterized as the third-person effect. Originally, the concept of "underachiever" indicated a specific disparity between inferior academic performance, on the one hand, and (relatively) superior aptitude-test performance, on the other. However, with the reification of aptitude, the "underachiever" epithet started to be attributed to and/or adopted by those whose intellectual performances fell short of some *un*specified, but, *nonetheless*, anticipated or desired level. Increasingly, students (and their parents) have reportedly characterized bad academic performance (low grades), *alone*, as "underachieving." Ironically, perhaps, "overachiever" has correspondingly transformed into a compliment—in tribute to someone who excels as a result of fierce determination and hard work. Of course, overachieving, was traditionally bound to the same comparative criteria as underachieving, and, as such, would hardly serve as an accolade; it describes the process of living beyond one's intellectual means. Tradition aside, "overachievement" has, apparently, joined the ranks of "pro-active"—words with (morphemically) redundant prefixes that function only as reinforcement or emphasis, like an exclamation point. Additional characterizations—sometimes focusing on an intervening variable—might similarly be offered without demonstrable evidence of competence-based performance, e.g., rationalizing low exam scores as a product of "bad test-taking."

Before aptitude's reification, students' inferior academic performances were also of concern and, typically, addressed by a different repertoire of explanations, i.e., including failure to study, inefficient work habits, and/or relative *intellectual weakness*. More recently, though, these explanations—especially intellectual weakness or *in*sufficiency—are increasingly uncommon. In fact, editorials on the subject have lamented that the idea or option of intellectual limitations, that are not simultaneously classified as pathological, have become ideas-*non-gratis*—inexpressible ideas in polite company. Some have even included this turn-of-events as one contribution to grade inflation; i.e., the increasing inappropriateness of characterizing someone as having even a poor, intellectual performance. Rather, with reified aptitude, it may be increasingly incumbent upon instructors to see students as vessels of intellectual *potential* awaiting stimulation and realization. This orientation may also have been influential in the administrative revision and growth of "learning disabilities."

This admittedly quick gloss is not at all intended to dismiss, *categorically*, discrepancies between measuring competence and performance; however, it would seem extremely important to distinguish those cases concretely including performance indices from those that don't. Moreover, concern has been voiced about this problem going one step further: in addition to *divorcing* aptitude from concrete performance, are people actually *defying* performance evidence in claiming aptitude? In the recent popular wave of reality-based TV competitions, examples of such defiance seem to be abundant—particularly as dramatized in what seems to be ritualized elimination segments. Some competitions, like *American Idol* or *Project Runway* dedicate whole episodes to showcasing auditions, and particularly those of untalented applicants who are rejected by the judges. What is of interest here is not the rejection, but how rejection is met by the loser. On *American Idol*, rejection after rejection is followed by the losing contestant's emotional soliloquy in which defeat is putatively *eschewed* because, according to the defeated auditioner, he or she is nonetheless in possession of unquestionable talent. While many will dismiss such responses as false bravado or defensiveness, an alternative theory may be postulated. People born after 1960 have been indoctrinated into a different philosophy of personal aptitude. This is *not* meant to suggest that only this generation was ideologically indoctrinated (which, of course, is ludicrous), but rather that post-1960s philosophy was somewhat altered on this matter of "judging" people. The unswerving dogma of post-1960s philosophy seems to affirm powerfully that self-assessment is the most accurate and important index of potential, and that self-esteem is the most fundamental dynamic of human happiness. If nothing else, this proposition has dominated the didacticism of popular-entertainment stories. A network, like *Lifetime*, has come virtually to specialize in dramatizing this tenet. Acceptance speeches at televised-award ceremonies are replete with sermons on this principle. Celebrity biographies, too, echo these beliefs when offering prescriptions for success. Popular lyrics in every genre sing or rap this message. Indeed, mainstream illustrations of this code are seemingly endless.

Another way to consider this philosophy is through what is, perhaps, a quirky, but complementary observation. Specifically, although media activists complain about the over-, under-, or mis-representation of all sorts of categories and conditions, one unequivocally under-represented issue that, nonetheless, virtually *never* provokes such protest is the under-representation of inferiority and failure.[10] Hypothetically, if we *were* to see inferiority and failure in proportion to their everyday occurrence, audience backlash would probably be monstrous; media producers would be assailed as reactionaries attempting to suppress people's (or *children*'s developing) aspirations. As our research has repeatedly detailed, the cultivation of unlikely, perhaps Pollyann-ish, dreams—fostering hope *despite* overwhelming evidence to the contrary—is a critical element of culture-maintenance. Belief in (unlikely) success—hope—can quell immediate hostility against the established order. Although our references here could refer to any type of inferiority or failure, the specifically *intellectual* dimension is no less a part of this general cultivation of probability-defying "optimism." That most people are deluded about their relative intelligence—as demonstrated by the third-person effect—is congruent with the larger systemic restraints to limit the spread of intellectuality and erudition in mainstream society. Specifically, delusion of superiority—or even adequacy—breeds comfort, and comfort prevents concern, contemplation and political action. That people are likely to judge their own intellectual potential favorably, then, ironically serves *to impede* rather than foster intellectual growth.

Learning Disabilities and Revising the Distribution of Intelligence

— "Her IQ is barely above room temperature"

A normal distribution is the most common means of statistically characterizing the dispersal of a phenomenon throughout a population. Historically, intelligence has principally been summarized in terms of the normal distribution. Most people are familiar with how this model allows approximate differentiation of "levels" of intellectual strength.[11] If nothing else, the technique offers a range for what is *statistically* average or normal, and by extension, a quantitative map indicating how each person's intelligence deviates from the central tendency. Thus, the normal distribution of intelligence has, traditionally, played a prominent role in determining what may, broadly, be considered boundaries between ranges of (relative) intellectual "conditions." We discuss this matter on the proverbial heels of aptitude because aptitude provides individual intellectual boundaries—or, at least, a baseline point-of-departure from which an individual's intelligence is constructed. Aptitude *also* figures importantly, *but obliquely*, into the normal scheme of distribution inasmuch as it was instrumental role in formulating the notion of intellectual "*disability*"—a term we use as an aggregate reference to conditions variously labeled disabilities, impairments, disorders, pathology, etc.).[12] By most relevant accounts, in the 1960s, Samuel Kirk coined the term "learning disability," identifying it as *a discrepancy* between one's aptitude and the learning actually accomplished. The notion of "learning disability," in turn, would seem to be, fundamentally, a boundary marker—distinguishing "competent" from "insufficiently-skilled" learning. Yet, precision will obviously be a problem in drawing this border. In this section, we explore whether or not the historical and conceptual development of "learning disabilities"—with the vast amount of definitions and descriptions proposed in its literature—either clarified these boundaries or, conversely, furthered their conceptual pliability. While, as an organizational component of institutionalized education, learning disabilities might be expected to contribute importantly to an exacting catalogue of states or levels of academic readiness (or some intellectual conditions along these lines), we will explore how and why institutionalized learning disabilities might, instead, have strategically served both to preserve the general *option* of intellectual "disability" while, at the same time, to ambiguate a defined normal distribution of intellectual skills. As might be expected, we attend to how such ambiguation may, on another level, contribute to anti-intellectualism's mythological uniformity, i.e., how learning disabilities complements popular imaging, intellectual quarantine, and other anti-intellectual mechanisms we have or will isolate. Indeed, in this context, we attend especially to those systems and institutions formally connected to learning and education. When mythological uniformity is achieved, in part, by contributions from social forces expected *to preserve* traditional intellectuality, its effect is more poignant.

Organizational and Conceptual Background of Learning Disabilities

As intimated above, one ostensible and major objective of learning disabilities, as an administrative division of educational institutions, is to help remediate problems associated with academic skills. We consider this objective here in light of a fuller contextualization. Clearly, educators have always been concerned about students' intellectual progress; however, the organizational development of "learning disabilities"—both in terms of (1) the political and scholarly climate in which it developed and (2) its departure from its historical "predecessor," educational and guidance counseling, has, arguably brought to the social construction of intelligence some new and compelling revisions regarding the social distribution of intelligence.

As noted earlier, we employ "learning disabilities"—the currently favored term—as an aggregate concept. We do so not only for stylistic economy, but because lexical specificity would not better represent the empirical conditions of this organizational unit. Two consistent features of this auxiliary-educational field have been (1) rapidly-shifting (or favored) lexicon, and (2) the *absence* of universally-adopted concepts or codes. It might be said that, since a Samuel Kirk's initial equation of disability with underachievement (itself a definition virtually defying standardization), there seems to have been a *cultivated* conceptual ambiguity . . . an avoidance of painting clear boundaries between that which is "able" and that which is "disabled." Still, later in its history, when the issue rose to the prominent level of government regulation (in the Disabilities Education Act), boundaries were still vague. In the legislation, itself, disability was classified as an "*im*perfect ability" [emphasis ours] to accomplish various intellectual activities including thinking and listening. Even dismissing "imperfect" as a flagrantly bad word choice, (it seems unimaginable to consider perfection as the norm), the general identifications of learning-disability issues have been consistently fuzzy if only because, as just intimated, the disability was always constructed to stand in comparison—but to some implicit and *tacit* condition. Thus, "perfection" is problematic not only because it lacks reasonable "frequency," but also because the delineation of criteria for intellectual perfection will likely remain elusive. In fact, the comparative standard that increasingly appeared in the literature was, instead, "normal" or "normalcy"—invoked in it colloquial, non-statistical sense. While the "abnormal" disability is probably far less provocative than the "imperfect" one, it remains conceptually ambiguous. Into this ambiguous framework, though, a later-twentieth-century development was very importantly introduced: (the popular advent of) *neuroscience*. The importance of this affiliation should not be underestimated; the evolution of "learning disabilities" as an extant administrative force is intertwined with and beholden to neurophysiological research. However, this connection may be characterized as anything but predictable.

The Imprimatur of Science

That the emerging area of learning disabilities was increasingly anchored to developments in neuroscience proved especially useful in marshaling greater authority and privilege than formerly extended to earlier educational-counseling programs. What was this research on which such a powerful, scientific bond was secured? Fortunately for readers and for ourselves, specificity will be unnecessary; our salient issues are not to be found in the methodological details or even the precise correlations obtained. More specifically, our concern is with the interpretation and application of the central findings. Thus, we only recap the simplest contours of the work, itself. The scientific studies that have inspired the current configuration of learning-disabilities demonstrate, in aggregate, significant correlations between neurophysiological activity, on the one hand, and intellectual conditions identified as learning disabilities, on the other. Principally through applications of scanning and imaging technology, researchers were able to find significant connections between anatomic and/or metabolic conduct, on the one hand, and certain identified intellectual conditions, on the other. We suggest that this research can be variously interpreted—even on this limited basis—and we shall discuss two distinct readings. Of course, one interpretation of this research was instrumental in developing the recent institutionalization and organizational management of learning disabilities. For this reason, we refer to this perspective as the "administrative" interpretation. This administrative interpretation of the neurophysiological research provided an important, *new* way for most people to think about and treat disabilities. From the administrative perspective, the overriding value of the research reveals why we evaded the methodological details; administratively, the most compelling finding has been that *there is* a physiological basis of (learning) disabilities (the PBD). We might point out that similar neurophysiological research (revealing a PBD-like premise) exists with respect to the definition and treatment of several *other* conduct areas, including gambling or sexual "addiction" as well as the motivation for news-seeking and/or the effects of media violence. Both of our interpretational approaches apply to these other areas as well as to learning disabilities. Thus, in all such cases, the administrative reading of the research was that it *proved* that the given disabilities or conduct were not merely "psychological" manifestations. Rather, the real hardware (scanning, imaging and/or other technology) indexically registered *real, physiological dimension*! The monumental revelation, then, is that the disability is physiological or physical. This is a premise that shapes all subsequent understandings and interpretations of the disabilities. Indeed, we will now use administrative logic to examine one such sequence of extensions.

Perhaps the most obvious place to start accounting for the PBD imprint is in connection with the question of origins: how is the learning disability caused? The administrative interpretation deduces that these real, physiological disabilities cannot, obviously, arise from personal, mental or emotional causes. Rather, real, physical conditions can only arise from internal dysfunction—genetic, congenital, or acquired physiological influences.[13] So, not only do physiological data (discovered through technological means) reveal the PBD, but also ready the organization for the *imprimatur of science*. Additionally, though, the inference of this *within-the-body*-causation polishes the scientific patina into a narrower, glossier *medical* finish. The reasoning here is equally straightforward: the locus of the disability is physically inside the body. Thus, disability *behaves like* a *medical* issue and may be treated as one. This medical option was has been widely understood and has, in turn, produced a parallel, medically-oriented narrowing of the otherwise popular "abnormality" criterion. In this administrative interpretation, the medical "translation" of abnormality" was "pathology"—although the term might be articulated or implied and, more importantly, was also invoked colloquially, i.e., indicating something undesirably dysfunctional, but not as measured by indices of morbidity or mortality. Consequently, the substitution of "pathology" eliminated virtually all conceivable relativism from diagnoses. The "pathologizing" of learning disabilities reinforced, more strongly than ever, that diagnostics was a binary system. The threshold criterion was *whether or not* a learning disability was present. More subtle evaluations—ratings such as "moderate" or "severe"—obviously only applied to the already-diagnosed disability.

The administrative (extended) reading of the neurophysiological research represents the empowered, mainstream understanding of learning-disabilities. From just the above summary, we should be able to understand how the neurophysiological research is thought to "reveal" the "physiological base," and how that PBD, in turn, was further seen *as license to construct* a scientific, indeed, a *medical* environment around learning disabilities. The whole reading and its extensions must also be considered an administrative *achievement* in that this line of thinking served to make learning disabilities a more credible and reckonable force. The scientific imprimatur—although mocked as "geeky" in other contexts—is tied here to the more applied (and less intellectual) medicine which, especially, provides traditional associations with compassion and healing. Moreover, the medical adaptation has been especially useful in establishing parity between learning disabilities and those other obviously real and credible sensory and motor disabilities with which learning disabilities now share an institutional platform.

As indicated, this administrative interpretation is an option—and the one option that importantly contributed to the general delineation of intelligence. By contrasting the administrative logic with the remaining—*structural*—perspective, we try to place in sharpest relief, those implicit *choices* through which conduct is defined by its undesirable or dysfunctional aspects. The obvious first point of contrast must be how PBD is understood. We know that from the administrative position, PBD is singularly critical insofar as the whole, recent development of learning disabilities is driven by the belief that neurophysiological research revealed that disabilities are, in fact, physiologically-based phenomena. (In fact, post-neuroscientific developments in other conduct areas completely parallel what we see in the evolution of learning disabilities.) If it is anticipated that the challenging, alternative interpretation will debate the premise of a "physiological basis," that would be wholly wrong. No known position challenges that. However, what *is* challenged is whether or not the PBD *is* what, in fact, was critically *revealed*, and, if not, what that unfounded premise means further down the line of application.

All types of structural theories have presumed that *all* behavioral conditions—conduct, ideas, emotional states, etc.—must, minimally occur (or co-occur) at the neurophysiological level.[14] A structural interpretation would never reject the PBD, but, *as a premise*, the PBD could never be the central "finding," much less a new revelation. What may seem, at first, like semantic quibbling, in fact, contradicts the whole, interpretive trajectory of the administrative position—the scientization, medicalization, pathologization, and all the organizational revisions emanating from the significance (*not* the accuracy) of the PBD. In other words, from a structural position, the administrative interpretation misunderstands the explanatory scope (or limits) of the correlation obtained. This misunderstanding can also be articulated in terms of what does *not follow* from the neurophysiological research. For example, the patterns of neurophysiological activity obtained do *not*, in and of themselves, provide grounds for diagnosing individual cases of learning-disability. Moreover, the fact that physiological patterns significantly match behavioral conditions identified as socially-undesirable doesn't establish the distributional, let alone pathological character of those patterns. Finally, and most broadly, these unquestionable physiological correlations don't even recommend that the recorded activities be interpreted *as* learning disability.

Whatever these caveats imply vis-à-vis the administrative inferences, they don't at all suggest that the research has no *structural* value. To the contrary, research of this type *importantly* refines what is, otherwise, an *abstract* physiological condition; the technological records allow specific types and specific sites of reactions (relative to studied conduct) to be known. For but one example, knowledge of specific areas and activities is requisite in developing pharmacological means to effect meta-intellectual conduct (e.g., through enhancement deterrence, reversal, etc.). On the grander conceptual level, *all* specificity contributes to the construction of a normal distribution of related patterns, from which even more exacting correlations between neurophysiological activity and intellectual conduct can be sought. The rigorous completion of a general model would, from a structural perspective, create the proper scientific preparation for conceptualizing, diagnosing and treating learning disabilities. For example, when the normal distribution of various "conditions' are reliably charted, one can validly discuss if physiological conditions or junctures consistently distinguish our most-common social assessments, or, if conversely, other patterns exist. We may discover, for instance, that many conditions we presently conceive of (and test for) on the basis of the aforementioned binary distinction, are actually distributed evenly along a continuum that may (or may not) parallel other, conduct continuums. Indeed, even before such knowledge should effect specific diagnoses, it should be considered in terms of general theory, particularly as certain conduct is typified as "abnormal." The translation of abnormality into dysfunction is probably something that can't even be addressed by this type of research. Unless the conduct in question is described by standard measurements concerning morbidity and mortality, it is probably more in the domain of logicians, ethicists, and social anthropologists to determine *why* variation in certain social characteristics is differentially meaningful and politicized. In our culture, for instance, it is illegal—at least within educational institutions—for students to consume pharmaceuticals that will enhance their physical strength and endurance, and thus, their athletic skills. However, we shall consider the comparative treatment of intelligence shortly.

The Myth of Optional Physicality and the Conversion to Physiological Determinism

Because the case of learning disabilities is not *sui generis*—because empirical identification of correlative, neurophysiological elements has become grounds for the wholesale, but inappropriate, conversion to physical/medical/biological premises for the study of numerous types of conduct (as suggested earlier), we should, perhaps, suggest the fundamental problem. Because it is such a commonplace misunderstanding, we offer an archetypal instance from television and from what is, arguably, the *most* popular spawn of neuroscience research.

> A TV commercial for an anti-depressant shows an animated stick figure moping around while a kindly voice-over explains that depression and its symptoms (sadness, listlessness, etc.) might very well be . . . *physical* (i.e., "real," *not* "mental"!) Affected viewers are urged to consult a physician to determine whether or not *their* symptoms are physical, too. If the viewer also has real, physical symptoms, he is urged to consider this particular product which, by the end of the commercial, has revitalized our stick figure.

The commercial misrepresents depression along the same lines with which learning disabilities are misrepresented in the administrative interpretation. The commercial is misleading because it suggests that there exists *another* kind of depression—one that is *not* physiologically rooted.[15] It is misleading because the very symptoms of depression, alone, incontrovertibly mean that *some*thing is physical—even so-called psychosomatic symptoms. As organisms, our only means *of* experiencing things is—at least as a threshold transformation—neurophysiologically; whatever the behavior—depression, learning limitations, responses to media, etc.—none may occur *without* neurophysiological dimensions. Depression is physical. Period. What makes for our conceptual problem is that to which the commercial, nonetheless appeals: the *myth of optional physicality*.

If, following Descartes, one imagines behavior as governed by either of two, irreducible, dimensions—mind/body, mental/physical—then, it, next, follows that the appearance of physiological correlates will be both revelatory *and*, then, (mis)identified as evidence of a physically-governed condition. Moreover, popular ontology translates that physical causality to mean that the condition is "real." In truth, though, this attribution packs more than mere ontological status—it also imposes value. Specifically, when conduct is revealed to be material and "real," any stigma attached to being thought (metaphysically) mental and "not real" is alleviated. With respect to depression, for instance, the TV commercial (and, generally, all those in the industry of treating depression) celebrate its PBD because it removes the stigma of being an "unreal" psychiatric, or emotional disorder. This general valence system involving physicality, reality and credibility now applies to various behavior issues.

Ironically, perhaps, the pharmaceutical focus has been included in administrative arguments to *substantiate* the *bona fide* pathological nature of disabilities. In other words, it has been argued that we know (indirectly or deductively) that disabilities are legitimately pathological when they are effectively remediated by pharmacological treatment.[16] However, this, too, is a spurious inference because default remediation hardly indicates that the effect actually requires physiological dysfunction. This principle is common enough to have been memorialized—with specific reference to learning disorders—in popular fiction. However, although it was fiction, the outcome dramatized followed mundane biochemical laws. Specifically, consider an episode of the eminently popular TV show, *Desperate Housewives*, showing overworked moms dipping into the prescriptions of their ADHD-diagnosed children to accomplish more successfully all their familial and organizational responsibilities. There are, in fact, pharmaceuticals that can be relied upon for such oblique diagnoses. However, when stimulants, such as Ritalin, apparently turned all the pill-popping moms into whirling dynamos of labor, they experienced a unilateral effect. Most people will similarly respond with intensified concentration and perseverance. The "unilateralness" of the effect, then, suggests it's a systemic reaction, not at all dependent on a pathological precondition.

Eradicating Intellectual Distinction

We expanded an earlier version of this section because, to our surprise, many otherwise knowledgeable readers were unfamiliar with recent learning-disabilities theory and, especially, with how that theory is operationalized in institutions. Of course, we apologize to those for whom the basic review was unnecessary and ask for indulgence in this one, additional instance where we spend a few sentences outlining policy and procedures in one major, institutional context. Specifically, our observations are heretofore confined to students in higher education whose disabilities are designated as wholly learning-related. Although learning disabilities are often diagnosed in grade school, there is no standardized program of evaluation, and thus, the "learning-disabled" designation may be assigned at any point in an individual student's career, depending on whether the student, or someone involved in his or her education, seeks assessment. In most college settings, students designated as learning-disabled are integrated with the non-diagnosed student population. Generally, unless they identify *themselves* as learning-disabled, students are indistinguishable in this context. Students *do* often choose to reveal their learning disabilities so as to avail themselves of certain services to which they are entitled. These services are typically administered out of a university's central disabilities office, which, in addition to a staff, typically maintains an environment of support facilities. These provisions include (but are not limited to) such things as environmentally-controlled rooms, special electronic tools such as spellers, audio books, sound recorders, text-to-speech (TTS) software, and so forth. These provisions allow for such services as private space and extended time for examinations, or equipment allowing for specially-adapted exams and assignments. Additionally, learning-disabled students are often provided with less instructor-collaborative assistance, such as special readers, note-takers, or proofreaders. Clearly, these techniques are *specifically* geared to help with problems in various aspects of language use and comprehension, writing, attention, memory, and generally, requisite cognitive processing. From what is clearly articulated in the literature, and from what is apparent in the very structure of the available assistance, the general objective of these treatment techniques is *compensation*. So, for example, a student with "dysfunctional" concentration and writing-anxiety might be assisted with taped lectures (to *compensate* for instances when concentration has experienced interference) and/or with a special office, additional hours, and TTS software for exam-taking (to *compensate* for classroom and writing tension and distractions).

There can be no doubt that this institutionalization of learning disorders has been undertaken with the most equitable and benevolent intentions; however, the illusory correlations and the subsequent inappropriate extensions of scope would seem invariably to compromise those intentions. The conceptualization of learning disabilities has a special role in the social construction of intelligence because it approaches the problem both obliquely (by identifying the "negative") and counter-intuitively (by seemingly *promoting* intellectual development). Of course, we say it is "*counter*-intuitive" because we argue here that the evolution of learning disabilities corresponds to other historically-parallel intellectual developments described in this chapter, i.e., they can be seen as part and parcel of the overall architecture of (anti-) intellectuality. However, if there is something akin to a division of labor in the social construction of intelligence, learning disabilities is a "specialized" mechanism by virtue of *how* it revises intelligence, i.e., through direct and implied measurement and statistics. We can derive from our overview of the popular theory and practice of learning disabilities a cluster of provocative conceptual tendencies—ambiguation, subjectivism, and compensatory-leveling—that can be seen as specially effecting how we imagine intelligence *to be distributed*. As the discussion that follows indicates, these elements are highly interrelated and not easily or aptly approached as exclusive characteristics.

We have already discussed a general "ambiguation" of intellectual commodities and boundaries. Most talk about competence and ability has become personal assessments unattached to external standards. We have also discussed how the anchoring of learning disabilities to neuroscience, was not "matched" by the anchoring of diagnostic concepts to explicit, public standards, e.g., those involving normalcy, pathology, etc. As we looked further into the practical administration of learning disabilities, there remained no explicit standards to use in drawing the diagnostic line of the binary division. This becomes especially poignant *vis-a-vis* the compensatory objective: to what "point" is educational assistance supposed to improve a learning disability? If, in abstract principle, compensation is supposed to be equal to the problem, then, is perfection reintroduced into the equation, i.e., compensatory to the extent that the assistance *wholly* offsets the problems? And how is *that* point revealed? Without explicit standards for cognitive skills, the best that can be achieved is a subjective assessment: checking the extent to which a student's performance is problematic to him or her or relevant others involved in the measurement and management of his or her education. Such subjective determinations don't at all correspond to the scientific tenor presumably introduced by the PDB.

It is further interesting to consider the precise *forms of assistance* offered to the learning disabled because they are, arguably, of at least two different types. One type is "tutorial" and is the kind of assistance traditionally provided to all students in need of help. It consists of subject repetition and technique-practice—routines that review unilaterally-employable information and methods e.g., note-taking, tapes, actual tutorial review, etc. Of course, the disabilities office may make such opportunities more obtainable to the learning-disabled student, but the assistance isn't calibrated to level overall classroom performance. Handicapping, though, is a different strategy—one that is *not* unilaterally available among students; indeed, the very notion of "handicap" suggests the imposition of an advantage on certain students *in the interest of* equalizing the chance of success. While handicapping is individually, compensatory, too, it works toward that balancing effect precisely by allowing "weaker" students to exercise options unavailable to those who are, presumably, stronger. From a more pedagogical perspective, handicapping is advantage-giving rather than traditionally instructive because the service does not contribute to strengthening intellectual skill or erudition. As we shall consider shortly, the amended tasks could very well possess the same unilateral effectiveness of, for instance, certain attention-disorder drugs, i.e., all students (regardless of diagnosis) might show improved performance were the handicap imposed in their favor. In all, then, we can see that the techniques of administering to learning disabilities serve, individually to compensate, but at the class level, to create a leveling effect on academic assessment.

There are, of course, other ways we may consider the ramifications of compensation and balance. As noted earlier, we might compare our treatment of intellectual talent to our treatment of *other s*ocial values or skills. For instance, both athletics and art are skills taught in schools and which also possess demonstrable PDBs. In fact, both athletic and artistic talent are obviously "physically"-related in terms of the demands they place on such things as hand/eye coordination, visual acuity, manual dexterity, etc. Yet, athletic and artistic problems or failures (unconnected to a wholly separate motor or neurological disease) are, typically, attributed to "being all thumbs," clumsiness, not being able to draw a straight line, and similar rationales. There is little doubt that these attributions *could be* reconceptualized . . . that the continuum on which they are now charted could be translated into some binary code differentiating dysfunction. (And here, we refer to revision beyond the notion of "intellectual pluralism" which we discuss later.) As we've seen, though, we don't imagine academic-intellectual problems in this vein. The idea of "normal" deficiencies—non-pathological or non-dysfunctional cognitive "imperfections" equivalent to being intellectually "clumsy" is hardly an option. If this parallel were stretched to exclude the PBD-factor, we might even compare intellectuality and wealth as (sometimes) valued social *characteristics*—and we would still find both more ambiguity and more leveling attached to the intellectual dimension. Metaphorically, at least, we could argue that "materially disabled" is one way to characterize poverty. And, in terms of an established and legal program, the materially disabled also receive assistance. However, with respect to economic values, the standards determining material disability are explicit and public; citizens are easily measured according to several unilateral, unambiguous, economic indices. Moreover, diagnoses and assistance for the economic condition are not only standardized, but also, confined to those severely affected. In practical terms, what this means is that material assistance is hardly calibrated to be *compensatory;* certainly, it is hardly intended to compress a normal curve so as to diminish interpersonal deviation. Indeed, such a compensatory/normalizing scheme might constitute an oblique definition of socialism. On the one hand, our cultural mythology, in many ways, impugns a socialized economy, while positively dramatizing the *laissez-faire*, competitive, marketplace enabling (putative) opportunities for social mobility. On the other hand, although it's rarely articulated in these terms, the notion of a "socialized" intellectuality is, by implication, far more compatible with our general worldview.

Especially in terms of leveling—constructing the understanding of a condensed intellectual scope—one looks particularly at the ends of the metaphoric spectrum. Learning disabilities would most obviously be connected with management at the "lower" end—performances that might be assessed as inferior without the application of the previously-discussed "correctives." And, of course, we have seen how a concept such as "pathology" or "dysfunction" has the power to provide a conveniently *ad hoc* and sympathetic rationalization of this (otherwise) "less capable" intellectual performance. This rationalization is not simply of personal use to the bearer of the disability, but by purging the general distribution of what was understood as "subnormal" performance, the social statistics of intelligence are obviously changed. These sorts of changes cultivate a conceptual environment that is hospitable to the development of popular phenomena such as intelligence's third-person effect or beliefs in unlimited, universal aptitude.

Although learning disabilities logically connect to the intellectually imperfect or inferior, its general leveling effect contributes to closing *all* ranks of intellectual variation. Certainly, the "remedial" technique of handicapping may especially influence the significance of *un*enhanced, better-than-average performances, that is, at least, to the extent that higher education functions as a comparative, competitive undertaking. Also in this context, unenhanced performances are comparatively weakened *beyond* the issue of relative academic standing. For plausible reasons, diagnosed students who have received remedial/compensatory assistance generally *remain* undistinguished from classmates, i.e., their transcripts for graduate schools or employers do not suggest that their grades reflect this assistance. In this way, the compression of the "curve" is sustained beyond academics . . . beyond learning. Of course, weak records also remain with students beyond academe and, here, we must revisit the idea that the reconceptualization of learning disabilities has significantly purged that hypothetical intellectual-performance curve of the non-dysfunctional, non-disabled below-normal performers. While learning disabilities has successfully reconfigured this category, it has not been purged. Given repeated reports that learning disabilities are significantly more likely to be diagnosed among the materially-advantaged, it doesn't take higher mathematics to consider that the constituency of those remaining as "non-disabled and below-normal" reflects far more about the cultural management of intelligence than it does about academic norms.

While, from a methodological perspective, subjectivity and ambiguity may be surprising and distressing, it can simultaneously be interpreted as serving a non-methodological social function. Ambiguity and indeterminism allows—indeed, almost demands—that the margins of intelligence be set by the practitioner or diagnostician. The dissolution of its standards further allows "intelligence" to be subject to the political convenience of *ad hoc* assessments. Leveling and the compression of intellectual variation helps to diminish the appearance of intellectual superiority. These revisions, along with reconceptualizations of aptitude and learning disabilities, permit more people to participate and succeed in the educational process. While we will later discuss the specific implications of educational participation, we can presently understand that the subjectivity, ambiguity, and leveling go toward "egalitizing" intelligence—extinguishing its traditional character as an elite and extraordinary asset. Increasingly, then, the existence of the third-person effect with respect to intelligence, and even the notion of self-determined aptitude makes cultural, if not empirical sense. As we will see in the following pages, there have been a few more strategic reformulations of important intelligence-related distinctions that contribute further to the seeming "liberation" of intelligence.

Dividing Intelligence

The notion of "divide and conquer" is a long-established strategy to gain or sustain power. Over the course of the twentieth century, there *has* been a strategy of both overt and implicit division in the conceptual construction of intelligence. Here we will consider the more literal "partitionings" of the general concept. All can be seen as intersecting with the issues already considered—the third-person effect, aptitude, and learning disabilities. While sometimes the various divisional mechanisms can be thought to divide the labor—to focus on one or another aspect of intelligence, all can be understood as reformulating and contesting the traditional power of intellectuality and erudition.

Cultural Bias

Most of the antipathy toward standardized tests responds to the "cultural bias" of "traditional" intelligence. Any position (such as functional constructivism) that sees culture as the real and dynamic force of accumulated social experience, will have conceptual difficulty with the colloquial notion of "cultural bias." In a socially-constructed world, the notion of cultural bias seems tautological and certainly not a value judgment. Thus, we need to consider the difference between the tautology emerging from cultural determinism, on the one hand, and the popular notion of "cultural bias," on the other.

There is no question that the social traditions upon which America was built entitled only white, property-owning males. Anthropologically, this tradition was, among other things, racist and sexist because the entitled men presumed *biological* superiority to all living things. Intellectual subjects and techniques were forged at universities to which only these entitled men had access. Thus, formatively, our intellectual tradition is symptomatic of all our traditions—steeped in and supportive of white, male, Christian, heterosexual power. Intellectual "bias" was not merely a by-product of generic cultural orientation; rather, the harnessing *of* intelligence and education (in all cultures) is, itself, a compelling force through which tradition (and oppression) is maintained. Throughout this chapter we have seen (and will continue to see) numerous contexts and means by which traditional, intellectual themes, methods, and standards have, in various ways, been distanced and disassociated from the "under"classes. Much of this separation has been (rationalistically) understood as the voluntary, popular *rejection* of "establishment" tradition—a rejection supposedly undertaken to censure and reformulate the conservative foundations of social power. Structural theory reframes, not this history, but both its meaning and method and asks: to what end and/or to whose benefit is it for the non-elite classes to eschew and/or be detached from that (old-fashioned) intellectual tradition invented for and bonded to the empowered classes? Is it simply sufficient to reject the standards of elite culture, or does it not make an enormous difference, if the traditional standards to be rejected have, ironically, long been practices and rituals unavailable to the underclasses? In this context, what does the multi-culturalization of intelligence imply? What would it effect? In sum, is it conceivable that the very idea of *characterizing* intellectuality *as* culturally biased (so that normal people eschew it) is, itself, a quarantine technique, the goal of which is to make traditional knowledge and intellectual skills inaccessible to the *hoi polloi*. The following anecdote from a popular television comedy helps to articulate these questions in more practical terms.

Years ago, a segment of the then-very-popular TV comedy, *Good Times*, featured the youngest and most studious of the characters, Michael, refusing IQ-testing at his school because of the test's cultural, or more specifically, racial bias. As he was an excellent student, his parents initially saw this as needless troublemaking. But Michael persisted and his beleaguered parents accompanied him to school where, in the presence of a smarmy, sanctimonious, white test-giver, Michael demonstrated the "issues." Pointing to a test question—an analogy problem beginning with "cup : saucer," Michael argued that this question, for example, was biased against any sub-culture in which saucers were uncommon. The parents realized the rectitude of Michael's plaint and supported his stance. Indeed, some retaliatory catharsis was achieved when the tables were turned and the smarmy test-taker was, instead, questioned *by Michael*, who asked for the definition of "get on down." The laugh track roared at the outsider's obvious ignorance and mortification. The family left the consultation proud and unified. While the ending scene was gratifying for the characters, the studio audience, and presumably the home-audience, does that gratification represent movement toward political equality? Or do such treatments of cultural bias serve as a placebo having unintended reactionary consequences? What if *there were* two, three or four I.Q. tests designed to accommodate different "versions" of intelligence?

Intellectual Pluralism

Arguments of cultural bias led to the notion of *intellectual pluralism*—the belief that there isn't (or *shouldn't be*) *a* "traditional" intelligence, but rather, a concept embracing many different *kinds* of intelligence or intelligence*s*. It's not a brand new idea—nor does it directly address variations in sub-cultures—but it has, nonetheless, since the last third of the twentieth century, remained powerfully revitalized, leaving in its wake some questions interweaving, logic, language, symbolic economics, and ideology.

It could be argued that, in English (as well as many other languages), we already have terminology generically referencing performance ability, e.g., "talent," "skill," "facility," etc. English even provides for more specificity within such broad rubrics, e.g., "emotional sensitivity" or "manual dexterity." The fact is that our language (as well as others) always allowed for such distinctions. Even the standard college-entrance exams have provided for at least two or more measurements of specific skills. And, if there is a skill for which the language presently lacks terminology, our vocabulary is hardly "full-up." After all, we are relatively big-brained and neurophysiologically sophisticated organisms; we may develop and manipulate pretty extensive lexicons. Despite what is or what could be, we sometimes use the *same* word or term to mean *different* things. At the very least, this lexical piggy-backing is inefficient—it makes vague what could be precise. Why, then, "borrow" or re-use terms for which familiar referents already exist ?[17] There is no question that people may be talented in innumerable ways—athletically, graphically, mathematically, musically, and so on-but it does *not* enhance, simplify or fine-tune our understanding of anything to stipulate that, henceforth, all these "talents" will be called "intelligence." However, this sort of cost-benefit analysis doesn't consider the social value of the word.

On certain occasions, intellectual failure or sloppy scholarship may be to blame for "misusing" a term, but in most cases, the language is not misused, but rather, co-opted—words with well-established referents are given one or many new references to compete with and to subvert the words' traditional meaning. Lexical co-option is inextricably rooted in linguistic determinism—the idea that language structures reality. As classically articulated by Saussure, and later by Whorf, a culture's lexicon and grammar must be understood *not* as empty, technical mechanisms by which to articulate universal ideas, but rather, as a framework molding *what* we know and *how* we know it. In the case of lexical co-option, then, one way to modify the power of an extant concept is by appropriating its lexical designation. With intellectual pluralism, the word "intelligence" has unquestionably been co-opted for ideological reasons. These reasons related to inclusion and tradition; intellectual pluralism allows the widest population to claim (some form of) "intelligence." It also allows for an intelligence devoid of traditional intellectuality and erudition.

Plebian Pluralism: Street Smarts Versus Book Smarts

Carefully-differentiated lists of intelligences (often coordinated with neuroscientific issues) are now reasonably well-recognized, but for a much longer time, there existed a popular means by which to construct an intelligence *without* intellectual components: "street smarts." Comparatively, intelligence based on traditional intellect and erudition, has been "book smarts." Clearly, the very notion of "street smarts" has provided the same egalitarian patina or balance sought in our culture. The third-person effect is likely realized because "street smarts" is the intelligence with which most people identify. However, the book-versus-street-distinction should not be understood as *merely* an older version of intellectual pluralism; unlike newer distinctions, "street smarts" can't be dismissed as some lesser or less-known form of intelligence. Quite obviously, the skills supposedly required for negotiating everyday life would seem to be the most necessary intelligence. For this reason, the book-versus-street-distinction constitutes one of the most powerful and broadly appealing of our intelligence myths. And just as we can see anti-intellectualism dramatized throughout media representations of intellectuality, we can, similarly, observe that there is no personal rebate demanded for street smarts—no relinquishing of fashion sense, of sexuality, of composure. In fact, it is a *rewarded* or positively portrayed intelligence. The street-smart character is typically cleverer than others. He is also glib—particularly if created as an action hero such as any Bond or (most) Bruce Willis roles. Street-smarties *rarely* lack sexual appeal, and it may very well *be* the street savvy that *is* the attractant. Street-smart characters are very adaptable and can usually conquer contexts and people unfamiliar to them. Consider the immensely popular, long-running and much syndicated series *Friends* featuring one major character for whom unusual detail *was* disclosed—Ross, who held a Ph.D. in anthropology with an emphasis on paleontology. Of course, a number of jokes were written at the expense of his education/profession. However, beyond that, can faithful fans tell us *where* Ross studied or, if anyone other than Ross *graduated* college? If so, from where? With what major? One could be a *Friends* zealot and not know these answers simply because such details may never have been invented, much less publicized. In fact, even vague information about characters' *general* intellectual level is typically avoided unless specifically cultivated for narrative or dramatic use, e.g., Joey's denseness on *Friends*.

The Frequency and Combinations of Fictional Smartness: As a rule, having both types of intelligence is really rare. Only ultra-heroic types are so characterized—and even then, infrequently (e.g., Indiana Jones or Batman). And, it is even rarer that book smarts are necessary to the "super" incarnations. Superman, if he desired, could devour all literature instantly, but he evinces no intellectual polish. The most recent Spiderman seems more good-natured than even street-clever. Similarly, the *Charmed* sisters don't seem interested in any book but the one in their attic containing potions and spells. Other than craven wizards, serious erudition is typically reserved for regular mortals, perhaps because they lack super powers; however, if erudite, the mortal is virtually never street smart, *too*. Even when science-fictionalized within the same body (e.g., Jekyll/Hyde, the nutty professor, the Incredible Hulk), the twain can't seem to meet simultaneously. This, of course, corresponds to the formula we have already examined—intellectuality demanding the forfeiture of all interpersonal cleverness.

While it's relatively easy to locate *educated* characters who are sociably stunted, it is, intriguingly, unlikely for the inverse to be created: *un*educated characters who are conspicuously *less* capable. Compare, for example, the children of *The Fresh Prince of Bel-Air*; their cleverness is unrelated to their varied educations—and, predictably, the most educationally advantaged is hardly the cleverest. Even in serious, supposedly realistic dramas, like *The West Wing,* there rarely is shown to be a "cleverness differential" among characters who, presumably, have widely-various educations. *The West Wing* is hardly the only series with professional characters, but regardless of the vehicle, characters are rarely exclusive or quicker because of their educational pedigrees. On medical series, for instance, the relationships between physicians and nurses don't typically manifest hierarchical characteristics (unless pecking order is part of the comic or didactic point). On a show like *Scrubs*, for example, administrators, pharmaceutical salespersons, physicians, janitors and nursing staff interact without a sense of effective intellectual division. And it should be noted that such observations emerge from comparison to media portrayals of *other* class divisions. Movies and TV stories, for instance, commonly reflect class variations based on both demographic indices (like economics, gender, race, sexual preference, ethnicity, etc.) or cultural castes (group divisions based on social standards like looks, popularity, etc.). Among all these culturally-acknowledged divisions, intellectuality/education seems the *least* powerful barrier in media-represented interpersonal interaction. The failure of intellectuality/erudition to function as a significant variable in "class"-division emerges in virtually all genres. In the many *Law and Order* casts, for instance, the frequent interaction between detectives, psychiatrists, city workers, coroners, judges and attorneys can be both friendly and/or contentious, but it is infrequently marked by an *intellectual* differential. Similarly, on the many forensic shows (such as *Crossing Jordan*), the M.D.s, the technicians, the counselors and the police interact—sociably and romantically without *intellectual-class* consciousness and/or interactional meaning. Even series that aren't profession-based still tend to neutralize potential educational hierarchy, largely by revealing very little about characters' intellectual lives, but also, by creating stories and dialogue that intellectually equalize the participants. When hierarchy *is* implied, it is dubiously both demanded by, and to the narrative detriment of, the intellectual—as dramatized in any of the previously-discussed myths. In a sense, then, the introduction of intellectual hierarchy typically has the function of revealing intellectual pretense and, in fact, the "real" intellectual equality among characters—if not the actual superiority of the non-intellectual. A very common narrative device in TV fiction, for example, is to positions the (seemingly) least powerful cast member as the wisest. A common variant of this device occurs in the dramatization of wealthy families and their servants. From *Hazel* through *Maude*, *Soap*, *The Jeffersons*, *Mr. Belvedere*, or *Fresh Prince*, the maids and butlers are those in possession of the necessary wisdom. Interestingly, American Anglophilia is particularly overt in these situations; if a servant is English, he may be shown to possess superior erudition and wisdom—such as Mr. French in the TV classic *Family Affair* or Arthur's valet in the movie of the same name. In non-domestic-based shows, service workers sometimes assume roles similar to the smart family servants, e.g., the janitor on *Scrubs*. What might be considered a corollary here is the portrayal of educated characters relying on their "common sense"—*not* on their education—even for *professional* success. For instance, the successful females of *Legally Blonde* or *Working Girl* ultimately prevail, not because they *have* gone to school to study law and business, respectively, but because they know about hair care and society gossip. Tom Cruise's studious good-boy in *Risky Business* wins the approval of the Princeton interviewer not through his academic achievement, but as a result of his intuitive (and hooker-assisted) entrepreneurial savvy manifested in converting his temporarily-vacant parents' house to a brothel.

This conflict—the value of erudition versus sociable skills—also became the premise of some competition-based "reality TV" series. On *Beauty and the Geek*, for instance, particularly nerdy, educated boys have been teamed with attractive, outgoing girls. The teams enter contests (dubiously) reflecting either intellectual skills (the males' forte) or "street" graces (related, for instance, to fashion), and the skilled member has to coach the non-skilled partner for each competition. As the competition is also gender-coded for intellect (female contestants are consistently the "beauties" and *not* intellectually-accomplished), female contestants would have difficulty with their assignments (e.g., memorizing the legal moves of chess pieces), not only because of their intellectual limitations, but because of a plausible lack of interest. In other words, that which the boys might teach the girls is, typically, not shown to be as vital as the mundane social skills the boys (more aptly) learned. Although the teams competed against each other, it was clear that the female contestants had only the prize money to win. The male contestants, though, were represented as winners despite the central competition.

One season of an even more popular reality show was ostensibly based on an overt book-versus-street competition. The general premise of the third season of *The Apprentice* (to Donald Trump) was the entrepreneurial competition of two teams of contestants, explicitly divided as "book-smarts versus street-smarts." As it happened, a member of the "*book smart*" team won, which would normally seem to go against the predictable pattern. However, closer examination of the competition does not suggest that a counter-example was actually proffered. To start, Trumps's original team assignments were determined by a single criterion—whether or not a contestant graduated college. Indeed, a player might have been deemed "street smart" although only shy a few credits for graduation. Conversely, a graduate of a middling institution with a barely-passing grade-point-average would have been assigned to the book-smart group. Thus, by avoiding rigorous distinctions in terms of educational institutions, degrees, majors, grades, and knowledge possessed, there, arguably, may have been little-to-no difference between the competitors' education or intellectuality. After every competition, Trump and his staff held review-and-discussion sessions with contestants, but, *even then*, educational backgrounds were rarely invoked to account for success or failure. Over the course of the season, the competing teams were often neck-and-neck, demonstrating, at least, that the street-smart team held its proverbial own against the college graduates (although, in fairness, the book-smart team was *not* branded as otherwise sociably retarded). Finally, inasmuch as one from each team constituted the two finalists, it was hardly a runaway victory for college graduates whose team member managed to survive. Probably, to the extent that the season exemplified anything, it was that *education makes little difference*. Viewers might be recalling, though, that in the show's successful premiere season, a personable Harvard MBA lost to a young man with a simpler education but more entrepreneurial experience—and, in that case, Trump cited *the experience* (as opposed to whatever's symbolized by a Harvard degree) as the scale-tipping criterion. Thus, *The Apprentice* has implicitly reinforced traditional intelligence mythology more than establishing a counter-example to it.

Street Smarts as an Imaginary Construction: There is, though, a radically different way of accounting for the events represented: what if the reason Trump's teams were not dissimilar is that "street smarts" and "book smarts," as isolatable categories, *are not commonly found?* In practice, when an attribution of either of these "smarts" is made, it is with the implication that the person in question owns one, but not the other. Rarely, if ever, is it asked how many Carla Tortelli's exist *without* a professionally-written script? Similarly, how common are very erudite people who are, otherwise, *in*capable of navigating a cocktail party, filing tax forms, or falling in love? Yet, these mutually exclusive categories are so profoundly bound and ingrained in our mythology, they defy what logic and everyday evidence would otherwise suggest: the cases are empirically unlikely.

Another question that is never asked, but to which the answer might be useful involves the implied definition of "practical" as meaningful and/or useful—in other words, those things that book-smart, apparently is not. While it is probably the case that few people are *unilaterally* skillful (i.e., considering, say, all the options on a list of pluralized intelligences), it is interesting to contemplate *why* we believe that those who, for instance, can fix things in the house are labeled as being more "practical" than those who can analyze the variables in a complex negotiation. There is remarkably little scholarship on this distinction but, hypothetically, we imagine that such assessments come from the same worldview which stipulates that the value of studying a foreign language is measured by whether or not one will need ever to speak it. We pause to question why related questions are not asked, e.g., who is more practical, the person who can do his own home repairs or the one whose work allows him not to know? Clearly, these options are not mutually exclusive in principle. However, men (especially), in popular fiction, who are not capable of these sorts of chores are, typically, characterized as effete and, often, sexually "dubious." One knows that a character is portrayed as especially laudable, if he can afford to hire someone, but can—and chooses to—do it himself. While our mythology on the basic book-versus-street distinction is common and overt, its larger text is subtle, layered, and largely unrecognized.

Before mass literacy, before mandatory public education and the omnipresence of media products—public libraries, cable TV, or the Internet—perhaps it was possible to speak of individuals who, despite a lack of knowledge *sources*, were more apt than others. However, *that* street-educated person is as much an historical artifact as Daniel Boorstin's "hero" (someone whose fame is accomplished *without* media celebrity). Nonetheless, the notion of the street-savvy person persists well into social periods during which most everyone has *physical* accessibility to intellectual information. To the extent that the original social conditions producing the book-versus-street comparison no longer exist, we see, instead, an excellent example of the *functional* promulgation of "unrealistic" or inaccurate knowledge.

The Paradox of Anti-Intellectualism Redux

At the outset, we considered a seeming paradox: Why would people be intensely concerned about their intelligence, if our popular mythology (which is hardly benevolent toward intellectuality) is an important acculturative force? After reviewing details of the conceptualization of intelligence, we should recognize how that potential paradox has been side-stepped. By allowing people the opportunity to identify with *an* "intelligence" that had personal resonance, the third-person effect is reasonably achieved, i.e., 88 percent of the population who claim to be smarter-than-average might be right, *if* the population for each individual assessment is different. Similarly, if aptitude can be assigned *without* performance correlates, on what basis can we deny *anyone's* claim to intelligence? Most importantly, perhaps, is that, while the ostensible reason for revising the concept is the promotion of egalitarianism, the practical effect may be quite the opposite.

It is often difficult for scholars or social critics to question egalitarian ideological ends. When intellectual pluralism (or any position) forged in the interest equality is *challenged*, it often engenders at least two common, interrelated and erroneous reactions. First, *intention is confused with rectitude*; it is often incorrectly assumed that if motives are pure, then the plan of action chosen follows the motivation—as if ethically-laid plans cannot produce an (inadvertent) unethical outcome. Second, *rationo*-logic is confused with *ideo*-logic—it is often incorrectly assumed that any challenge to an earnest and morally well-*intentioned* strategy, is *de facto*, in opposition to the ethics and principles that strategy supposedly represents. A multi-cultural, pluralistic intelligence *is*, for all intents and purposes, a paragon of political rectitude. What ***is*** problematic with such pluralistic strategies is their double agency; they appear to engender egalitarianism while proposing concepts and policies that *thwart* such objectives. This often happens when an illusion providing immediate "psychological" comfort is favored over a more complex, but appropriate and long-lasting solution. For instance, it might be comforting to alter language and standardized tests so they measure skills that are far more culturally inclusive, if not universal. However, such alterations would be senseless *if* these revised terms and tests provided measurements of characteristics that don't at all lead to empowerment. In that case, the alterations are cosmetic—without the practical outcomes formerly attached to the traditional terms and tests. Even worse, the revised concepts and language might, in fact, "set up" the disempowered by having them go through the motions without reaping the old benefits. Pluralizing intelligence is senseless to the extent that social systems and institutions continue to rely on or prefer those with traditional intellectual training.[18] One might argue that even just cosmetic changes might have a positive effect by virtue of the resultant self-esteem of *believing in* one's intelligence . . . by feeling justified in producing the third-person effect. Such arguments, though, must be tempered by considering the "common market" of everyday interaction—whether or not that in which someone believes will have any value beyond the belief, itself. Indeed, although it deserves a critical chapter of its own, *self-esteem* is a particularly deceptive delusion; not only may it lack currency outside the individual's self-assessment, but its very possession—the very real comfort it provides—can block vision and hinder advancement.

The Distribution of Intelligence: Is Access to Blame?

The social-science research literature does not address social class nearly as frequently as it considers other major demographic categories. Sometimes, it is argued that this avoidance reflects the conceptual slipperiness of "class." However, it might be otherwise (or, at least, additionally) suggested that social class is to demographic variables, what intelligence is to "social-asset" variables—the most important, and therefore, the most politically sensitive case. This sensitivity, in turn, engenders politically-correct treatment, one aspect of which is the plain circumvention of the topic. Arguably, the conceptual slipperiness is a function of this avoidance (through the absence of cumulative labor). Now, if we find scholars retreating from social-class research, we can imagine the increased sidestepping resulting from the incorporation of intellectual indices "in with" class division. It is hard to imagine a social-science question that's *more* governed by "don't ask/don't tell" policy. However, there is one seeming, but consistent exception: theorizing intellectual-variance-by-social-class *as a consequence of differential access* to intellectual materials and services.

The communication literature is replete with access-blaming positions, the most recent incarnation of which is the "digital divide" hypothesis. This argument interprets hierarchical class division *vis-à-vis* intellectual matters as a function of inequitable access to digital hardware, services, and/or training, and it largely revolves around two interrelated beliefs: 1. the classic/generic argument that *access* to technology or communication channels (techno-access) is an important means by which to sustain class division, because inequitable access accounts for intellectual variation, and 2. the immediate/temperocentric argument that) digital technology *particularly* exacerbates the divide because it demands not only access to (costly and rapidly obsolete) hardware, but also, the *continuing* requisites of literacy, software and ancillary services. Like campaigns for "intellectual pluralism," the digital-divide argument arose as earnest, politically progressive, social criticism. While there is no debating that resources differ along class lines, the question is whether these differences are significant (let alone, causal) in constructing intellectual inequality. Arguably, it might be said that the digital divide is more tautological than causal.

Corporate Endorsement and Functional Skepticism

Access solutions are appealing for too many reasons not directly relevant to our ultimate objective; unfortunately, then, we haven't the luxury to explore what are, otherwise, some intriguing issues, e.g., technocentrism, temperocentrism,[19] materialistic privileging of palpable "symbols," and altruistic hedonism. We can, very briefly, though, describe yet another appeal that invokes a type of functionalist deduction. Deductive logic, of course, will not refute the appropriateness of the digital-divide hypothesis; it might, however, prompt a type of skepticism not addressed in the techno-access literature.

The digital-divide hypothesis has the distinction of being a seemingly highly-progressive strategy that's, nonetheless, *supported* by the empowered, corporate sector. Major technology corporations and communications entrepreneurs exercise grand largesse through substantial endowments to educational institutions. Virtually all major manufacturers of digital hard- and software reward schools and academic organizations with special discounts and legal waivers. It is *not* new to be skeptical of such generosity—at least not on the grounds that these major charitable efforts, more than supplying tax hedges, serve to insinuate *the benefactor's* product-line into the educational infrastructure. We don't dispute this interpretation. However, one might dilute (or, some might say "rationalize") its power by considering that, one way or another, *some* corporation will ultimately profit, and that the existence of a particular profiteer doesn't *minimize the educational access gained* through the contributions. Yet, another layer of skepticism attaches to this collaboration with the corporate sector—independent of *whose* software and services will, ultimately, be endorsed. Specifically, given that the digital divide's overt objective is a form of social "*leveling*," is it not *un*characteristic of corporate powers to invest in any endeavor leading seriously to equalizing class division . . . to invest in any endeavor contributing to the vulnerability of empowered classes? If universal access powerfully contributes to effecting such equalization (as opposed, simply, to defining it) then these cooperative corporate agencies would, essentially, be contributing to their own dispossession. Do those whose activism is directed toward universal access and leveling believe that the super-powered sponsors *underestimate* the political value of their charity? This latter option is something we *would* protest. As noted at the outset, the contradictory element that corporate capitalists do, in this specific effort, join with those who would, perhaps, undo capitalism is, at best, the proverbial food-for-thought to which we shall return briefly.

Studying Access: Research and Social Class

Many years ago, in cultural anthropology, a theoretical argument was launched about the advantages of studying "up"—of analyzing materially-*advantaged* subcultures. Unfortunately, this scholarly orientation never quite caught on. More than a century ago, the economist Veblen famously compared consumption patterns of "new" and "old" money, and much more recently, a few sociologists, most notably Baltzell, addressed traditions of patrician "society"; but, generally, this perspective is not much assumed—particularly in media studies. Perhaps the everyday interaction of the rich is overlooked by academics' progressive tendencies on behalf of the disenfranchised. They may discount the need to champion the materially-endowed. Yet, representations of wealth can be especially meaningful or even instructive to those who are not, themselves, familiar with the condition. Popular writers are hardly apologists for the rich. For example, as our running example has indicated, (mediated) wealth, repeatedly and almost exclusively, is *not* accompanied by happiness. Functional theory explains that myths about the serious disadvantages of money serve to subdue proletarian outrage resulting from "unfair" failures *vis-à-vis* the American dream. While *just* this function is undoubtedly essential, it may be that representations of wealth offer us *other* insights. One can imagine countless stories that do or can relate to techno-access, money, and intelligence.

A story in a parenting magazine tells about a child who must go to summer school after failing classes. Predictably, he is most unhappy about sacrificing his vacation. His parents' lament is different: "All he wanted was this super computer-docking station with a small laptop that could go from school to home and to our house in the Hamptons. And, of course, we gave him a card and told him to get everything he needed. *But* imagine our chagrin upon discovering that it didn't make one bit of difference!" Another similar story describes a wireless, college lecture hall in what is described as a "boutique" university. Of the thirty-eight students in attendance, eighteen have electronic notebooks open and operating. Throughout the lecture, ten of these screens overtly display games, messaging, and extraneous-websites. Three other students were gaming or text-messaging on cellular phones. The attentiveness of those *without* electronics or apparent diversions is unknown. A third set of examples comes from celebrity media-coverage. One piece regales singer/actress Jessica Simpson's putative lack of intelligence. If not her infamous confusion of seafood with poultry, she reportedly asks actress, Pamela Anderson (whose old *Baywatch* role Simpson is alleged to be reprising) how Anderson ran so slowly in *Baywatch*'s opening (slow-motion) credits. Also gleaned from following young celebrity-reality programs and related webpages (featuring the Gottis, Hilfiger, Hiltons/Ritchie, Reid, Osbournes, Simpsons, and the *Laguna Beach* teens), is that the majority of these young, rich celebrities have neither expressed plans for nor interests in formal education. It also appears that the proportion of these reality stars who *didn't* graduate high school is actually *higher* than the national average (of high-school dropouts). Research tends to show us, though, that the poor aren't driven to access available education any more than these anecdotes show the rich doing so. This seems to be the case, historically—long before digital media. Consider, for instance, the development of public lending libraries. Apparently, these institutions did not have a profound direct or culture-lagged effect on literacy or erudition despite the fact that universal literacy—literacy for the masses—was among the central objectives of the early social reformers who established these libraries.

In media stories—"reality" or fiction—the rich are not especially prone to pursue educational opportunities. However, the same stories show that poor kids typically only fail because the world has, *economically*, failed *them*. The mass media generally do not tell stories about the poor forsaking opportunity, however, the intelligence mythology is not isomorphic with money myths. If a poor person grows wealthy, he also grows increasingly troubled—in fiction or nonfiction. The stories about, for example, big lottery winners—whether reported in tabloids (about Indiana trailer-dwellers) or shown on *Hill St Blues* or *Roseanne*—is the same story. However, with respect to *intellectual opportunity,* rich and poor, as we have suggested, are *not* shown to follow the same paths.

With respect to fiction, virtually a whole American genre is devoted to the story of *poor* kids who are intellectually adopted by someone who gives them serious educational "opportunity"; whether it's Alger's *Mark, the Match Boy*, Will Hunting, or Forrester's student, opportunities are virtually never forsaken. Intriguingly, except for the story of the lending-library research, academic theories offer the same structural alternatives as those produced in Hollywood. The digital-divide hypothesis (and all positions blaming access for class division) implicitly characterizes the disenfranchised as *predispositionally different* from the empowered population—*different with respect to something other than the tautological economic sufficiency.* It would seem that the disenfranchised have somehow remained *un*affected by or immune to the culture's reigning anti-intellectualism. It is not clear what the energy source or the mechanics of such drive would be, but one might infer from theory that the deprived are *more* intellectually driven as a result of their deprivation. Perhaps it is assumed that the poor experience such gratitude upon receiving access to digital goods and services, that the charity becomes intellectually coercive? We certainly don't proffer any of these explanations, but we don't know the basis on which any theory should assume that, upon gaining access, the poor will respond radically differently from, say, Paris Hilton or that wealthy child who had to go to summer school, or plainly, most *anyone* else? Carried to its logical conclusion, we would suggest that the digital-divide hypothesis places an *unduly heavy* burden on those for whom access is demanded. What is the long-term effect when such burdens are *not* met—when access is not intellectually effective? To what extent do campaigns for access have a narcotizing effect? Ironically, we would predict that narcotization is probably beside the point. Perhaps one of the main reasons *there is* so much corporate collaboration on behalf of equalizing access is because people already know that any intellectual promises *beyond access* are undeliverable . . . that the access divide does *not* significantly alter class structure? This possibility is unlikely intentional—we don't propose it as conspiracy theory, but as a functional alternative. In other words, it's another interpretation of observing energy and resources unintentionally channeled *away from* the root causes of inequity. It's an interpretation because, as we shall see, it is hardly the most egregious example of substituting conservative-function-for-progressive-action.

How to Address Social Structure

We do not pretend to be even *among* the first to see problems with the digital-divide hypothesis—although the problems we identify tend to be different from those commonly found in the extant critical literature. We also mean to be clear that our criticism of techno-access-based theories of social structure is by no mean offered to rationalize a moratorium on or suspension of efforts to achieve universal access. To whatever use it's put, digital technology is so increasingly presumed in daily life, everyone should be connected for the sheer integrity of our society. In all, we don't criticize access, just the faith developed around it.

Ironically, perhaps, it is *on behalf* of the disenfranchised, that the alternative should be considered—that access *not* be the cornerstone of attempts to redress problems of social structure. However, philosophically, it is an almost-impossible debate to recommend serious revision of very-entrenched ontology. To ignore the *consequences of "the ideas of the thing"* (to use materialist language) is *to feed* rather than undermine that problem. Elements such as those discussed in the foregoing sections—storytelling, institutional geography, or semiotic dimensions—are typically wholly overlooked, despite their critical role in the development of anti-intellectualism. The anti-intellectualism to which we refer is no obscure abstraction—at least we hope we have, thus far, shown that it is a constant in the mundane world. In this sense, it is hugely relevant to educational motivation. In the next section we examine one last instance of its mundane manifestation and, also, perhaps, that more egregious example of (functionally) conservative conduct posing as "progressive" strategy.

Divided Intelligences: Social Class and Education

We consider last what is commonly regarded as both the last stage of formal intellectual training, as well as the (idiomatic) last place one would expect to find anti-intellectualism systematically disseminated: the academy—particularly the "ivory tower" of higher education and scholarship. Earlier, we explored the routinely-cultivated *image* of academe, but here we focus on the academy, itself—its structure and practices—as an important source of social-order maintenance *vis-à-vis* intelligence. In the discussions that follow, we explore how education is the site at which relevant public consciousness about learning and "necessary knowledge has been reshaped in recent history. To this analysis, we specifically introduce and focus on social class. Specifically, we examine how the conduct of educational systems (combined with popular ideology *about* learning) impinges on social order maintenance and, particularly, establish the relationship between intelligence and upward social mobility.

The Social-Class Diffusion of Education

At one time when the modern university was being conceived, it offered education only to those men who were "to the manner born." That these institutions catered exclusively to the empowered is hardly surprising. Except for an historically atypical period in ancient Greece (when literacy was forced upon slaves so that they might read to their illiterate masters) formal education had long been the province of the gentry. Nonetheless, education was ultimately extended beyond its historically-exclusive, social-class (and gender) boundaries. Just as it is inaccurate to imagine that the advent of mass communication merely involved the extension of human communication through new means to a broader audience, it is similarly inappropriate to characterize the advent of mass education as simply bringing classical instruction to a larger, more diverse audience. In the following pages, we will outline how the diffusion of education created new versions *of* education. However, a threshold question precedes that overview: *Why* was education diffused into the general population in the first place?

The gradual "diffusion" of formal education to young, *non*-noble men has often been chronicled as the product of progressive social reform—campaigns waged on behalf of societal welfare and/or of redressing inequities of birth. While reformist campaigns were, in fact, launched, many historians have (rightfully) argued that social reform, in and of itself, is rarely the major or effective cause of social change. For instance, the American Civil War and the Lincoln presidency will forever be historicized as a fight against slavery—which no doubt they were. However, history also reveals other (notably economic) reasons for the emergent opposition to both slavery and the secession of the southern states. Other less progressive motives for the spread of education have also been tendered, for instance, the *nouveau riche* conspicuous consumption of "genteel" education for their children. One problem with many accounts of education history is that this history is presumed always (or even mainly) to be *about* education. As we shall see, education can and has been used to (help) satisfy *other* systemic needs. Particularly with respect to the construction of intelligence, attention must be paid to the less overt and more systemic influences. With such an orientation, it is most economical to confine our historical investigation to an era and area of, arguably, the most significant diffusion: the development of higher education in the second half of the twentieth century. While the issues to be addressed can be found throughout the educational process, the most noticeable indices of class variation occur *after* curricula no longer revolve around "indispensable" fundamentals. Moreover, the curricula in higher education are those most overtly linked to applied and vocational options which, as will be explained, are intimately tied to the relationship between intelligence and social class.

Education as Social Constraint

There are several important economic, medical, and technological conditions that either hadn't developed until or experienced significant development *during* the twentieth century—and particularly, onward from the second half of the century. Probably most important among these developments are the increased human lifespan, automation, the growth of multi-nationalism, the burgeoning cross-national use of workers in developing nations, the dwindling coffers of social security, the increasing costs of medical coverage, and the end of mandatory retirement. At first blush, all these conditions (which, obviously, are themselves rooted in other events) might not seem related to each other, much less to education. However, these are the more immediate *pre*conditions of social-class expansion in higher education. In this case, it is education—or rather, institutionalized education—that becomes, itself, a mechanism of containment. It has been often documented that the aforementioned preconditions, in confluence, generated socioeconomic problems relating to employment and related fiscal responsibilities of the government. Moreover, depending on the economic analyst, these preconditions either caused or were exacerbated by critical shifts in both the birth rate (particularly the post-war boom) and the potential, domestic employment pool. Overall, what the second half of the twentieth century saw, then, is the collision of two incompatible forces; an increasing pool of job-seeking unemployed (at *both* ends of the lifecycle) and dwindling government resources to tend to any types of overrun and ancillary problems. In this context, the education industry is used to ameliorate troubles involving unemployment and government assistance. If more people of working age *go* to school and attend school for *longer* periods, then the pressures exerted by these problems are, at least, shortened and deferred. University students defer employment and, typically, are disallowed various government monies, but for scholarships and other school-assistance. Even with rising tuition, academic assistance tends to be substantially lower than public assistance. Moreover, a burgeoning student population has, on many fronts, been deemed preferable to enlarging the bureaucracy of government-assistance or its dependents.

Edumobility and Diffusion through Social Classes

Given a nation of numerous representatives and almost-constant elections, deferral becomes a common administrative strategy. As a result, government and other empowered agencies provide incentives—or help provide the means—for less-advantaged people to partake in higher education. This general practice strikingly increased in the second half of the twentieth century—beginning most notably, perhaps, with the G.I Bill. However, given the culture's long-standing anti-intellectualism, these benefits are, in and of themselves, insufficient motivation for people to volunteer for education beyond the amount mandated. Along with this assistance, other conditions and ideas had to emerge—among these, a mythology of educational advocacy—the mythology of "edumobility." Edumobility myths, collectively and generally, recommend education and academic degrees as a means of upward social mobility. After all, educated people before the diffusion *were* more successful than others.

Edumobility is a distinct subset of the longer-standing American, upward-mobility mythology. The general mythology of social-mobility grew especially strong after the American Civil War, and, as immigration numbers and stories substantiate, these representations were sufficiently powerful to reach far beyond the nation's borders. However, now in the twenty-first century—with a society inured to the likes of Bill Gates, $27 million Powerball lottos, or the "real" lives of Laguna Beach teens—it may be difficult to imagine the earlier versions of class-breaking success. Popular artifacts, such as Horatio Alger novels or Thomas Edison's published career advice, stipulated that the road to economic achievement was one lined in hard and determined work that, in turn, could lead to a respectable and contented life (of comparatively modest success). It really was not until well into the twentieth century, that increased *education* started to be understood as a detour, if not a shortcut, off the path of serious hustle and labor.

Eduinflation and Depreciation

Along with the social-class expansion of higher education, the ascendance-of-faith in edumobility responded to (largely post-war) socioeconomic problems; the significant expansion in student-hood allowed for some postponement and, possible, deterrence of the spiraling growth of unemployed and publicly-assisted. Problem *deferral*, though, is invariably a short-sighted remedy, and often induces its own complications, requiring further social adjustment, and so on. In this case, cause-effect linearity is difficult to discern because the original action—the social-class diffusion in education—is never reversed, and so, the consecutive adjustments overlap.

Throughout time, education is class "coded"—differentially constructed as a function of social class. Throughout most of Western history, the coding is simply its literal *availability*, i.e., education was generally available only to the rich and empowered. In the beginning of the period we currently analyze, we see a more elaborate coding pattern. The beginnings of this pattern can be observed in the differential promotion of education. After WWII and rapidly increasing through the 1960s, higher education was advocated to those youth for whom, *prior* to the 1940s, only high school graduation was anticipated. Consisting of those neither subsidized by inherited wealth nor those "ranking" below "lower-middle-class," the males would likely assume mid-level/non-executive management positions, small-business ownership, or like-income employment both prior to or after education's diffusion." The working-class population—an even larger group—was also "targeted" for extended education. Their pre-diffusion record of public-school completion was less regular than that held by members of the middle class; the high-school diploma, short-term trade-school, or apprenticeship became the elevated "terminal" standard for working-class students. Thus, although different, increased educational goals were promoted to all non-wealthy workers.

This promotion was unquestionably successful to the extent success is measured by enrollments; it was successful because this promotion resulted in an historically *disproportionate increase* in degrees, certification, and diplomas, as well as "time served" educationally. However, if, instead, "success" were measured by "structural" employment—an older economics term indicating the degree to which employment opportunities befitting or substantively dependent on this more-educated labor force are available—the result was not so profitable. As the average-time-spent-*in*-education increased, so, too, did structural unemployment or "misemployment," as it were. Indeed, in the second half of the past century, the condition of "structural unemployment" became so familiar that it served as a cultural irony, e.g., "armed with a degree in fine arts and a bus pass, he was able to secure a position stocking shelves at the supermarket." Such drollery notwithstanding, the *greater* relief of problems stemming from *general* unemployment and government-assistance has been sufficiently compelling to have maintained, and even intensified, broader and longer educational terms *despite* the ensuing structural un/misemployment. Moreover, the structural unsuitability of the employment market has been addressed with correctives (which we will soon detail). Beyond structural unemployment, there have been other problematic outgrowths of diffused and inflated education. For instance, the substantial increase in degrees, certification, and diplomas has also caused their (seemingly inevitable) *depreciation*; as their number grows, their relative value, consequently, decreases. Although different conversion techniques have derived varying figures, virtually all economic analysts agree that a late-twentieth-century baccalaureate, (despite the much inflated cost of pursuing it), became less valuable than one held as recently as the 1970s. Of course, the depreciation of a specific education must, ultimately, consider the degree level, the discipline, and also, the general distinction of the granting institution. Degrees awarded by elite institutions, for instance, have been strongest at holding their value. Importantly, though, these expansions and extensions of education have commonly been understood as progress rather than inflated "growth." One generation in a family rejoices in the following generation's increased education (and higher salaries), often without regard for inflationary-conversion (on either front).

Although, we must necessarily limit the specific examples we outline, as suggested above, there arose different forms of *adjustments* to redress problems ensuing from the diffusion of education. Also as indicated, some developments were likely reactions *to* reactions, e.g., some strategies arose to address structural unemployment and/or "education"-depreciation—themselves reactions to the eduinflation (in turn) resulting from the social-class diffusion of education. Indeed, since advocacy of extended education continued despite these reactions, it is sometimes simpler to view ensuing adjustments as simply redressing extant problems in the sequential complex of eduinflation. One type of adjustment that responds to more than one issue has been the subsequent inflation of educational *pre*requisites. This inflation could clearly be seen in terms of programs for professional preparation, college-degree programs, and interaction between the two. For instance, nurses' training was often transformed from three-year, in-hospital, certification (R.N.) to a baccalaureate program. Similarly, three-year "normal schools" for training teachers (grades 1-12) were also phased out in exchange for four-year baccalaureate degrees. In many locations, these normal schools were literally converted to four-year state colleges—still expressly for the training of teachers. The additional, "preparatory" year was typically filled with theretofore unavailable electives (outside of pedagogy) and/or new, generic, institutional requirements. This sort of elongation through prerequisites can also be observed as redefining graduate study. As late as the 1960s, medical- and dental-schools did *not*, uniformly, require a baccalaureate degree for admission; however, probably within a decade from that point, these professional schools became exclusively graduate institutions, stringently demanding an undergraduate degree for admission. Not irrelevantly, the professional schools' curricula were *not* uniformly revised to reflect the imposition of this new prerequisite. Similar protractions, lengthier reorganizations, and/or increasing requirements affected many other areas of professional training such as architecture, engineering, or law, to name just a few. As the inflation compounded, some professional-training programs (e.g., social work or public-school teaching) metamorphosed within a few decades from not requiring a completed baccalaureate to, ultimately, mandating graduate-study! In some areas, public-school-teaching "certification" consisted of a given number of semester hours in relatively unrestricted graduate study. Other certification programs were also rooted in "additional" credit-hour generation—based on compounding the institutional (or economic) unit of measurement. Not coincidentally, after a cycle or two of prerequisite-adjustments for eduinflation, dozens of new graduate programs (including those for Ph.D.s) were established at universities. Many of these new programs established graduate degrees in areas for which advanced study had never before been created—thus, expanding institutional structure for additional eduinflation.

All these added educational factors (commonly demanding additional time spent in the academy) were, typically, met with complimentary adjustments on the part of institutions for which education was a screening criterion, i.e., graduate and professional schools, and employers. For instance, job opportunities once requiring a high school diploma would—fewer than two decades later—demand college completion. It is in this period that the baccalaureate became the minimum credential for professional employment. While logic might suggest that extensions of educational programs and degrees should follow changes in employment criteria, quite the opposite occurred. Education did not rise to the demands of the employment market; labor, instead, reordered its priorities in response to extended education. This "reverse logic" reconfirms the central argument that these changes in academe have no objective greater than constraining larger, societal problems (not, themselves, connected to education). In turn, the complementary adjustments in the professional and business world can be said to offer pragmatic rationalization of the eduinflation. Or, put another way, the eduinflation that helps defer certain economic problems requires the cooperation of consumers (who *will* send their children to—more and longer—school). The corresponding changes in professional requirements legitimate the added time and expense of this schooling. That the edumobility mythology *precedes* the conditions it promotes and/or that the educational "reforms" are virtually independent of either intellectual/erudition values and/or the foreseeable availability of structurally-appropriate employment are hardly trifling social conditions.

We have now outlined inflationary adjustments to the social-class diffusion of education. As we have seen, these inflationary elements fall into two broad classes: (1) an increased "number" of events (e.g., students, schools, programs, degrees, high grades, etc.), and (2) increased time (e.g., length of revised program, years in standard program, etc.). In many cases, the net result has been an intersecting inflation, e.g., stiffer job prerequisites—increased credentials as well as time served in academe. Before examining two other types of adjustments, we add briefly the only educational issue with which "inflation" has been popularly associated—"grade inflation." This problem has received so much attention that it requires no definition; however, since its place among other edu-inflationary indices is not self-evident, we analyze specifically how higher grades *do*, in fact, follow from social-class diffusion. Now, the literature cites lamentations about grade inflation as far back as the nineteenth century, however, it has been commonly suggested that the current grade inflation is an artifact of American involvement in Southeast Asia—a military engagement which relied on Selective Service conscription. Because of that conscription (coupled with the unpopularity of that undeclared war), higher grades were allegedly awarded to protect students' draft exemptions. Although there seems no reason to doubt this allegation, there *are* a number of reasons to discount "Vietnam" as the point of origin. First, the period of conscription (1969-1972) lasted but three years, nonetheless, rising grades were, in Keynesian terms, "sticky," i.e., they persisted, and some say even hyper-inflated, *after* the withdrawal of US troops in 1973.[20]

Also, analyses do not seem to report significant gender variation in inflated grades even though only men might benefit from draft exemption. But, even if we imagine a deliberate attempt not to penalize (comparatively) female students—although *popular* feminism and political-correctness movements proceeded Vietnam—we would still have to believe either that (1) the (genuine) academic performances of those eligible for student-deferment were, comparatively and significantly inferior, or that (2) inflated curves were instituted despite earned grades not in need of inflation. Absent these initial conditions, the older patterns would not have shown inflation. While other, even *less* likely, conditions might be invented, it seems more profitable to propose that the modern trend of grade inflation is rooted *earlier* in the social-class diffusion of education that spurred eduinflation, in general. It is not simply the rubric of "inflation" that's persuasive, but rather the logic of the precipitating conditions."

Metaphorically, academe had also been drafted—enlisted to help balance some serious difficulties emerging from the cumulative effect of WWII on the heels of the Great Depression, i.e., problems associated with unemployment and government assistance. As higher education was more widely diffused, the ivory-tower transformed from a sheltering edifice protecting *research and erudi-*

tion to a corporate entity devoted to maintaining and cultivating *enrollments*. Educational standards and expectations had, correspondingly, transformed. To maintain and cultivate enrollments in a world of proliferating and competing programs, degrees, etc., marketing and retention had to become institutional priorities. Such competition also reversed basic conditions of market demand. Universities conducted business as service-oriented sellers with significant effort directed toward mainstream publicity and customer relations. This revised orientation had to accompany the edumobility mythology because, in addition to the influence of a competitive market, most of the targeted consumers did not possess long or strong familiarity with or preparation for academic tradition. In other words, the promotion needed to be persuasive *to this new population*. University administrations, then, exerted both formal and informal pressures to "discourage" any conduct (such as awarding low or failing grades) that might *inhibit* attracting (as well as retaining) students. From the "other" side, these adjustments were hardly *sub rosa*; students have increasingly become aware of the consumer-side economics. Although edumobility mythology has, presumably, convinced many consumers to endure extended education, it also encourages consumers to see student-hood *as* time invested and served. And as suggested, since motivation was principally connected to the "destination" (the credentials) rather than the journey (the learning), it is understandable that travelers would prefer itineraries featuring minimal labor and stress. Surely, this is a complex matter with several more, unaccounted for variables. Another variable that might have stimulated grade inflation, for example, was the mainstream introduction of "intellectual-pluralism." After all, one salient feature of pluralism is its skepticism of unilateral evaluatory standards . . . like, for instance, grades. The point here, of course, is not to account precisely for grade inflation, but rather to show how it, very plausibly, is explicable as part of the eduinflationary complex.

The Paradox of Modern Education?

Our review of eduinflation, along with most histories of modern American education, suggests that accessibility and growth were defining characteristics of higher education in the second half of the twentieth century. This typification is eminently reasonable *apropos* the history of eduinflation, in isolation. However, is this theme of "accessibility and growth" contradicted when examined against the backdrop of all preceding sections in which major, cultural conduct has been characterized as *anti*-intellectual? How does the growth and accessibility of education fit with repeated demonstrations of intellectuality and erudition being either discredited, obscured, diminished by conceptual division, or somehow *disempowered* in, or (physically and/or conceptually) *estranged* from mainstream culture? Our analytic perspective is not so Procrustean as to reject contradictions. Indeed, functionalism is oriented to *aggregate* assessment; it anticipates exceptions. But here, the contradiction is no mere exception-to-the-rule; it is, rather, an incompatible conclusion. Education is too central to the construction of intelligence to interpret it against the grain of all other systems and institutions. Thus, either all our earlier assessments about anti-intellectualism are wrong, or the social-class diffusion and inflation of education does *not* indicate the paradoxical promotion of intellectuality. Inasmuch as we stand behind all earlier sections—and the general existence of cultural anti-intellectualism—the answer is that the growing surge of education after the Industrial Revolution is most definitely *not* a testament to the culture's respect for and allegiance to serious education. Indeed, we would claim that the development of modern education—particularly that education provided for those working-to-middle-class people historically, disenfranchised from the academy—is anything *but* supportive of intellectuality and/or erudition. In other words, modern education is every bit as anti-intellectual as our popular stories. Through the next few subsections, we explore additional repercussions of the social-class diffusion of education where we can see how the seeming paradox of *anti-intellectual* education may be ironic, but not at all impossible.

Applying the Brakes: Instructional Deceleration

Here, we examine instruction by uncommon means: *volume* and *rate*, i.e., the length, breadth and depth of material covered and the speed at which it is taught. The corporatization of academe has been criticized for, among other things, measuring "education" by semester hour; ironically, though, people familiar with academe also know that semester hours are often the sole means by which learning *is* measured. For instance, a university might demand twelve semester hours of a foreign language to satisfy pedagogical objectives regarding the acquisition of translation skills and/or a multi-cultural perspective. Comparative analysis of volume and rate certainly does not assess all the nuances of instruction, but it can address whether eduinflationary extension and expansion actually do extend and expand the learning process, or whether they merely enlarge the institutional "sentence." Educational deceleration occurs either when (1) an inverse correlation is obtained between intellectual progress and educational-time served, or, less dramatically, (2) two educational terms are held constant, but intellectual progress is reduced. Thus, analyzing volume and rate assesses both the equation of education and semester hours, as well as the seeming paradox of eduinflation in an anti-intellectual world.

While colleges and universities typically require tests for admissions as well as student evaluations of courses, comparative analysis of instructional volume is something hardly ever acknowledged, much less measured. While much of our university system was, originally, borrowed from the classical, British institutions, the employment of mass-administered comprehensive (level) exams is not something that prevailed here (or, for that matter, throughout the contemporary British system, either). It has been suggested by some social critics that the failure to monitor volume and rate, both in terms of curriculum and successful instruction, was responsible for the later-twentieth century "discovery" of (or previous failure to find) the increasing proportion of functionally-illiterate students among American high school graduates. Critics suggest that an institutional aversion to empirical attempts to analyze standardized objectives continues to thwart discovery and remediation of instructional problems. Many of these plaints were offered during the 2000 presidential campaign for national-resource allocation based on such standardized measurement. Regardless of resistance, nothing has thwarted deceleration from being casually observed.[21] Such observations can only be made by those who have access to the comparative information relevant to deceleration—and, anecdotally, these observations are pretty much limited to people literally old enough to have witnessed different educational eras and, presumably, the lost educational ground between them. This is probably why such anecdotal observations are typically dismissed as senile dementia. Consider how many films or TV programs—particularly comedy skits—feature a stereotypical grumpy, older person engaged in a "when-I-was-a-boy"-type tirade. What is ironic is that the relevant comparative data suggest that such tirades are *not* indicative of a romantically-distorted or ravaged memory.

Educational deceleration may be accomplished through several means, some of which more easily submit to comparative analysis; as is often the case, the most compelling indices may also be the most subtle. The less overt indices involve three, broad issues: (1) the *depth* of the instructional material, (2) the degree to which the instruction mandates ancillary "intellectual resources" (as defined earlier), and (3) the nature and depth of material for which students are responsible. Evaluating these variables may present methodological problems because the more available instructional records, like syllabi, may not fully reflect instructional depth or intensity. Educational research has, nevertheless, shed some light on these indices. For example, tracking of undergraduate and graduate reading assignments—especially the comparative use of textbooks versus original, issue-specific readings—has provided evidence of deceleration of the depth of both instructional materials as well ancillary resources these materials might call into play. Before the 1970s, anthologies of original readings by the seminal scholars were far more popular in the social and behavioral sciences, in history, philosophy, and religion, and to a less-documented extent, in certain business-related subjects. Reading assignments in undergraduate courses also show a huge reduction in the amount of supplementary readings (of any type); in fact, most commonly, the number of readings—*other than* one main text—dropped to zero in state universities and community colleges. Publishers suggest that, just decades ago, assigned readings in over half of advanced, undergraduate courses would include specialty readings in addition to a text (if one were assigned). These sorts of trends don't *merely* reflect a decreased amount of required reading; the differences between texts and other readings, are important for other reasons.

Arguably, the elimination of original, individualized scholarship eliminates an important type of cognitive integration that is rarely required by instructors using a single textbook. In other words, supplementary readings, not simplified or otherwise revised to fit into a semester's worth of material, demand that student-readers (with, perhaps, the instructor's rhetorical-modeling) reread and analyze work—moving between lecture material, text, and readings—until a synthesized interpretation is cognitively formed. On a textbook-only diet, the tome typically assumes all of this intellectual burden by predigesting and organizing material into a series of simplified packages. Amplifying this problem is another major criticism directed, specifically, at college texts, to wit, that even textbooks themselves have become increasingly *more* simplistic. For instance, some critics observe that, in addition to more explanation of fewer ideas [22], most texts now include bullet-pointed, chapter summaries.

Changes in academic publications and reading assignments are, of course, only one way to examine deceleration. The popular advent of digital technology has also been especially cited as changing the breadth and depth of instruction. Some of this criticism seems to assume a Luddite'ish, generic anti-technology posture, but other more reasoned discourse acknowledges that digital technology and the Internet may, potentially, enhance instructional possibilities, too. Most often cited in this context is either the use of bulletin boards for interactive discussion or the increased accessibility of hard-to-find scholarship. Still, many have argued that such obvious benefits are wholly undone by other, deleterious practices inherently encouraged by the structure of much digital technology. For example, one plaint is that students have become increasingly less likely ever to have entered and used a campus library. This plaint, at first blush, seems speciously romantic, even if it's entirely true. Unless the special benefits of a library's architecture or environment is put forth, the importance of conducting work in a library proper has not been established. However, beyond architecture, perhaps one can hypothesize some more serious implications. For instance, perhaps it's not the absence of the library, but the substitution of the Internet and special educational websites (instead of older-fashioned library research-resources) that amputate certain key intellectual exercises? Links, for instance, that allow students to research "efficiently"—to go directly to specifically-isolated content as indicated by key words—may eliminate sifting through research and theory to detect what may or may not be applicable. Is that lost sifting, interconnecting, and prioritizing a skill that's important in intellectual development? Is it a skill that only library work helps develop? What is the price for search-efficiency that locates only that information with which the searcher is already familiar (i.e., familiar enough to enter it as a search criterion)? Probably the most heralded criticism of the Internet is that the huge and easily accessible array of esoteric sources has led to an increase in plagiarism—the intellectual disadvantages of which need no discussion. Or, a more recently-cited problem relates to the use of digital technology *in* the classroom. The so-called "smart" classroom has, it is suggested, contributed to the decline of critical thinking. Even more than over-simplified textbooks, it has been argued, certain instructional software (most notably, PowerPoint) has supposedly "freed" students from having *to process* complex verbal information. Instead, even lecture material is increasingly expected in synthesized and "memorizable" bullet-points. This seemingly dustless and streamlined technique has, it is debated, not only depreciated students ability to listen, to extract, and to prioritize information, but also, lessened the degree to which full comprehension is even required—especially to the extent that examinations befit the highly-condensed summaries. If most classroom information comes as a list of discrete data, and that list is the basis of examinations, what *is* it we are teaching students to learn?

In addition to reading assignments and the use of classroom technology, there is also considerable discussion of deceleration-through-evaluation apart from grade inflation. It has often been argued that the inflation of student populations made it increasingly impossible for instructors to work with and assess students personally. Thus, eduinflation, it is claimed, made a virtue of so-called "objective" (e.g., multiple choice) testing which, in turn, reduced greatly both the complexity of information for which students were responsible as well as the necessity *to be* intelligible, let alone convincing, through written expression. Finally, there are a range of simpler indices, that might be used to examine deceleration, e.g., changes in credit-hour generation, changes in course-load expectations, the assigned credit-hours to courses, and/or the total number of credits required in various subjects, to graduate, etc. Literature suggests that there has been a general deceleration with respect to *all* these types of indices—or rephrased, increasingly, more college credit is given for relatively less work or achievement. Studies of college credits or "semester hours" suggest an earlier history in which credits were not only tied to a uniform standard within the institution, but across institutions. However, although the specifics vary according to report, the uniform rules for awarding semester-hour credits has, it is argued, increasingly varied by institution. And even *within* an institution, different colleges and departments have employed different standards to arrive at what a unit of instruction is worth.[23] In the end, students, it is argued, have increasingly been required to taken fewer courses to graduate, wherein the courses taken are neither more comprehensive nor more rigorous—just more credits. Other studies of "core" courses in various disciplines have found that introductory prerequisite courses have also been elongated to two, and in some cases three, semesters, without comparable enlargement of the syllabus or assigned readings. Like the changes in grading criteria, some have credited the growing student population with the need to increase teaching time.

This section has offered few and abbreviated examples that are meant mainly to suggest how deceleration is specifically accomplished in the structure of instruction itself. The importance of deceleration is that it is one means by which the extensions and expansions of education are, otherwise, offset so as to suggest, perhaps, that *plus ça change, plus c'est la même chose*. In all, then, it might be said that while eduinflation lowers the market value of educational credentials, *deceleration* diminishes the intellectual value of knowledge obtained in the course of that education.

Social-Class Coding

Social assets are, by definition, relatively scarce, yet desirable, and their cultural management is complex. Zealous interest in or pursuit of the asset itself cannot occur too frequently or too profoundly. With wealth, for instance, social mobility is sufficiently "dangled" to induce most potential workers to continue culturally-necessary labor; but everyday economics along with economic mythology all go *to constrain* fervent capitalist obsession among the have-nots. To achieve such contradictory ends simultaneously, special mechanisms evolved for the unobtrusive encouragement, discouragement, or silent maintenance of related conduct. Many such mechanisms may have a "duplicitous" effect, i.e., promoting one outcome while creating its antithesis. In any case, these processes are, typically, covert because, if detected, they reveal how social mobility is manipulated by factors *other than* merit, luck, or hard work. In other words, their detection shows the American dream to be an impossible one. And, despite social criticism over a rise in cultural cynicism, the American dream seems not hardly in the final throes of existence.

Decelerated education is one example of such a covert practice; it lengthens the educational process while awarding more academic credentials, but it does so *without* commensurate adjustment to instruction and/or *relative* earning power. Deceleration is covert and duplicitous; no one typically imagines "more is less" with respect to education. Yet, more complicated and insidious than decelerations is a mechanism mentioned earlier: *social-class coding*. Social class coding is the differential (and, usually, covert) formulation and/or dissemination of education as a function of *social class.*[24] Class coding began as soon as education infiltrated the population at-large. We approach it here in terms of the twentieth century, social-class diffusion of higher education, but in all cases, class coding is cloaked by the wrapping of "inclusion." More specifically, elite education is believed to have spread to the underclasses—to have become more inclusive. Social-class coding adds the sense of "included-but-different," i.e., while the underclasses are indeed provided with many of the same subjects, diplomas, grades, and so forth, they are *not* incorporated *into* and/or don't receive all the important benefits of those of the elite tradition.[25]

Typically, the relationship between social class and education has been limited to such indices as the median income of students' families, the size of endowments, the number of affiliated renowned scholars, or even the social-mobility potential embedded in school ties or socially-extended "nepotism." Our introduction of class coding deals with an additional set of categories: (1) overall organizational structure (colleges, schools, programs, degrees offered, faculty degrees, possible majors and concentrations)—in terms of what types of study and knowledge are available and constructed, (2) requirements for completion/graduation, i.e., what types of knowledge are designed to be *in*escapable in a given institution, and (3) course curricula—what is the breadth and depth of knowledge presented and how is learning breadth and depth assessed? In sum, the class coding that most impacts the construction of intellectuality concerns the extent to which classical instruction (an emphasis on traditional intellectuality and erudition) is a curriculum requisite. While it is impossible to report all the individual variables relevant to this level of social-class coding, we, instead, organize and summarize the two most important effects of their variation. Or, more simply, variations in intellectual class-coding permit us to isolate mechanisms by which decidedly non-intellectual material is offered as either traditional instruction or as equally good as (or superior to) traditional instruction. Predictably, the non-traditional variant is taught to the less-advantaged student populations. Clearly, there are subtleties and caveats relating to this coded instruction, and we try to attend to them below in the more specific descriptions of (1) knock-off and (2) "useful" (or pragmatic) instruction.

Knock-Off Intellectuality

Etymologically, "knock off" is, interestingly, part of mass society terminology; it emerges from mass production in the fashion industry, and refers, specifically, to churning out or "knocking off" facsimiles of original, master-tailored garments (correspondingly, the *haute couture*). Thus, knock-offs bear a resemblance to the original, but lack its quality and integrity. While the term is still popular in the fashion world, it's also eminently applicable to other modes of production having an elite tradition that is counterfeited for popular sale. In the fashion world, illegal designer labels and logos are often attached to the knock-off, so that the garment might "pass" for the genuine article.[26] The notion of counterfeit or forgery—as opposed to an honest alternative or even *homage*—is critical.

Virtually all forms of social-class coded instruction emerge without disclosure or acknowledgement. These mechanisms are unobtrusive precisely because the distinctive codes are *not* to be observed; the simulated version is intended "to pass." As knock-off intellectuality, then, it is more commonly taught in non-elite institutions. As we've indicated, the deception (in material coded for dissemination predominantly to working-and middle-class students) is located in the appearance of inclusion—the appearance that this education made available to most working- and middle-class students is that which has been offered all along to the sons of rich men. However, the instructional material is coded—differentially formulated and disseminated for the less privileged.

As noted earlier, the patterns of class coding became very clear while examining curricula, syllabi and reading assignments *for deceleration*. However, this was no serendipitous discovery. Deceleration and class coding share many data bases, e.g., institutional structure, curricula, reading assignments, syllabi, evaluations, etc. Secondly, decelerated material *may also be* class coded, and, when it is, it's invariably a decelerated knock-off. More specifically, eduinflation affected all institutions of higher learning, and deceleration was, correspondingly, a broad, systemic response to eduinflation—generally serving to offset the extensions, expansions, and other inflated aspects of education, so as to preclude a net intellectual gain. However, *beyond* this general and unilateral restraint, deceleration has been particularly enlisted to impede intellectual growth where it would be the most dangerous, i.e., in that context where it would be, historically, the most unusual, and, sociologically, the most damaging: within the non-elite institutions and among the predominantly working- and middle-class students. Literally, dozens of examples may be cited here and our illustrations are merely representative of distinctions in instructional material and conduct. Indeed, we tried to select specific illustrations to indicate the very smallest variables we analyzed in inducing class coding. For instance, basic studies or core requirements are part of the curricula in most undergraduate institutions—elite or not. Just comparing these sets of academic minima, as it were, arguably tells more about the institutions' intellectual mission than their formally-composed statements-of-mission. One commonality among *all* types of institutions with a general-studies curriculum is the inclusion of at least one mathematics-type course, i.e., typically, at least three semester hours. Often a limited menu of possibilities is indicated, so that choices in some places include, for example, statistics and logic in addition to courses offered by mathematics departments (and, in non-elite institutions, sometimes, by "general studies" programs). In examining core studies across institutions, then, we looked, both at the general areas deemed "core" *and* at the specific selections available for each area. Continuing with our specific example, the menu of courses fulfilling a math requirement at community colleges and state universities is almost four times more likely to include at least one course described as a review of geometry, algebra, and trigonometry principles. High-prestige institutions did not significantly require *more* mathematics credits than working-class or middle-class schools; however, as implied, it is comparatively rare to find elite institutions offering any credit for coursework covering high school-level, remedial mathematics; when geometry was included in an introductory course, it was invariably non-Euclidean geometry and, often, mixed with an introduction to calculus. As suggested earlier, we don't mean to call attention away from the larger scheme, but class coding is often located in such details. In fact, this variation in mathematics requirements is mirrored throughout a comparison of general-studies courses. For example, in comparing the common "science" requirement, we find that science departments in community colleges and, particularly, state universities often offer a "special introduction" for *non*-majors—a course described in catalogues as one-semester, "theoretical" overviews of the given discipline. Relevantly, these non-major courses may *not be taken* for credit by science or pre-medicine majors, but they can fulfill non-majors' basic-studies requirement. Similarly, in the humanities, the vast majority of all schools offer introductory literature/writing courses—the particular section often determined by preliminary placement exams. What distinguishes a surprisingly high number of community colleges and state universities in terms of a general "composition" course is the option of fulfilling the standard writing requirement with a *for*-credit *remedial* course. Overall, then, one can find in the "details" a tendency for non-elite institutions to develop "softer," diluted, and often non-advancing or non-collegiate instruction in *core* curricula. These instructional inventions can be interpreted as demonstrating class-coded, impostering (knock-off), as well as decelerated instruction.

Arguably, what is coded in the details is not—in the scheme of social-order maintenance—itself a detail. And, it must be remembered that the details of core curricula, for example, are just one set from dozens of class-differentiated codes. These variations can be found in the most prosaic contexts—such as the average number of books assigned per class (elite schools' averages significantly more books), or in more subtle, organizational frames. For instance, elite universities are more likely to have courses in—or, indeed, support whole departments of—arcane, academic subjects. In the late twentieth century, foreign-language departments, for but one example, started to be collapsed and/or eliminated in many institutions of higher education. Yet, these departments have remained significantly more intact within elite institutions.[27] This maintenance of "knowledge-for-it's-own-sake," though, may be more relevant to our next type of social-class coding.

The Pedagogy of Pragmatism

Knock-off education is a simulation of traditional instruction because it is similarly packaged, labeled and promoted. In most instances, though, the distinctions can't be detected simply because it is taught largely in non-elite institutions where consumers don't, and probably will never, have any basis of comparison. It would be too simplistic and, frankly, inaccurate to suggest that traditional instruction is not available at state or community-sponsored schools. However, in terms of both empirical availability as well as cultivated motivation, it is fair to say that it is exceedingly *easier* for students at non-elite universities to have a fairly regular curriculum of knock-off offerings; for the same reasons, it is far easier for the student enrolled at an elite institution to acquire a classical education. In fact, in this latter instance, it is much harder to escape traditional instruction or to fail to see it modeled.

In earlier sections we have, in various contexts, seen intellectuality counter-positioned with a number of qualities or characteristics, e.g., "street smarts," emotionalism, or even, by extension, sexuality. To counter traditional intellectuality and erudition *in the very context of instruction* is to introduce a "new" opposition: the practical, pragmatic, useful, concrete, viable, and so forth. This parallel also provides a "negative" identity to intellectuality, one that degrades traditional knowledge—especially "knowledge-for-its-own-sake—as having currency only within the ivory tower. It takes little imagination, then, also to imagine why consumers vitally concerned with employment and wherewithal could be easily sold an alternative education—one that does materially translate to the "real" world. That is the alleged *sine qua non* of the knock-off instruction, but it should be remembered that, in the overall balance, consumers are simply *more attuned* to the advantages of that which they receive. In other words, those patronizing non-elite institutions likely have strong opinions about the relative value of types of courses and majors: those that are theoretical, offering no overt, applied technique are, typically, those deemed useless by students inculcated to assume the pragmatism of applied-skills training. We don't suggest, though, that just because students recognize such academic distinctions, they also connect them to social class. The knock-off character, then, emerges inasmuch as it is different, and that it is marketed as the same, equal to, or possibly even better than, any other type of instruction offered anywhere.

Educational missions and objectives as expressed in catalogues and publicity materials reveal how institutions differentially emphasize practical/applied subjects, degrees, experience and "connections." Institutions' social-class status could, on average, be accurately identified by observing the extent to which these materials emphasize elements such as "professional training," internships, practicums, work-study, or the extent to which the language includes ample references to "real" worlds, employment, "hands-on," and so forth. Similarly, by examining an institution's academic structure—its colleges, schools, departments and programs—one can make similar inferences about the school's dedication to the pedagogy of pragmatism and, in turn, its social-class composition. Often, disciplines and majors that are the most powerful in an academically-inferior school don't even *appear* in the curricula of elite institutions. Of course, just as more traditional academic education is available at non-elite universities, practical training is also available at institutions principally servicing the privileged; the difference is in the proportions. The pedagogy of pragmatism is unquestionably both pitched *and* desired far more powerfully among the less advantaged. What remains to be considered is the promise of such "pragmatism"—particularly in terms of social mobility.

It is probably true that *no* amount of "practical" instruction will ever be half as compelling as the school ties, extended nepotism, connections or networking derived from personal relationships. Elite institutions are understood to be vastly superior (directly and indirectly) in developing such relationships. Indeed, elite school ties are understood *to be a perquisite* of attending such schools. Such advantages are also understood to follow graduates beyond, perhaps, any other force. One might even argue that the very meaning of "prestigious school" is that its mere imprimatur, alone, is influential. But these real and pragmatic influences are not directly encoded in academic instruction. The prestige operates by titular association and, in some contexts, it might be argued that the credentials and ties from an elite institution obviate (in whole or in part) the need for particular training. Thus, the practical training in non-prestigious schools is seemingly believed to be the *second*-best option—the best a student *without* "connections" can do.

Regardless of social class, traditional intellectuality is probably not regarded as particularly useful (unless one plans to become a professional scholar.) This would be an illustration of how ideology is dominant *because* it overrides class. In any case, neither the reliance on school ties nor learning applied skills capitalizes on traditional intellectuality and erudition. While virtually no opinion research on this matter could be located, it would seem that the classical instruction accompanying school ties—if it's recognized as a special form of education in the first place—is probably understood as incidental to the interpersonal network. That intellectual substance may, in fact, *be* a school tie seems almost silly. Yet, if boardrooms are dominated by people who suffered the same type of esoteric and, otherwise, useless erudition—otherwise interspersed with football games, eating clubs, etc.—this seemingly "incidental" education offers the unity of shared knowledge and experience. Yet, even if we accept this reading, would it then be fair to say that traditional instruction is otherwise effete and irrelevant?

Students, their parents, along with many faculty members and university administrators have, increasingly, approached higher education as if its value increases to the extent that it equips students with specific job skills. The pedagogy of pragmatism virtually blinds people to the direct consequences of "specific" education. First, there are relatively few "how-to" skills specifically sought or taught that make students eminently and immediately employable. Ironically, the few *academic* subjects offering such employable skills—e.g., accounting, statistics, and foreign-language fluency—are, reportedly, *not* courses often chosen by students wishing to avoid intellectualized and/or theoretic approaches. Much material emerging from the corporate sector suggests that many other applied skills have considerably less applicability that promised. Many corporate representatives have suggested that, to the extent that the "academic" preparation is for a very specific (and often hierarchically "low") job, the applied skills may be suitable; however, students even with entry-level jobs, students are urged to consider how many other students have also developed those skills. More to the point, business professionals have claimed that in considering new candidates for executive training, those with a highly-focused, but narrow range of practical skills will be overlooked in favor of candidates who project more general potential—particularly in terms of traditional subjects like literature, art history, and history. Similarly, representatives of certain professional schools (most notably law) are also on record for preferring—given parity in test scores and records—students with esoteric and eclectic backgrounds over those whose undergraduate transcripts have limited scope. In this context, corporate representatives suggest that many employers would prefer to know that a prospective hire is simply an apt *learner* rather than one who has, unfortunately, spent time mastering antiquated, unprofessional, or simply alternative versions of supposedly usable skills.

This last caveat refocuses discussion on what is probably the most important issue concerning the pedagogy of pragmatism, to wit, the extent to which this training significantly correlates to social-mobility. *Even if* specific-labor skills enhance a student's employability, a relevant question is how a student's education prepares her for her last job, not her first. Of course, *some* employment is a threshold criterion for social mobility—at least the kind that doesn't come through marriage, luck, or crime. However, what is rarely considered in constructing an education is the extent to which specific "employability" comes with its own "cap" or ceiling. It is in this context that the *practical* value of intellectuality and erudition shouldn't be overlooked. While very many education critics have created a stir about the, often implicitly-taught, intellectual resources that come with "non-decelerated," traditional, "elitist" instruction, they are often dismissed as conservators of stale educational values that, supposedly go against the interests of poorer students in need of practical employment skills. In state- and community-supported institutions, for example, supporters of more classical instruction for a largely less-advantaged student body have been criticized for being inappropriately superior to and out-of-touch with the needs of the relatively poor. Thus, in the interest of their material disadvantage, poorer students are benevolently provided with educations befitting their obvious need for regular work. This might be called the mythology of practical pedagogy. Intellectual methodology, probably as much as anything, is capable of providing an unbound capacity to reason inductively and systematically, and to integrate knowledge beyond simple learning. In this sense, it provides *adaptive* resources that *are practical* skills for upward mobility—at least the skills that probably should come in second, after nepotism and school ties.

Such analyses of conditions for social mobility remain largely hypothetical insofar as we can neither manipulate nor control for all the variables involved in the general process. To the extent that what we (and others) have suggested *has* merit, though, it is a serious matter, indeed, exemplifying a mass proposal of reactionary ends as progressive goals. However, such debate must be restrained from grasping at utopian principles. Specifically, how different could educational institutions—and, indeed, all cultural institutions (including popular media) and mythology—be vis-à-vis traditional intellectuality and erudition? The class coding described—as well as all other mechanisms we've outlined in this chapter—are functional. Despite popular mythology, in our hierarchical class structure, limitless social mobility is wholly impossible. People cannot *be* educated for unlimited "upward" travel.

Conclusion

In this chapter, we have attempted to introduce some of the mediated and non-mediated processes and mechanisms through which myths about intelligence, specifically those regarding anti-intellectualism, are constructed. Grounded in the functional-constructivist perspective, we have provided a framework for the critical analysis of intelligence mythologies along with extensive, but by no means comprehensive, evidence of such mythologies in various contexts; as stated earlier, uncovering significant socially-constructed patterns among and across various contexts is vital to the process of verifying the accuracy of an argument. Here, we have demonstrated how ideas about intelligence and intellectuality are socially constructed through aphorisms and popular stories (especially those told in popular media), within the Ivory Tower of academe, by public discourse about science and research, through the reification of aptitude, via conceptions of learning disabilities and distributions of intelligence, and by way of educational systems as they are connected to social class and instructional deceleration. Through these mechanisms, among others, mythologies about intelligence are disseminated and, ultimately, function to maintain stability within the social order.

Notes

1. This chapter is a much summarized and collaborated version of lectures we have given for the last three decades. Because this chapter was to be introductory, and because so little has been written about the construction and representation of intelligence, we wished to provide a summary of the many research and theoretical issues we have covered, rather than isolate one particular study. To provide this summary in combination with its scholarly history would, itself, be book length. One of the major reasons is that such a combination would require a critical review of the vast and vastly important sociology of knowledge literature. Inasmuch as more elementary lectures had been given along introductory lines, it was decided at the outset to make this chapter almost entirely discursive. Sari wishes to express her gratitude to several, one-time graduate students (now established scholars, themselves) who took up the cause and dedicated their dissertations and further research to this area so much in need of systematic exploration—in many cases, putting her once-armchair theories to the empirical test: Bill Evans, Amy Franzini, Susan Kahlenberg, and Selcan Kaynak as well as the numerous other students who served as coders, respondents and library assistants in the process. Also, thanks are given to Michael Krippendorf with whom the first "beauty and brains" research was undertaken, and Carolyn Blake and Craig Shank at Indiana University who helped tie up loose ends of this project. Paul Messaris remains the most trusted critic. Lisa would like to express much appreciation to Kevin Warner for his continuous dialog and assistance with all things scientific. This chapter is dedicated to the memories of George Gerbner and Robert Merton, who first provided the critical direction for this interest.

2. A very clear marker of cultural bias is found in identifications marked "*not* something else," e.g., non-Euclidean. Obviously, what it is not is the privileged, cultural default. Surely, we could suggest Gaussian or Riemannian geometry, but their general obscurity, ultimately demands identification in terms of Euclid or plane geometry. Similarly, even though research has advanced substantially with respect to various forms of communication—e.g., visual, body-motion, musical, proximal, temporal and a host of other

types of interaction—all these other forms (many we actually use *more* than verbal language) are persistently lumped together as "non-verbal," thereby reflecting the cultural privileging of (verbal) language.

3. Of course, in a culture dominated by rationalism and capitalism, it is inevitably unpopular to infer that there are regular forces more powerful than the self-motivated intentions of individuals. This is one very big reason that structural functionalism (as well as other wholly structural perspectives) is rejected while rationalism-based theories are embraced. There has long been some confusion about whether or not scholarship may determine who is empowered by virtue of who is *theorized* as empowered. Obviously, this introduction is geared only toward minimal description, not proselytizing. However, we would suggest that, to the extent that scholars gets to compose their own epistemological orientation (i.e., to the extent they learn of optional configurations), it is not something that should be decided on the basis of "psychological" comfort and/or (immediate) wish fulfillment. This is discussed, again, in a later note. In any case, the value of particular orientations should be assessed in reaction to comparative analysis of the philosophical *products.* In other words, it is not whether or not the initial premises fit with an ideologically-comforting view of the universe, but whether or not a given orientation can offer insights leading to social change and *the construction* of an ideologically-comfortable universe.

4. Although structural concepts do receive explicit scholarly attention (presumably because of their status as "challenger" notions), it can't be argued with any confidence that this attention is meticulously coherent. In particular, although textbooks may define key terms, one won't necessarily find consistent nor cogent definitions across sources. These definitions, *en masse*, often manifest a fluid, elusive character because "different" keywords seem to draw definitions from the same terminological reserve, e.g., agency, complex, function, institution, interdependent, interrelated, norm, order, organization, pattern, process, role, structure, system. Sometimes, these terms seem casually shuffled sufficient to produce a case of lexical vertigo. Strangely, perhaps, the words are more ambiguous than the overall theory because they are interconnected, if not synonymous, and tend to circumlocute the basic perspective. As we also use some of these terms, we try to differentiate and define them in a very loose, theoretical narrative. In any case, we apologize in advance for any violence we seem to inflict on other definitions and stipulate that the definitions are provided to serve the chapter.

5. At least one, classical American functionalist has suggested that functionalism is only a method, not a paradigm or approach. We are probably less circumspect about meta-theoretical terms and could understand an argument suggesting that functionalism is just structural theory applied to the analysis of social function (as described). This "method-only" attribution is appealing because it allows certain associations to be disentangled, e.g., it permits functionalists to eschew seeing society as a *literal* analog to the biological organism. However, given that that social-functional analysis is so particular, it seems evasive to suggest that it has no theoretical premises. Unfortunately, such evasive tactics might have been the fearful response of functionalists who, for whatever reasons, didn't thoroughly contest the illogical and perverse conflation of (unpopular) empirically-based inferences and dysteleogical theory. We suggest that when social stability is clearly distinguished from a research *objective*, numerous canonical theorists, such as Marx, Weber, or, more recently Bourdieu, to name but a few, can clearly be identified as performing functional analyses.

6. We point out, without elaboration, that functionalists wholly recognize and admit biological cause. In traditional functionalism, the social supervenes directly on the biological. What is more to the point, perhaps, is that the perspective that we, at least, endorse is invariably phylagenetic and never ontogenetic. It should not be surprising, then, that the most recent renaissance of functionalist theory has been in sociobiology.

7. To be fair, systematic analysis might make us modify this hypothesis which we'd say is accurate *in theory*. However, in practice, it may be the case that, the more an approach is tied to progressive, prescriptive, ideological objectives, the more it is likely to characterize big industry as one-dimensional. This unidimensionality, in itself, does not formally remove independent, cognitive behavior from the individuals involved, but, as described, it generally provides little *other* explanation for a singular corporate mentality. In a postmodern theory, for instance, there doesn't seem to be any reason that workers should have the propensity to exercise their individuality more than privileged owners of the same culture. Yet, it is the variations in textual readings and not the (inevitable) inclusion of "dominant ideology" in those texts that is, consistently, presumed.

8. The ultimate irony of academic publishing is that textbook writing is the one genre most capable of generating sales. What's ironic is that textbooks (which summarize others' research) typically do *not* count as scholarly production in most academic institutions. So, we see an inversion of worldly value and reward.

9. It must be mentioned that, of course, it is only a matter of linguistic convenience to refer to this the journalism as "covering the controversy." In fact, the coverage, arguably, *creates* as much as covers the debate. Whatever grassroots reaction may have resulted from the publication of the original research, it is likely minimal in comparison to the agenda set by giving it public attention.

10. The closest argument that comes close to such protest was made by Jhally and Lewis many years ago concerning the then-very-popular *The Cosby Show*. They argued that rather than delivering inspirational role models with an upper-middle-class, intact, nuclear family helmed by a professionally successful physician and attorney, the show, instead provided *unrealistic* models that best served an empowered white culture that might turn such fictional characterizations into unachievable standards for a disempowered, minority population. This reading is an inverted corollary of the notion of "underrepresented inferiority." In any case, such pronouncements are rare and, generally, unpopular—especially in a discipline in which scientific research has literally been chastised for being "pessimistic."

11. Here and throughout the chapter, the reader is reminded that our unmodified use of "intelligence" refers to the default definition stipulated earlier, or more generally, to traditional intellectual indices. We will, predominantly in this section, also consider other characterizations.

12. To be precise, we limit our use of these terms to their non-overlapping application to *students in higher education*. By this we mean diagnoses that are not coextant with other medical conditions (of which they may be a part). This limitation corresponds to those applied in our later section on education, and is not meant to compromise our theoretical arguments at all. We choose this context for two reasons: First, university policies and other information are handily available online. Second, our argument here pertains to the *normal* distribution of "intelligence" in the general population, and questions the degree to which our culture allows for variation within the range deemed "normal." At the level of higher education, students' learning issues are most likely to have been determined and, thus, we have the ideal population for our examination: people who likely

fall, at least, into the normal range of intelligence and who are distinguished as having learning disabilities *specifically*.

13. Of course, the details of *how* one acquires physiological conditions, when carried to *their* logical extension, reintroduce the very environmental issues that were believed to have been eliminated as a result of the neurophysiological findings. Virtually none of the educational literature on the matter proceeds this far back in inquiry about acquisition of disabilities.

14. As a unilateral, pre-existing hypothesis, it is akin, for instance, to Darwin's "inference" of genetics before Father Mendel's inductive substantiation with bean plants. While, of course, we certainly do *not* suggest that this analogy applies in scale, it is (structurally) isomorphic. Specifically, structural scholars have always assumed that *all* behavior either (1) *stems* from physiological conditions (the position of physicalists and materialists), or (2) has two explanatorily-exhaustive, interactive but mutually exclusive, formative levels, the physiological and the social (the position held by non-reductive physicalists, emergentists, and, structural functionalists, for examples).

15. On at least one meta-level removed from the theoretical plane of this commercial, on might actually consider physicality in terms of whether a person's depression was preceded by socially-provocative events, or, if, instead, he developed symptoms seemingly without social prompting. *However*, this level of inquiry does not remotely attend to the same question raised *either* in the commercial or in the learning-disabilities literature.

16. Problematically, perhaps, the particular illustration provided pertained to ADD/ADHD which is often included in a miscellaneous-type category, but not specifically listed as a disability. Given, though, that a substantial amount of learning disabilities registered in higher education are attention-linked, and also that we wish to use a popular-media example—again to call upon something that is culturally commonplace—we stay with the ADD/ADHD example despite this caveat. Clearly, the structure of the proof could apply to other disabilities and other drug treatments.

17. One reason sometimes given for pluralization schemes is that the reconstructed categories are based on distinctions recently provided by neuroscience. However, given the myth of "optional physicality," we also know that such correlations, in and of themselves, are sometimes interpreted to mean more than they do. However, even when appropriate distinctions are codified, it is possible that *even finer* neurophysiologic variations might be distinguished, e.g., what we now collapse under "math and logic" might be otherwise divided; logarithms, for instance, may involve different processes from algebraic word problems. Unless *everything* is both physiologically identified *and, then,* distinguished by *systemic calibration*, it is pointless to develop a scheme in which categories are named on an *ad hoc* basis. Indeed, even if such systematic calibration *were* devised, the specific expansion of the "intelligence" rubric—as opposed to the terminology we have or could coin—is not rationalized by anatomic or metabolic connections.

18. A segment of Comedy Central's, *The Daily Show*, offered a satire of a parallel discrepancy. Host, Jon Stewart, was conducting a mock Q and A about capital-gains tax cuts with an actor playing an economic apologist for President Bush. Stewart asked the interviewee to admit that these cuts were only going to exacerbate the gap between the poor and the rich. "Yes," said the mock analyst, "but *only if you define rich and poor in traditional ways*; but you really can't define rich by how much wealth you have."

19. We reference here an adaptation of Lasch's cultural narcissism. In this case, the narcissism is manifested as forms of "centricism"—as in ego-, ethno-, or anthro- specifically applied to methodological conceits of time, technology and vocation. Temperocen-

trism, is the tendency to identify one's own lifespan as *that* period in which the past is resolved and the future determined. People seem to have much difficulty imagining events in *their* lifetime as, ultimately, insignificant blips on the grand scan of historical radar. Arguably, it is temperocentrism that, for instance, possesses seemingly contemplative people to label the study of *recent* technology as "*new* media"—as if the word "new" should be retired, like an old football jersey, because, surely, no other compelling technologies (from which we will want to distinguish present systems) are henceforth imaginable. Thus, with temperocentrism, everything in the analyst's lifetime is seen as historically unique. Media and popular culture scholars, especially, are prone to developing a similar historical orientation, but as "uniquely" identified by the reigning *media* or *technology*—technocentrism. Considering the present discussion, one might argue that the reduction of generic techno-access to a "digital divide" is illustrative of technocentrism—*or* that reducing a confluence of social issues *as* techno-access, is more fundamentally technocentric. Finally, mediacentrism could also be called an example of what we have called industriocentrism—the deductive positioning of one's work or vocation as the determinant of phenomena studied. That media scholars periodize history in terms of dominant media is an obvious illustration, but one could also claim that no less mediacentric and industriocentric is a prominent body of media scholarship that structure all forms of research so that the media may be blamed (or not) for societal problems. We don't refer to it all as "mediacentrism" because, of course, such centrism isn't confined to communication. A scholar devoted to the study of social class, for instance, may see *that* variable as responsible for anything theorized. When this is done purposely, we conceive of it as determinism. Arguably, formal determinist positions may be found more methodologically acceptable because it is assumed that the determinant has been inductively discovered.

20. This argument may recall for the reader the persistence of eduinflation beyond what can indisputably be characterized as the post-war period—the general extension of eduinflation, in many respects, into the twenty-first century. Economists have, in fact, provided ample explanation for this extension—current and continuing explanation, rather than the persistence (stickiness) of an historical reaction. Of course, social organization, itself, is integrated and cumulative, so the dialectic of cause and effect is always a factor. Still, many of the issues mentioned at the outset, such as, increase in birth rate, extension of average life-span, retirement policies, etc., are, actually, responsive to the whole period in question and *not* necessarily causal conditions in place in 1945.

21. Inasmuch as we included "grade inflation," and now, propose eduinflation's inverse intellectual effect, we, perhaps unnecessarily, anticipate a question concerning the other, conceptually-related inflation—that uncovered by Flynn. Since our earlier use of Herrnstein and Murray's *The Bell Curve* was specifically *illustrative* of rhetorical modeling, and we don't, anyplace, otherwise invoke IQ tests as anything but part of the social construction of intelligence, we question the relevance to our discussion of any IQ data use interchangeably as "intelligence" data. With that said, Flynn's findings (secondary analyses undertaken to rebut Herrnstein and Murray) showed that later twentieth-century IQ scores significantly inflated throughout the world. We do not find that these data, or their interpretation as the Flynn Effect, at all contradict our suggestion that, in this same period, increased education did not, comparatively, yield increased traditional intelligence. There are several factors supporting this inference. First, what accounted for Flynn's inflated scores was specific: a significant rise in the "non-verbal"-skills scores. He found no parallel inflation with respect to the "cognitive"-skills questions, which, according to many neuroscientists, are those most intimately associated with what our

culture traditionally regards as "intellectual." Moreover, the inflation Flynn discovered is unilateral—global, and not accounted for by a rise among scores in any one demographic category. Thus, although higher, IQ scores have been *proportionately* stable. In this context, the tests have always indicated significant differences between various demographic categories, including—given our interests here—social class. Thus, as we propose a social-class differential in education, this, too, is not contradicted by the Flynn's findings. Indeed, inasmuch as the Flynn Effect *establishes* the very strong influence of social patterning in determining intelligence, the findings are, if anything, *supportive* of our overarching position. As an ironic footnote to a note, it might be added that Flynn's thesis is also an inflationary effect of post-WWII conditions. The changes in nutrition and prenatal healthcare, the emphasis on (reunited) families and child-rearing would all, logically, contribute to the numbers Flynn uncovered. And while Flynn's inflated scores appeared millennia too soon to be interpreted as a genetic outcome, their timing is quite appropriate to a technological-determinist interpretation, to wit, the rapid growth of televisual and, then, digital media throughout this period could certainly account for newer generations improved skills in pattern recognition and processing spatial relationships. While one may hypothesize further—that people's increasing familiarity with standardized testing, itself might make them more adept—this is not borne out. Scores have not inflated in other, popular "aptitude" tests, such as the SAT's. Of course, that very contradiction reinforces our basic hypothesis.

22. For but one example, a comparison (over time) of texts on the history of major, world civilization showed that more recent publications actually included longer discussions of significantly *fewer* variables in accounting for nineteenth-century civil uprisings and revolutions! Interestingly, to the extent that these introductory histories are taught by inexperienced teachers who, presumably, follow the assigned texts more closely, there is an increasingly smaller chance that students will be familiarized with the complexities of historical evolution. Unfortunately, research that actually attends to such seemingly minor variations is relatively rare, but it should be apparent that it is within these very focused orientations that much broader instructional problems are revealed.

23. Interestingly, some critics lay blame for the unevenness of credit-hour assignments on the later-twentieth-century wholesale introduction of unconventional laboratory and/or conservatory-like courses-of-study within the university. In other words, this criticism suggests that the failure of higher education in the twentieth century is the boarding of anything remotely "studiable" within the academy—despite the fact that the new entries had no academic tradition and, thus, no conventional way to measure units of instruction. Shortly, we will examine this issue with respect to something more profound than credit-hour generation.

24. As we begin seriously to discuss social-class in this section, we need to provide two methodological or rhetorical qualifications: (1) our "unexpressed" systemic assumption, and (2) our use of binary polarity. In the case, of the systemic orientation, it goes without saying that, in any functionalist approach, parts of the whole are not understood to manifest behavior or consequences independent of other constituent elements. This pertains to how we talk about social class because we adopt a rhetorical style that may, superficially, suggest otherwise: specifically, we sometimes borrow the common style of framing discussion in terms of workers or the lesser-privileged. This framing is not simply a bow to political rectitude, but reflects a complex theoretical statement that we distill here in very abbreviated fashion: It might be said that, despite an oft-cited bit of Marxist pith, something that most analysts of social class realize (perhaps, *including* Marx) is that

the ideas *of the working class* are the ideas that rule society. Few, if any, would argue that the ideas that *serve* the working class are *not* the ideas that rule. However, contrary to the notion that there are in circulation two distinct worldviews (a duality that would be devilish to sustain), we only suggest that there is a dominant ideology insofar as it represents the ideas—the mythology—that conserves social structure. In this sense, the only "false" or altered consciousness that is ever necessary is that attached to the workers or lesser-privileged. Whether or not the empowered also learn these popular ideas (which evidence suggests they absolutely do) is not all that material—except, perhaps, to the extent that a single, dominant worldview is inordinately more efficient. Thus, to the extent that one frames analysis in terms of the disempowered, one *is* framing popular worldview and prescribed conduct. Secondly, we, generally, confine our social-class analysis to the broad distinction between "empowered" and "disempowered," rich versus poor, advantaged versus disadvantaged, etc. We don't, typically, find any need to employ "finer" discriminations. This matter clearly extends beyond simple taxonomy and is far too complicated to be resolved here. To prevent grievous misrepresentation, we try to limit our analysis to fundamental issues for which evidence does *not* indicate *multi*-level behavior or consequences. However, to the extent that a working-class-perspective does not adequately imply the overall pattern of class-coding, we will shift perspective until the representation matches our understanding. Finally, we indicate here that class coding is "covert." We do so as to avoid the tautology trap, i.e., anything that varies by social class is *de facto* class coded. Of those class-separated elements, we wish to address here only those that are, specifically, *neither* widely-recognized (nor meant to be recognized) as class-related.

25. We recognize that no system remains the absolute same through time—and especially after the more dramatic economic geopolitical, religious, and technological changes that precipitated the institutionalized diffusion of education. Nonetheless, evidence points to the continuity of the elite, classical tradition. In fact, many of the original institutions once serving aristocratic men remain in operation today—changed, of course, but still within the recognizable tradition—particularly in comparison to newer schools established for educating "the masses." Coincidentally and conveniently, the very current issue of *economic diversity* in higher education has produced a literature that confirms our assignment of "class." This very topical issue involves studies indicating that more and more wealthy students are increasingly consuming positions at elite or prestigious American institutions. By simply raising this issue, cognizance of the value of an elite, traditional education is demonstrated. Nevertheless, that awareness is never connected to issues of learning or intellectuality. Thus, it's hard to derive much theoretical support from this movement. However, inasmuch as it is necessary to assign class ranks to universities in order to examine economic diversity, this literature has supported our important methodological decisions. First, the economic diversity literature indicates that there are a few different concepts that can be used interchangeably with respect to institutions of higher education: "prestige," *academic* ranking, and the percentage of students from upper-income families all correlate positively with each other. When we started our exploratory research, involving only 90 universities, we found that our assignments matched entirely with those provided in this literature. Still, given the heuristic nature of our class-coding discoveries, we discuss these variations in very broad terms.

26. Interestingly, recent, popular references to *smaller*, elite institutions as "boutique" schools, seem to complement this central metaphor.

27. There is no denying *other* interpretations of these patterns, e.g., one might argue that wealthier schools assign more books because they know their students can afford to

buy them. Our starting point is that, *for whatever reasons,* the nature and amount of required readings *are* class coded. When most of the class-coded elements can "also" be related to the differential cultivation of intelligence, we suggest that the pattern is not incidental.

Chapter Two

Book, Street, and Techno Smarts: The Representation of Intelligence on Prime-Time Television

Susan G. Kahlenberg

Representations of the intellectual abound in popular culture. Although there is confusion and dissent when describing the quintessential intellectual, the image often conjured up is the professor, teacher, expert, and scientist (Claussen, 2004; Gerbner, 1974; Holderman, 2003). Kadushin (1982) suggests that "there are about as many definitions of the term intellectual as people who write about intellectuals" (p. 255). Posner (2001) describes the intellectual as a "person who, drawing on his intellectual resources, addresses a broad though educated public on issues with a political or ideological dimension" (p. 170). This definition, like the trend in scholarship, connects intellectuals with higher education, and in turn, academic knowledge.

Brown (1980) argues that the definition of an intellectual may be problematic for the simple reason that every person has intelligence. Extant literature in social science and philosophy reveals that there are various manifestations of knowledge, including academic, technical and popular or "common-sense" knowledge (see Foucault, 1977; Habermas, 1989; Hegel, 1977; Kant, 1992; Konrad and Szelenyi, 1979). Scholarship has also raised the question of whether these various levels of knowledge have equal value and usefulness within the public sphere.

For instance, Gramsci (1971) claims that all manifestations of knowledge are legitimate, yet not every person functions as an intellectual. When comparing academic knowledge with that of technical and lay knowledge, Konrad and Szelenyi (1979) contend that academic knowledge does not differ in quantity or complexity, but rather in the quality and character of the knowledge. Conversely, Hegel (1977) finds no legitimacy in common knowledge, stating that "to be independent of public opinion is the first formal condition of achieving anything great or rational whether in life or in science" (p. 205). Additionally, Marx and

Engels (1970) recognize public opinion as false consciousness and therefore do not consider it to be equal with that of the academy, the state and the nation. Public opinion is considered false, reflecting the ideology of the ruling class and contradictory to class interests:

> The ideas of the ruling class are in every epoch the ruling ideas, i.e. the class which is the ruling *material* force of society, is at the same time its ruling *intellectual* force. The class which has the means of material production at its disposal, has control at the same time over the means of mental production, so that thereby, generally speaking, the ideas of those who lack the means of mental production are subject to it (Marx and Engels, 1970, p. 64).

Thus, public opinion could not be equal with the academy because its ideology was regulated, in a sense, by the academy, state, and the nation. As this review of literature will suggest, the dominant ideology in mass media functions to reinforce the status quo that emanate from other social, political, religious, and economic institutions.

The theoretical incentive to examine the portrayal of intelligence in media is threefold. First, as suggested by Foucault (1977), knowledge is essential to obtaining social power, as the concepts are inextricably linked. Second, it is also a general sociological principle that the social structure does not establish an equitable distribution of power; thus, if power is not evenly distributed, it can be argued that intelligence must be differentially available (Foucault, 1977; Lamont & Fournier, 1992; Levine, 1988; Squibb, 1973; Thomas & Krippendorff, 1988). At issue here is not the public availability to knowledge, but the degradation of intelligence in the public sphere. This leads, thirdly, to the imperative of determining whether varying levels of knowledge are perceived as equal in terms of their value and usefulness within myriad systems of representation (Claussen, 2004; Mills, 1963; Said, 1994).

Cultivation theorists have long established reciprocity between media and society. Gerbner (1990) describes the process by which media, through its topics, themes, persons and actions, shape common attitudes, perceptions, preferences and behaviors. Since the 1960s, content analyses have coded for themes of violence, sexuality, gender, marriage, age and the aging, religion, politics, and labor, determining that the "television" world cultivates shared conceptions of reality about the "real world," particularly for the heavy television viewer (Gerbner, Gross, Morgan, Signorielli, & Shanahan, 2002). It has also been established that common themes cut across television programs, and these themes present an aggregate and conventional image that maintains the status quo (Gerbner, Gross, Morgan & Signorielli, 1994). Similarly, Schiller (1973) and McChesney (1996) maintain that it is a mistake to think that within a multi-channel communication landscape that there is a diversification of content:

> The fact of the matter is that, except for a rather small and highly selective segment of the population who know what they are looking for and can therefore

> take advantage of the massive communications flow, most Americans are basically, though unconsciously, trapped in what amounts to a no-choice information bind (Schiller, 1973, p. 10).

A political economy approach is crucial to recognize, as it allows us to consider media representations of intelligence within a wider political, economic and social system and the potential implications if knowledge is portrayed as undesirable.

Several scholars have explored how message production within the media is often dominated by commercial interests (see Gerbner, 1958; McChesney, 1996; Mosco, 1996; Schiller, 1973). As case in point, the Cultural Indicators Project has, for decades, recognized television as a powerful socialization agent, based on its ability to produce and reproduce symbols and images that shape public consciousness, all while reflecting the interests of capitalism. Given the industry's reliance on conventional formulas that are employed as part of this capitalistic enterprise, with the audience as primary commodity, it can be surmised that media content teach audiences, through its characterizations, what attributes they should value, prioritize and aspire for (see Gerbner, 1990; McChesney, 1996). If "we live in a world that is created by the stories we tell" (Morgan, 2002, p. 7), and the bulk of the stories reflect social relationships that bear the assumptions and interests of the media conglomerates of the few, wealthy media owners who control and own most media fare, then it is imperative to determine if there is a degradation of certain kinds of knowledge within these mediated stories. This will determine if media content contribute to the anti-intellectualism that pervades our culture.

Thus, it can be deduced that how knowledge is portrayed and valued on fictionally-based television programming can serve as effecting social-order maintenance. Several scholars have explored the social function of the intellectual, resulting in a range of opinions as to whether intellectuals function to legitimate or challenge the existing social order (Benda, 1928; Gouldner, 1979; Gramsci, 1971; Konrad & Szelenyi, 1979). A review of this extant literature reveals that there has been public resentment toward intellectuals, and specifically, academic knowledge. Yet few examinations of intelligence in media situate within the history of American anti-intellectualism (see Claussen, 2004; Rigney, 1991). Thus, it can be surmised that the way intelligence is portrayed in popular fiction might contribute importantly to the public's desire and evaluation for particular types of knowledge (Evans, 1995; Ross, 1989; Thomas & Krippendorff, 1988). For example, Claussen's (2004) analysis of the representation of higher education in five popular national magazines reveals that media contribute to anti-intellectualism by routinely marginalizing intellectuals. He suggests that:

> News media are capable of transforming modern societies in positive ways; government, corporations, labor unions, religious groups, voluntary associations, and

> other institutions are controlled by elites who generally are unable and/or unwilling to initiate or facilitate movements that both affect the entire culture and benefit the general public (p. 200).

Certainly, there are several scholars who concur that media fall short of their initial goal of participatory democracy (see McChesney, 1996; Mosco, 1996; Schiller, 1973) yet it becomes imperative to determine whether other forms of popular fiction reflect and promote anti-intellectualism, since critical social theory reviewed here already establishes that media function to help perpetuate and solidify the status quo.

While intelligence connects to important indices of vitality in our culture, relatively few scholars have explored the portrayal of intellectuals on prime-time television programming. Scholarship of this nature typically centers on the representation of certain occupational roles, especially professionals. When compared to their percentages in the U.S. labor force, Signorielli and Kahlenberg (2001) find that professionals (e.g., doctors, lawyers, entertainers) are over-represented, while white-collar (e.g., managerial, clerical) and blue-collar (e.g., service) workers are under-represented on television. This pattern persists through race and gender, suggesting that television programs place a high emphasis on professionals because their jobs serve as a dramatic function in the storyline. It's also been argued that television programs focus on characterizations that are associated with status, prestige and respect in relation to work—attributes often associated with professional occupations—as this provides for more glamorous, adventurous, and exciting storylines (Jeffries-Fox & Signorielli, 1979; Signorielli, 1993; Signorielli & Kahlenberg, 2001). As an aside, perhaps the trait of intelligence is most likely assigned to professionals, since many of these jobs require degrees in higher education and, in turn, intellectual knowledge. For decades, content analyses have examined occupational roles of men and women to determine whether prime-time television programming disseminated gender role stereotypes of men and women (see Dominick, 1979; Elasmar, Hasegawa, & Brain, 1999; Glascock, 2001; Greenberg, 1982; Haskell, 1979; Kahlenberg, 1995; Kanigua, Scott & Gade, 1974; McNeil, 1975; Miller & Reeves, 1976; Seggar, 1975; Signorielli, 2004; Signorielli & Kahlenberg, 2001; Tedesco, 1974). More recent content analyses suggest that women and minorities are portrayed with stronger recognition and respect on television than determined in earlier years. For instance, Signorielli and Bacue (1999) determine that since the 1970s, more women are portrayed as working outside the home, in a greater variety of occupations that are classified as traditional male or gender-neutral, rather than merely relegated to indeterminate or traditional female occupational roles. This suggests that women are afforded more prestige and respect on the screen. Yet the research does not explore the power associated with male and female characters per se, beyond the natural associations one might be inclined to make based on the status of occupation, nor the perceived intelligence of television characters.

Interestingly, Gitlin's (1977) examination of professionals finds that television programs "draw their energy from the social world, but transform reality—including its unconscious wishes, desires, and fears—into salable fictions" (p. 54). Perhaps these representations exist routinely insofar that the public desires that intellectuals be denigrated on television; this certainly contributes to anti-intellectualism sentiments that pervade our culture. He claims that when teachers are presented on television, they are depicted as buffoons. With few exceptions, Gitlin states that teachers are the *object* of their hostile class, spending their time at best struggling to maintain control.

There are previous studies that focus exclusively on teachers and scientists, occupations directly linked to academic knowledge within extant literature and popular culture. Gerbner (1974) indicates that when compared to their U.S. proportions, there are relatively few teachers presented in media. In fact, throughout the years, studies have consistently reported the under-representation of teachers on television (DeFleur, 1964; Kahlenberg, 1995; Signorielli, 1993; VandeBerg & Strekfuss, 1992). Moreover, there are often negative depictions of those relatively few teachers. For instance, Gerbner (1974) claims that teachers in the media are "depicted as tyrannical, brutal, pedantic, dull, awkward, queer, and depressed" (p. 482). Smythe (1954) also indicates that there are relatively few teachers portrayed in television dramas; he recognizes teachers as benevolent, but also as weak and slow minded.

Foucault (1977) describes Oppenheimer, a man who functioned as an atomic scientist and more importantly, as the transition from the universal intellectual to the specific intellectual. Gerbner, Gross, Morgan and Signorielli (1981) also examine the scientist, a quintessential representation of the intellectual in popular lore. They find that "the scientist is a relatively rare and specialized dramatic character" (p. 42); as case in point, they determine that less than one percent of TV characters are scientists. Additionally, Gerbner et al. (1981) determine that when compared to other professionals, scientists are less attractive, fair, sociable, warm, tall and young, though very smart.

Extant literature also investigates the relationship between televised occupational roles and gender differences in problem solving abilities (Downs, 1981), advising and ordering (Greenberg, Richards, & Henderson, 1980; Henderson, Greenberg, & Atkin, 1980; Turow, 1974), dominance (Barbatsis, Wong, & Herek, 1983; Jeffrey & Durkin, 1989; Lemon, 1974; McNeil, 1975; Seggar, 1975; Sternglanz & Serbin, 1974; Tedesco, 1974; VandeBerg & Streckfuss, 1992), and achievement (Sternglanz & Serbin, 1974). A review of this scholarship determines that although female characters are often rated positive on the aforementioned traits, male characters, for the most part, are portrayed with higher ratings of power, achievement and dominance. Wober and Gunter (1988) argue that many of these traits are central when determining the overall competence and prestige for specific occupations. It can be inferred, then, that if female characterizations are routinely denied access to power, achievement, and dominance, they may also be portrayed as less competent and intelligent when compared to their male counterparts.

Jeffries-Fox and Signorielli (1979) analyzes the relationship between personality traits and occupation for certain major characters on prime-time television. More specifically, their study examines those characters appearing as doctors, lawyers, police officers, psychiatrists, paramedics, and judges on television. Overall, these characters are rated positively on most of the personality scales. Of particular interest, perhaps, is the higher rating on intelligence, stability and power as well as the lower ratings on happiness and youthfulness. This substantiates the findings of other research studies that have found intellectual fictional characters to be balanced (or offset) by serious weaknesses in other areas of their lives, particularly romance and sex (Gerbner, 1974; Thomas & Krippendorff, 1988). This may explain, for example, the stereotyped "schoolmarm" and "mad scientist" images that permeate popular culture. In other words, if intellectuality is shown routinely to make characters less fortunate in other coveted aspects of life, such as love and sex, then the "message" may be that intelligence is not all that desirable.

Bell (1996) concludes that the following messages are embedded in television programs that focus on education: "anybody can teach" and "all teachers have an antagonistic relationship with their principal." These messages not only dismiss teachers as unintelligent, but they indicate that teachers have problems with authority. Moreover, he finds that several teachers, in particularly, science and physical education, are portrayed as incompetent. When describing the intellectual as unrepresented, weak, subaltern, and exiled beyond the mainstream, Said (1994) states that the "intellectual always stands between loneliness and alignment" (p. 22); his sentiments are certainly concomitant with the public resentment toward intelligence that has a long historical background within general social theory.

A critical examination by Hofstadter (1964) finds that many Americans have negative and resentful attitudes toward intellectuals, and these "anti-intellectual" attitudes have become stronger as intellectuals gain prominence in the social structure. He states that "the common strain that binds together the attitudes and ideas which I call anti-intellectual is a resentment and suspicion of the life of the mind and of those who are considered to represent it; and a disposition constantly to minimize the value of that life" (p. 7). In other words, as the intellectual gains increased access to power, privilege and luxury, so too does the resentment the layperson has toward the intellectual. Although no one questions the value of intelligence, the "intellectual" per se is looked upon with resentment or suspicion. Hofstadter's examination of anti-intellectual tendencies extended to business, political, religious, and scientific realms, though he believed that it was in educational institutions that anti-intellectualism coalesces and cultivates.

More specifically, Hofstadter (1964) provides several examples from American literature to support his claims of anti-intellectualism in America: the 1952 proposal of American novelist Louis Bromfield that a definition of "egghead" be included in the dictionary, President Eisenhower's claim that a definition of an intellectual as "a man who takes more words than are necessary to tell more than he knows" (p. 9), and even the dominant stereotype of the schoolmaster epitomized by Ichabod Crane, a man who was extremely lanky, unattractive, unkempt, and an "odd mixture of small shrewdness and simple credulity" (p. 315).

Neisser (1976) discusses how anti-intellectual tendencies exist in public discourse. He suggests that academic intelligence is not, for the most part, revered outside of the academy. Neisser (1976) describes how intellectuals are "often regarded as ineffectual, devoid of good sense, and preoccupied with trivia: stupid despite being intelligent" (p. 138). Durkheim (1951) maintains that in the pursuit of erudition, intellectuals were often detached from the center of society—it was this detachment that contributed to the frequency of suicide among those with the most active intellectual life. Benda (1928) is pessimistic about the social role of intellectuals, suggesting that they betray the layperson by bringing politics and religion into their poetry, art, philosophy and other activities. Rather than supporting the status quo in their activities, Benda asserts that intellectuals should stay critically independent of political and national interests.

Ross (1989) suggests that the public opinion of intellectuals may be influenced by their portrayals in popular culture. Using the example of the movie *Back to School,* Ross (1989) demonstrates how popular disrespect for intellectual expertise is somehow justified by its depictions of professorial pretension and the dependence of academic freedom on philanthropy and, in particular, shady business deals. Also, academic intelligence is often seen as attempting to claim control of "what other people do naturally" (p. 2); common sense, then, is valued within popular culture more than academic intelligence, an idea that this content analysis empirically and systematically investigates.

Gerbner (1974) proposes that there is a "hidden curriculum" embedded in popular culture—"a lesson plan that no one teaches yet everyone learns. It consists of the symbolic contours of the social order" (p. 476). He implies that images of schools and scholars complement this hidden curriculum, and as a result, members of society learn lessons about "learning" itself. Unfortunately, these lessons may suggest that intellectual knowledge is not to be respected in the social structure. It is this premise that underpins the theoretical incentive of this content analysis.

Thus, it is evident that the literature pertaining to intelligence in the public domain is not, in general, vast and decreases in quantity as the literature moves from general theoretical concerns to media portrayals of intelligence. This research attempts to fill this void in the literature, with a content analytic design that centers on two research questions: (1) how are the three manifestations of intelligence (academic, practical and technical) portrayed on prime-time television programs? and (2) how is intelligence distributed among plot-functional characters along with other attributes, most notably, gender, social age, social class, education, labor, general life power, and humor? In other words, is intellectuality shown routinely to make characters less fortunate? Are there distinctions in the representations of academic, practical and technical intelligence that could cultivate the message that intelligence is not all that desirable, contributing to the degradation of intelligence within the public sphere?

Method

This content analysis analyzed a weeklong sample of prime-time fictional programming appearing on six different networks (ABC, CBS, FOX, NBC, UPN, and WB). In total, this sample contained 118 programs or 94 hours of programming. Most programs were taped from September 29, 1997 through October 5, 1997, or on October 6, 1997 (UPN and WB). During this time period, a few programs on NBC and FOX were pre-empted to show live coverage of the Baseball World Series. These pre-empted programs were taped the subsequent week, October 6, 1997 through October 12, 1997, to ensure a representative sample of a typical week of prime-time television programming.

The unit of analysis was each plot-functional character appearing in the sample; plot functional characters were those in major and supporting roles who contributed to narrative development. Each character was coded according to the following attributes: gender (male, female); social age (child, teen, young adult, middle aged, senior citizen); social class (upper class, middle class, lower class); education (elementary school, high school diploma, associates degree, bachelors degree, masters degree, professional degree, technical degree); occupation (unknown, not working, professional, white collar, blue collar, criminal, law enforcement, military); general life power (very powerless to very powerful); humor (straight, sarcastic, witty, joker); and intelligence (academic, practical, technical, and overall). Each of these categories of analysis also contained a mixed/other slot to account for all possible character attributes.

In terms of intelligence, each character was assessed in terms of academic, practical and technical intelligence. Academic intelligence referred to traditional scholarliness, sometimes colloquially referred to as *book smarts*. University degrees are often used as an important index of academic intelligence (Neisser, 1976; Sternberg, 1988, 1990; Sternberg, Wagner, Williams, & Horvath, 1995). Practical intelligence, common sense, or *street smarts*, assessed how the character handled the everyday business of life such as family or work problems, and was not dependent on formal education per se (Sternberg, 1988, 1990; Sternberg et al., 1995). Technical intelligence involved a character's ability to perform simple tasks without necessarily understanding the theory behind the performance. This type of ability is commonly referred to as applied knowledge or *techno smarts* and is manifested in, for instance, the ability to service cars (Sternberg, 1990). More specifically, characters coded for technical intelligence had manual skills in terms of some science, mechanics, art, craft and trade, but could not necessarily understand the physics of the internal combustion engine. Additionally, skills in courtroom summation or college lecturing were not considered to be demonstrative of technical intelligence. An overall measure of intelligence—one which designated the combined effects of the three specific indices—was assigned also to each character.

All types of intelligence were originally measured using a five-point ordinal scale, which was later collapsed into three-points to yield higher cell frequencies for multi-variable cross tabulations. Additionally, specific characters were omitted who could not be measured for the assigned variables (i.e., cannot code), allowing the statistical analyses to be more robust. All intelligence ratings were measured by a three-point ordinal scale for purposes of the statistical analysis.

A pre-test was conducted prior to coder training to identify any inconsistencies with the recording instruments, operational definitions, coding directions, and the design of the coding sheets. For the reliability analysis, the author served as primary coder, and two additional coders were employed to recode a total of 15 percent of the sample for reliability. Inter-coder agreement was determined using Krippendorff's (1980) alpha; there was .92 overall agreements among coders. Reliability was above .82 for each individual variable in this analysis. Chi-square analysis identified differences in the frequency distributions among all variables.

Results

Six-hundred and sixty four (644) plot-functional characters were analyzed. There were 401 (60.4%) males and 263 (39.6%) females (χ^2 =28.681, df=1, p=.000). Consistent with previous research (Signorielli & Kahlenberg, 2001), the world of television was male dominated with a 1.5 to 1 ratio of male to female characters.

An examination of *social age* revealed that the majority of characters were young adults (42.2%) and middle-aged adults (44.0%). Relatively few characters were senior citizens (3.3%), children (3.0%), teenagers (7.2%), or mixed/other (.3%)(χ^2 =843.735, df=5, p=.000, N=664). An examination of chronological age indicated that the mean age for all characters was 34.2 years old, and an ANOVA of chronological age by gender revealed that males (M=36) were older than females (M=31)(Sum of Squares =102555.307, df=662, F=20.986, p=.000). This is typical of previous research, which found that the majority of characters were between the ages of 25 and 45 years old, and that women were portrayed as younger than men (Signorielli, 2004).

With regard to *social class,* more than three-quarters (82.8%) of all characters on prime-time television were coded as middle class. The remaining characters were assessed as follows: 11.7% were upper class, 3.2% were lower class, 2.0% had no determinable social class, and less than one percent (.3%) were mixed/other (χ^2 =1664.298, df=4, p=.000, N=664). These findings were relatively consistent with previous research (Butsch, 1992; Thomas & Callahan, 1982).

In relation to *education*, a large percentage of characters could not be coded for their level of education (28.0%). Approximately one-half of the characters had or were working toward high school diplomas (25.3%), bachelor degrees (26.2%), and professional degrees (13.4%). The remaining characters were assessed as follows: 3.2% were in elementary school, 2.3% were mixed/other, and 1.7% had technical degrees (χ^2 =541.880, df=1, p=.000, N=664). The analysis of education also accounted for whether education was revealed explicitly or implicitly within TV programs; of those characters with discernible levels of education, 44.7% were implicit and 26.8% were explicit (χ^2 =39.075, df=2, p=.000, N=664).

With respect to *occupation,* one-quarter (25.5%) of all characters were employed as professionals and one-quarter in total were portrayed as not working (15.2%) or in unknown occupations (17.3%); this is consistent with findings from previous research (Signorielli, 1993). The remaining characters were assessed as follows: 12.0% were law enforcement officers, 9.3% were white-collar workers, 5.3% were blue-collar workers, 2.7% were in the military, 2.4% were criminals, and 10.2% were mixed/other (χ^2 =266.633, df=8, p=.000, N=664).

General life power was an overall assessment of the control and force that a given character had in an episode, regardless of social context. The majority of characters in this sample had *average power* (48.2%) or were *somewhat powerful* (27.4%). Less than one-quarter of the characters were *somewhat powerless* (11.9%), *very powerful* (9.6%), and *very powerless* (2.6%).

The variable *humor* assessed the general disposition of a character as straight, witty, joker, sarcastic and mixed. Almost three-fourths of all characters had a straight or serious demeanor (72.7%). The remaining characters were assessed as follows: 10.8% were sarcastic, 8.6% were jokers, 4.1% were mixed, and 3.8% were witty.

Table 2.1 identifies how the characters were distributed for academic, practical, technical and overall intelligence. *Academic-intelligence* measured traditional scholarliness or "book smarts." University degrees were important indicators of academic-intelligence, but college students, for instance, were not automatically assessed as more erudite than those characters not presented on a college campus. The majority of characters had average academic-intelligence (45.9%). The remaining characters were assessed as follows: 24.5% had *above-average academic-intelligence*, 8.1% had below-average academic-intelligence, 7.4% had high academic-intelligence, 1.5% had low academic-intelligence, and 12.5% could not be coded for academic-intelligence.

Table 2.1. Frequencies of Academic-, Practical-, Technical-, and Overall-Intelligence

	N	%
Academic-Intelligence		
Cannot code	83	12.5
Low intelligence	10	1.5
Below-average intelligence	54	8.1
Average intelligence	305	45.9
Above-average intelligence	163	24.5
High intelligence	49	7.4
Practical-Intelligence		
Cannot code	5	.8
Low intelligence	7	1.1
Below-average intelligence	93	14.0
Average intelligence	391	58.9
Above-average intelligence	155	23.3
High intelligence	13	2.0
Technical-Intelligence		
Cannot code	402	60.5
Low intelligence	8	1.2
Below-average intelligence	31	4.7
Average intelligence	79	11.9
Above-average intelligence	112	16.9
High intelligence	32	4.8
Overall-Intelligence		
Cannot code	1	.2
Low intelligence	6	.9
Below-average intelligence	62	9.3
Average intelligence	392	59.0
Above-average intelligence	178	26.8
High intelligence	25	3.8

Note: Academic (χ^2 =527.867, df=5, p=.000);
Practical (χ^2 =1014.898, df=5, p=.000);
Technical (χ^2 =984.536, df=5, p=.000);
Overall (χ^2 =1051.548, df=5, p=.000).

Practical-intelligence measured how characters handled everyday family and work problems and was not dependent on formal training or education. The majority of characters had average practical-intelligence (58.9%). The remaining characters were assessed as follows: 23.3% had above-average practical-intelligence, 14.0% had below-average practical-intelligence, 2.0% had high practical-intelligence, 1.1% had low practical-intelligence, and less than one percent (.8%) could not be coded for practical-intelligence.

Technical-intelligence measured a character's ability to perform specific tasks without necessarily understanding the theory behind the performance. For instance, characters could have high technical ratings because of their ability to service cars, but still not understand the physics of the internal combustion engine. More specifically, characters coded for technical-intelligence had skills or applied knowledge in terms of some scientific, mechanical, artistic, professional, and/or trade application. As suggested in the next section, a more nuanced measure of technical-intelligence should be refined for future research. In this study, the majority of TV characters could not be coded for technical-intelligence (60.5%). The remaining characters were assessed as follows: 16.9% had above-average technical-intelligence, 11.9% had average technical-intelligence, 4.8% had high technical-intelligence, and 1.2% had low technical-intelligence.

Overall-intelligence was an index of a more general impression of characters' knowledge. The majority of characters were measured as having average overall-intelligence (59.0%). The remaining characters were assessed as follows: 26.8% had above-average overall-intelligence, 9.3% had below-average overall-intelligence, 3.8% had high overall-intelligence, and approximately 1% had low overall-intelligence (.9%) and indeterminable overall-intelligence (.2%).

It may be salient that fewer characters could be coded for academic-intelligence than practical-intelligence simply because TV programs very often center on everyday family and work problems. Technical situations also appeared less frequently on TV programs, so there were a large percentage of characters that could not be coded for technical-intelligence. It might also, at first, seem surprising that characters were generally shown to be academically rather than practically able (i.e., more characters in the sample were below-average in practical-intelligence than below-average in academic-intelligence). However, if characters were adept in solving interpersonal problems it might eliminate many script possibilities, and thus, this finding may be an artifact of production requirements. Correspondingly, more characters were placed in the high range for both academic- and technical-intelligence, and not for practical-intelligence. This representation would seem to correspond to the economics of everyday life—people are more likely to hire others to do things either based on applied knowledge and/or erudition (e.g., school tuition) than for assistance with mundane life. Moreover, it might be somewhat intimidating were viewers made to watch characters extremely skillful in negotiating interpersonal issues.

Cross-tabulations of intelligence with gender revealed significant chi-square tests for male versus female characters, and illuminated several findings of interest. A shown in Table 2.2, with the exception of technical intelligence, the majority of male and female characters had average levels of intelligence. More specifically, it was found that males (35.4%) were more likely to be portrayed with high academic-intelligence than females (26.6%). Similarly, more males (29.8%) than females (19.8%) were presented as having high practical-intelligence. Interestingly, males (55.1%) were portrayed in more situations codeable for technical-intelligence than females (68.0%). Finally, it was found that males (35.0%) were more likely to have high overall-intelligence than females (24.0%). Therefore, with respect to all kinds of intelligence, men were more likely than women to be highly intelligent.

Table 2.2 Chi-Square Analyses of Intelligence and Gender

	Gender			
	Males		**Females**	
Intelligence	R%	C%	R%	C%
Academic (All N=)		401		263
Cannot code	51.8	10.7	48.2	15.2
Low intelligence	68.8	11.0	31.3	7.6
Average intelligence	56.4	42.9	43.6	50.6
High intelligence	67.0	35.4	33.0	26.6
Practical (All N=)		399		260
Low intelligence	59.0	14.8	41.0	15.8
Average intelligence	56.5	55.4	43.5	65.4
High intelligence	70.8	29.8	29.2	18.8
Technical (All N=)		401		263
Cannot code	55.0	55.1	45.0	68.0
Low intelligence	74.4	7.2	25.6	3.8
Average intelligence	59.5	11.7	40.5	12.2
High intelligence	72.2	25.9	27.8	15.2
Overall (All N=)		400		263
Low intelligence	63.2	10.8	36.8	9.5
Average intelligence	55.4	54.3	44.6	66.5
High intelligence	69.0	35.0	31.0	24.0

Note: R%=Row percentages; C%=Column percentages;
Academic (χ^2 =10.313, df=2, p=.016, N=664);
Practical (χ^2 =10.194, df=2, p=.006, N=659);
Technical (χ^2 =16.564, df=2, p=.001, N=664);
Overall (χ^2 =10.616, df=2, p=.005, N=663).

With regard to *social age* and *intelligence* (see Table 2.3), chi-square tests were significant for academic-, practical-, and overall-intelligence, but not for technical-intelligence (χ^2 =15.250, df=9, p=.084, N=662). The majority of TV characters, regardless of age, were portrayed with average intelligence. Consistent with previous findings (Signorielli, 2004), the elderly made up a small segment (3.3%) of TV characters; although they had less recognition on the screen, they were deemed potentially with more respect, as few were measured with low practical intelligence (9.1%), and none were measured as having low academic- or overall-intelligence (Signorielli & Bacue, 1999). Another interesting finding related to the familiar doctrine that "age brings wisdom" was that more middle-aged characters were presented with higher academic-, practical- and overall-intelligence than younger adults; for instance, an examination of high overall-intelligence found that 37.1% of the characters were young adults and 56.4% of the characters were middle aged. Interestingly, children and teens were 1.6 times more likely to have low practical-intelligence (27.9%) than low academic- (17.6%) or overall-intelligence (16.2%), suggesting further support for the aforementioned age doctrine.

Table 2.3 Chi-Square Analyses of Intelligence and Social Age

	Social Age							
	Child/Teens		Young adult		Middle-aged		Senior Citizen	
Intelligence	R%	C%	R%	C%	R%	C%	R%	C%
Academic		68		280		292		22
Cannot code	7.2	8.8	42.2	12.5	41.0	11.6	9.6	36.4
Low intelligence	18.8	17.6	53.1	12.1	28.1	6.2	0.0	0.0
Average intellect	10.9	48.5	41.1	44.6	44.1	45.9	3.9	54.5
High intelligence	8.1	25.0	40.8	30.7	50.2	36.3	.9	9.1
Practical		68		279		288		22
Low intelligence	19.0	27.9	52.0	18.6	27.0	9.4	2.0	9.1
Average intellect	10.8	61.8	41.6	58.1	43.4	58.7	4.1	72.7
High intelligence	4.2	10.3	38.7	23.3	54.8	31.9	2.4	18.2
Overall		68		279		292		22
Low intelligence	16.2	16.2	54.4	13.3	29.4	6.8	0.0	0.0
Average intellect	12.0	69.1	42.7	59.9	40.4	54.1	4.9	86.4
High intelligence	5.0	14.7	37.1	26.9	56.4	39.0	1.5	13.6

Note: R%=Row percentages; C%=Column percentages;
Academic (χ^2=30.044, df=9, p=.000, N=662);
Practical (χ^2=29.023, df=6, p=.000, N=657);
Overall (χ^2=31.247, df=6, p=.000, N=661).

In considering the relationship between *social class* and *intelligence* (see Table 2.4), chi-square tests were significant for academic-, practical- and overall-intelligence, but not for technical-intelligence (χ^2 =6.767, df =4, p=.149, N=258). There were two findings of note with respect to this relationship. Of lesser interest, perhaps, because it is more predictable, upper class characters were more likely to be assessed as having high academic-intelligence (60.9%), than average academic-intelligence (29.7%) or low academic-intelligence (9.4%). In contrast, lower class characters were more likely to be portrayed as having low academic-intelligence (46.2%) and average academic-intelligence (38.5%) than high academic-intelligence (15.4%). To wit, those with material wealth appeared to be more scholarly, and those lacking material wealth were less so.

Table 2. 4 Chi-Square Analyses of Intelligence and Social Class

	Social Class					
Intelligence	Upper class		Middle class		Lower class	
Academic	R%	C%	R%	C%	R%	C%
All N=	64	64	300	492	205	13
Low intelligence	9.4	9.4	81.3	10.6	9.4	46.2
Average intelligence	6.3	29.7	92.0	56.1	1.7	38.5
High intelligence	19.0	60.9	80.0	33.3	1.0	15.4
Practical						
All N=	98	78	383	545	163	21
Low intelligence	10.2	12.8	81.6	14.7	8.2	38.1
Average intelligence	9.7	47.4	87.5	61.5	2.9	52.4
High intelligence	19.0	39.7	79.8	23.9	1.2	9.5
Overall						
All N=	68	78	383	549	197	21
Low intelligence	8.8	7.7	82.4	10.2	8.8	28.6
Average intelligence	8.9	43.6	88.0	61.4	3.1	57.1
High intelligence	10.3	48.7	79.2	28.4	1.5	14.3

Note: R%=Row percentages; C%=Column percentages;
Academic (χ^2=36.040, df=4, p=.000, N=569);
Practical (χ^2=19.008, df=4, p=.001, N=644);
Overall (21.955, df=4, p=.000, N=648).

Second, previous research indicated that poorer characters are more "helpful" in solving every day work and family problems (Thomas & Callahan, 1982). However, in this study, more than one-third (38.1%) of the lower class characters were measured as having low practical-intelligence as compared to only 14.7% of middle class characters and 12.8% of upper-class characters.

In relation to *education* and *intelligence*, chi-square tests were significant for all types of intelligence. As shown in Table 2.5, characters with professional degrees were assessed as having high academic-intelligence (87.6%), whereas characters with elementary (22.2%) and high school (24.2%) education composed almost one-half of the characters with low academic-intelligence; this is consistent with previous research connecting intellectuals with higher education (Posner, 2001). It is also interesting to note that one-half (50.6%) of the characters with professional degrees were assessed as also having high practical-intelligence, with an additional one-third (37.9%) having average practical-intelligence. This may be contrasted with more common worldview as well as scholarly interpretations which suggests that intellectuals are commonly portrayed as lacking common sense and even "emotionally maladjusted" (see

Gerbner, 1974). In relation to technical-intelligence, high school students were assessed as most likely to have low technical-intelligence (46.2%); a college degree was most strongly linked to high technical-intelligence. More specifically, it was found that approximately one-third (38.5%) of those characters with college degrees had low technical-intelligence, and conversely, almost two-thirds (63.4%) had high technical-intelligence. In retrospect, the operational definition constructed for this variable considered computer skills as related to *techno smarts*, as portrayed on programs like *X Files, Diagnosis Murder,* and *Law and Order.* Programs like *Step by Step* and *Family Matters,* which featured youth, did not provide storylines that emphasized robust technical-intelligence in these particular episodes, despite what might be expected from their inclusion of supporting characters that epitomize the "geek" stereotype.

Table 2.5 Chi-Square Analyses of Education and Intelligence

			Academic-Intelligence			
	Low intelligence		Average intelligence		High intelligence	
Education	R%	C%	R%	C%	R%	C%
All N=		54		204		191
Elementary	22.2	7.4	50.0	4.4	27.8	2.6
High school	24.2	68.5	62.7	47.1	13.1	10.5
Bachelors	4.9	14.8	48.5	38.7	46.6	39.8
MA/Ph.D./MD	1.1	1.9	11.2	4.9	87.6	40.8
Tech/other	15.4	7.4	38.5	4.9	46.2	6.3
			Practical-Intelligence			
All N=		74		266		134
Elementary	33.3	9.5	57.1	4.5	9.5	1.5
High school	21.4	48.6	64.9	41.0	13.7	17.2
Bachelors	9.3	21.6	57.0	36.8	33.7	43.3
MA/Ph.D./MD	11.5	13.5	37.9	12.4	50.6	32.8
Tech/other	19.2	6.8	53.8	5.3	26.9	5.2
			Technical-Intelligence			
All N=		26		49		115
Elementary	33.3	3.8	66.7	4.1	0.0	0.0
High school	24.0	46.2	26.0	26.5	50.0	21.7
Bachelors	8.7	23.1	36.2	51.0	55.1	33.0
MA/Ph.D./MD	8.9	15.4	13.3	12.2	77.8	30.4
Tech/other	13.0	11.5	13.0	6.1	73.9	14.8
			Overall-Intelligence			
All N=	48		261		169	
Elementary	23.8	10.4	61.9	5.0	14.3	1.8
High school	20.2	70.8	66.7	42.9	13.1	13.0
Bachelors	3.4	12.5	56.9	37.9	39.7	40.8
MA/Ph.D./MD	0.0	0.0	28.1	9.6	71.9	37.9
Tech/other	11.5	6.3	46.2	4.6	42.3	6.5

Note: R%=Row percentages; C%=Column percentages; Academic (χ^2=145.386, df=8, p=.000, N=449); Practical (χ^2=53.269, df=8, p=.000, N=474); Technical (χ^2=21.735, df=8, p=.000, N=190); Overall (χ^2=115.618, df=8, p=.000, N=478).

The coding instrument also considered whether education was revealed explicitly or implicitly; chi-square tests were significant for *academic-*, *practical-*, and *overall-intelligence,* but not for *technical-intelligence* (χ^2=6.923, df=4, p.140, N=262). As shown in Table 2.6, characters with explicit references about education were more likely to have average- (51.4%) to high academic-intelligence (46.8%). In relation to high-practical intelligence, characters were just as likely to have educational references be explicit (28.0%) or implicit (24.0%). This implies that common sense may not be strongly linked to erudition, sustaining previous research findings claimed by Hofstadter (1964), who examined anti-intellectual tendencies of higher education via the anecdote of Ichabod Crane, a schoolmaster described in American literature. When Crane was frightened out of town with a pumpkin smashed on his head, the Pumpkin was symbolic of the defeat of intellectual knowledge by common sense knowledge.

Table 2.6. Chi-Square Analyses of Explicit/Implicit Education and Intelligence

			Academic-Intelligence			
Explicit/Implicit Education	Low intelligence		Average intelligence		High intelligence	
	R%	C%	R%	C%	R%	C%
All N=	135	64	273	305	173	212
Cannot code	18.8	8.9	51.6	12.1	29.7	11.0
Implied	33.1	74.8	43.0	48.0	23.9	42.2
Explicit	10.4	16.3	51.4	39.9	46.8	38.2
			Practical-Intelligence			
All N=	188	100	295	391	176	168
Cannot code	27.0	14.4	38.0	12.9	35.0	19.9
Implied	32.2	67.0	43.7	58.0	24.0	53.4
Explicit	20.8	18.6	51.2	29.2	28.0	26.7
			Overall-Intelligence			
All N=	188	68	297	392	178	203
Cannot code	30.9	11.2	47.1	10.8	22.1	8.4
Implied	33.7	70.2	41.1	54.2	25.3	55.6
Explicit	17.2	18.6	51.2	35.0	31.5	36.0

Note: R%=Row percentages; C%=Column percentages;
Academic (χ^2 =39.234, df=4, p=.000, N=581);
Practical (χ^2 =11.824, df=4, p=.019, N=659);
Overall (χ^2 =18.681, df=4, p=.001, N=663).

With regard to *occupation* and *intelligence*, chi-square tests were significant for all types of intelligence. There were three interesting findings of interest in relation to occupation and intelligence. As shown in Table 2.7, characters with high academic-intelligence were typically in professional (42.2%) or law enforcement (19.9%) occupations, such as physicians, teachers, lawyers, professors, scientists, police, private detectives and government agents. Less than one percent of blue-collar workers (.5%) were portrayed with book smarts, such as laborers, food server/preparers, and household workers. This suggests that erudition and intellectuality was linked, stereotypically, to those more likely to have degrees in higher education.

Of all the occupational categories, professionals (23.1%) and unemployed (22.0%) had the largest percentage of characters rating as low in practical-intelligence. At first these are not groups that would be expected to manifest themselves similarly; however, when it is explained that 78% of unemployed characters were, in fact, *students*, these results might then be interpreted as reflecting popular mythology. More specifically, when occupation as a variable is isolated (as opposed to the more general *social class*) these data show that blue- and white-collar (i.e., ordinary) workers were shown to be more practically-intelligent than those more prestigious (or professional) occupations. Thus, this may indicate another methodological subtlety underlying a discrepancy between present findings and earlier work and beliefs.

Also, one would expect to find a reasonably large number of blue-collar workers represented as having high technical-intelligence inasmuch as many blue-collar jobs involve the operations of tools and machines. However, the data also showed that professionals (30.1%) and law enforcement officers (23.1%) were more likely than blue- collar workers (4.9%) to be in possession of high technical-intelligence. This finding, though, may be more an artifact of coding definitions because this study included computers and weapon skills as demonstrative of technical-intelligence.

Table 2.7 Chi-Square Analyses of Occupation and Intelligence

	Academic-Intelligence					
Occupation	Low intelligence		Average intelligence		High intelligence	
	R%	C%	R%	C%	R%	C%
All N=		64		304		211
Unknown	11.5	14.1	79.5	20.4	9.0	3.3
Not working	18.0	25.0	57.3	16.8	24.7	10.4
Professional	6.8	17.2	38.3	20.4	54.9	42.2
White-collar	13.6	12.5	62.7	12.2	23.7	6.6
Blue-collar	35.7	15.6	60.7	5.6	3.6	.5
Criminal	9.1	1.6	36.4	1.3	54.5	2.8
Law enforcement	0.0	0.0	46.8	12.2	53.2	19.9
Military	0.0	0.0	33.3	1.6	66.7	4.7
Mixed/other	15.5	14.1	50.0	9.5	34.5	9.5

Table 2.7—continued

Occupation	Low intelligence		Average intelligence		High intelligence	
	R%	C%	R%	C%	R%	C%
			Practical-Intelligence			
All N=		91		370		164
Unknown	11.5	14.3	75.2	23.0	13.3	9.1
Not working	27.4	22.0	53.4	10.5	19.2	8.5
Professional	12.7	23.1	55.8	24.9	31.5	31.7
White-collar	14.5	9.9	66.1	11.1	19.4	7.3
Blue-collar	31.4	12.1	54.3	5.1	14.3	3.0
Criminal	26.7	4.4	46.7	1.9	26.7	2.4
Law enforcement	5.1	4.4	49.4	10.5	45.6	22.0
Military	11.1	2.2	55.6	2.7	33.3	3.7
Mixed/other	10.8	7.7	58.5	10.3	30.8	12.2
			Technical-Intelligence			
All N=		39		78		143
Unknown	24.1	17.9	55.2	20.5	20.7	4.2
Not working	11.5	7.7	50.0	16.7	38.5	7.0
Professional	14.9	25.6	20.9	17.9	64.2	30.1
White-collar	28.6	15.4	33.3	9.0	38.1	5.6
Blue-collar	29.4	12.8	29.4	6.4	41.2	4.9
Criminal	28.6	5.1	14.3	1.3	57.1	2.8
Law enforcement	0.0	0.0	26.7	15.4	73.3	23.1
Military	9.1	2.6	0.0	0.0	90.9	7.0
Mixed/other	11.4	20.6	78.9	24.3	9.6	5.6
			Overall-Intelligence			
All N=		63		370		196
Unknown	11.4	20.6	78.9	24.3	9.6	5.6
Not working	17.7	20.6	65.8	13.0	16.4	6.1
Professional	6.0	15.9	49.4	22.4	44.6	38.3
White-collar	11.5	11.1	68.9	11.4	19.7	6.1
Blue-collar	34.3	19.0	51.4	4.9	14.3	2.6
Criminal	13.3	3.2	53.3	2.2	33.3	2.6
Law enforcement	0.0	0.0	47.5	10.3	52.5	21.4
Military	5.6	1.6	38.9	1.9	55.6	5.1
Mixed/other	7.7	7.9	55.4	9.7	36.9	12.2

Note: R%= Row percentages; C%=Column percentages;
Academic (χ^2 =111.395, df=16, p=.000, N=579);
Practical (χ^2 =56.927, df=16, p=.000, N=625);
Technical (χ^2 =46.026, df=16, p=.000, N=260);
Overall (χ^2 =105.960, df=16, p=.000, N=629).
* Chronological age ≥ 16 for all characters

With respect to *general life power* and *intelligence*, chi-square tests were significant for all types of intelligence (see Table 2.8). Interesting relationships were found when comparing intelligence variables to general life power. Confirming the findings of Foucault (1977), power comes with more formal academic knowledge. In this study, the majority of very powerful characters were assessed as having high academic-intelligence (62.5%). Also, there was a positive correlation between practical-intelligence and general life power. Very powerful characters (61.3%) were more likely to have high practical-intelligence, and very powerless characters (64.7%) were more likely to have low practical-intelligence. In turn, characters with average power (54.1%) were more likely to have average practical-intelligence. This finding demonstrated the strong internal validity of this study; the variables of practical-intelligence and general life power, at some level, both considered the skills at which characters negotiated everyday situations and problems. Of course, one could argue in fact that certain measures of internal validity are tautological in this very way.

Table 2.8 Chi-Square Analyses of Intelligence and General Life Power

	General Life Power									
	Very powerless		Little power		Average power		Some power		Very powerful	
Intelligence	R%	C%	R%	C%	R%	C%	R%	C%	R%	C%
Academic										
All N=		17		79		320		182		64
Cannot code	4.8	23.5	10.8	11.4	56.5	14.7	20.5	9.3	7.2	9.4
Low	4.8	17.6	38.1	30.4	44.4	8.8	11.1	3.8	1.6	1.6
Average	1.6	29.4	11.5	44.3	56.7	54.1	24.6	41.2	5.6	26.6
High	2.4	29.4	5.2	13.9	34.1	22.5	39.3	45.6	19.0	62.5
Practical										
All N=		17		79		320		179		62
Low	11.1	64.7	33.3	41.8	46.5	14.4	8.1	4.5	1.0	1.6
Average	1.5	35.3	10.2	50.6	57.5	70.3	24.8	54.2	5.9	37.1
High	0.0	0.0	3.6	7.6	29.3	15.3	44.3	41.3	22.8	61.3
Technical										
All N=		17		79		320		182		64
Cannot code	3.5	82.4	11.7	59.5	49.0	61.6	26.1	57.7	9.7	60.9
Low	5.3	11.8	34.2	16.5	42.1	5.0	10.5	2.2	7.9	4.7
Average	1.3	5.9	15.2	15.2	58.2	14.4	20.3	8.8	5.1	6.3
High	0.0	0.0	4.9	8.9	42.7	19.1	39.9	31.3	12.6	28.1
Overall										
All N=		17		79		320		182		64
Low	7.4	29.4	41.2	35.4	42.6	9.1	5.9	2.2	2.9	3.1
Average	2.8	64.7	11.5	57.0	58.7	71.9	23.0	49.5	4.1	25.0
High	.5	5.9	3.0	7.6	30.2	19.1	43.6	48.4	22.8	71.9

Note: R%= Row percentages; C%= Column percentages; Academic (χ^2 =110.983, df=12, p=.000, N=662); Practical (χ^2 =177.451, df=8, p=.000, N=657); Technical (χ^2 =48.768, df=12, p=.000, N=662); Overall (χ^2 =179.971, df=8, p=.000, N=662).

In relation to general life power, the majority of characters could not be coded for technical-intelligence, with 82.4% for characters assessed as very powerless and 60.9% for characters assessed as very powerful. Yet it was evident that high techno smarts was reserved for the very powerful characters (28.1%) rather than the very powerless (0.0%) or somewhat powerless (8.9%).

In relation to *humor* and *intelligence*, chi-square tests were significant for all types of intelligence (see Table 2.9). Although the majority of characters were portrayed as having "straight" dispositions, different humor types were shown to correlate with different intellectual types. More specifically, two interesting findings pertain to the interrelationships of humor, academic-intelligence, and practical-intelligence.

Table 2.9 Chi-Square Analyses of Humor and Intelligence

	Academic-Intelligence					
	Low intelligence		Average intelligence		High intelligence	
Humor	R%	C%	R%	C%	R%	C%
All N=		64		305		212
Straight	8.0	53.1	52.0	72.8	40.0	80.7
Witty	22.7	7.8	54.5	3.9	22.7	2.4
Joker	30.6	23.4	61.2	9.8	8.2	1.9
Sarcastic	13.6	12.5	45.8	8.9	40.7	11.3
Mixed	8.3	3.1	58.3	4.6	33.3	3.8
	Practical-Intelligence					
All N=		100		391		168
Straight	14.8	71.0	58.8	72.1	26.5	75.6
Witty	8.0	2.0	56.0	3.6	36.0	5.4
Joker	28.1	16.0	63.2	9.2	8.8	3.0
Sarcastic	7.1	5.0	68.6	12.3	24.3	10.1
Mixed	22.2	6.0	40.7	2.8	37.0	6.0
	Technical-Intelligence					
All N=		39		79		144
Straight	10.0	48.7	27.9	67.1	62.1	81.9
Witty	15.4	5.1	53.8	8.9	30.8	2.8
Joker	45.8	28.2	33.3	10.1	20.8	3.5
Sarcastic	18.2	10.3	36.4	10.1	45.5	6.9
Mixed	23.1	7.7	23.1	3.8	53.8	4.9
	Overall-Intelligence					
All N=		68		392		203
Straight	8.5	60.3	58.5	71.9	33.0	78.3
Witty	16.0	5.9	52.0	3.3	32.0	3.9
Joker	28.1	23.5	64.9	9.4	7.0	2.0
Sarcastic	6.9	7.4	61.1	11.2	31.9	11.3
Mixed	7.4	2.9	59.3	4.1	33.3	4.4

Note: R%=Row percentages; C%=Column percentages;
Academic (χ^2=39.100, df=8, p=.000, N=581); Practical (χ^2=22.756, df=8, p=.004, N=659); Technical (χ^2=31.156, df=8, p=.000, N=262); Overall (χ^2=32.855, df=8, p=.000, N=663).

First, within academic-intelligence, more than three-fourths of the straight (92.0%) and sarcastic (86.5%) characters were assessed as having average to high academic-intelligence. Thus, the majority of "scholarly" characters were portrayed as serious. Additionally, when humor was compared to program genre, it was found that more than three-fourths of the sarcastic characters (79.2%) were portrayed in situation comedies, with relatively few characters portrayed in dramas (9.7%), whereas the majority of straight characters were portrayed in situation comedies (42.0%) and dramas (34.6%). The high distribution of sarcastic characters in situation comedies perhaps suggests that characters were sometimes ridiculed for their scholarliness by less erudite characters to fulfill production requirements and, in turn, social-order maintenance functions. In other words, the narrative of situation comedies incorporates certain characters for comic-relief, and the butt of their joke may, in fact, be scholarliness—or the "nerdiness" or priggishness that often characterizes fictional scholars. If not poked fun of as geeks, intellectual characters are often represented as too serious, caustic and unlikable.

Second, with practical-intelligence, given that humor arguably helps to negotiate everyday situations, it was interesting that more than one-third of the witty characters (36.0%) had high practical-intelligence; these characters were portrayed as having the ability to make lively, clever remarks in a sharp, amusing way. In other words, having a witty sense of humor would certainly help characters to better negotiate interpersonal situations. However, few jokers (8.8%) were assessed as having high practical-intelligence. For example, the title characters from the programs *Drew Carey* and *Ellen* were rarely successful in negotiating everyday situations with humor. More often than not, their attempts at humor were moderately successful, and only within their "inner circle" of friends was their sense of humor understood, though not always appreciated.

Discussion

Data related to the first research question (how are characters represented in terms of *academic-, practical-, technical- and overall-intelligence*) can best be summarized in terms of three issues: (a) the availability of codeable characterizations; (b) the distribution of intelligence "types" among characters; and (c) the convergence of intelligence types among characters. These three intelligence issues—availability, distribution, and convergence—are relevant to social-order maintenance theory. More specifically, based on both the extant literature and the popular mythology of intelligence, it was predicted that television programming would likely underplay the social significance of *book smarts* or academic-intelligence, recommending instead the relative greater importance of first *street smarts*—or what is called here practical-intelligence, the ability to negotiate the rigors and problems of everyday life, and secondly, *techno smarts* or technical-intelligence—the ability to perform specific tasks.

Thus, the first question that arises was whether the availability of data itself responds to this larger theoretical issue. The opportunity to determine whether or not characters possessed academic-, practical-, or technical-intelligence was not the same. While most characters could be coded for practical-intelligence, their academic and technical abilities were less often suggested in the scripts. More specifically, whereas almost all characters could be coded for practical-intelligence, the academic intelligence of relatively fewer characters could be discerned, and even fewer characters could be coded for technical-intelligence. The fact that television stories do not reveal information about characters' intellectual and technical capabilities suggests that this type of information is not essential for viewers to identify with the characters, that is, to be interested in their concerns and problems, or in order to consider the narrative of TV programs entertaining.

When the first research question was examined in terms of the distribution (as opposed to availability) of intelligence among characters, the vast majority of characters were coded as average for all types of intelligence. Additionally, more characters were placed in the high range for both academic- and technical-intelligence than for practical-intelligence. These findings as well would seem to be incompatible with extant thought. While it might be expected that TV characters would be richer and perhaps more gainfully employed than the "average" viewer, it would not seem quite so logical to create them as smarter, if only because their speech and behavior must be intelligible to, if not resonant with, the viewer. Moreover, inasmuch as each episode's story for both drama or comedy series typically involves some interpersonal or social (practical) "problem" with which the characters struggle, it would make sense that the storytelling requires characters who are neither too quick nor too slow to solve the weekly dilemma. Thus, average practical-intelligence among characters would seem to suit the narrative requirements of TV programming. Of course, many television programs rely on the presence of a particularly wiser character that facilitates problem solving for the other, often younger, characters—as quintessentially shown in *Seventh Heaven*. But even the presence of this stereotypical mentor character typically requires the collaboration of those less wise.

The third issue relating to the representation of intelligence—how the different types of intelligence (academic, practical, technical and overall) converge in characters—presented more intriguing results. As the data for this study indicated, by and large, characters with high academic-intelligence were generally presented as at least average in terms of codeable practical- and technical intelligence; in other words, intellectual characters were not assessed as incompetent in other aspects of their lives, particularly when considering the interrelationship among social class, education, labor and general life power, and humor.

Unlike those data summarized thus far, this particular finding is the first (and perhaps most fundamental) of several other convergences that is not compatible with expectation and does not resonate with the logic of earlier scholarship and/or popular worldview. More specifically, this finding was inconsistent with the expectation that more academically-intelligent characters would possess less practical- or technical-intelligence (other than specific laboratory or computer skills) because such inverse associations are typical in American storytelling. These more typical inverse correlations are thought to discredit intellectuality inasmuch as intellectuality, in the larger political context, is regarded as dangerous to the status quo if possessed by the masses. Moreover, when the data relevant to the second question are reviewed, we see other correlations that, on the one hand, contradict earlier scholarship and popular lore about intelligence, and on the other hand, are consistent with these findings.

The second research question examined how intelligence connected to other indices of vitality and power in our culture—gender, social age, social class, education, labor, general life power and humor—to determine if there was a traditional discreditation of intelligence. As suggested in the previous discussion of social-order maintenance theory, it was suggested that characters with high ratings of intelligence—particularly in academic intelligence—might have lower ratings on youthfulness, social class, and humor (Jeffries-Fox & Signorielli, 1979). It was also expected that intellectuals may be typecast into professional occupations that necessitate specific credentials or degrees in higher education.

The intent of this study was also to explore how a television character's intelligence connected to other important indices of vitality and power in our culture. As related to *gender*, whereas previous research indicated that women were generally outnumbered by men at about three-to-one (e.g., Tedesco, 1974), and more current research established almost a two-to-one ratio (e.g., Signorielli & Kahlenberg, 2001), this study found a 1.5 ratio of male to female characters. This suggests that in mere numbers, or recognition, that women are now more populous on the screen. When gender is analyzed in relation to this study's more important variable of intelligence, we see again what might be called a traditional, if not, reactionary connection.

First of all, if we return to the three sub-issues raised earlier—availability, distribution, and convergence—we see that it was simply easier to obtain intelligence ratings for male characters than for female characters. Thus, in television stories, the "minds of men" are apparently more significant to narrative development than those of women.

Secondly, in considering all types of intelligence measured, male characters were more likely than females to have *high intelligence* ratings. This means that male characters are not only more likely to be more scholarly and technically able than female characters, but also better at solving everyday, practical problems. Thus, it might even be said that these data not only conform stereotype, but in fact are *more* reactionary than what might be expected in popular modern thought. More specifically, certain scholarship (e.g., Tuchman, 1974) claimed that women on television are shown to be the emotional leaders in domestic problem solving. Of course, this study examined practical-intelligence in a variety of contexts (including home and work), and may not be directly comparable to these findings (If this study had examined specific domestic problems—e.g., helping children with their chores—it might also have found that women are better at it than men). So, this study, in general, found that if a character is going to be smart, that character is more likely to be male than female.

In relation to *social age,* Signorielli (2004) established that television programming may contribute to the maintenance of ideas about aging, and potentially, to the familiar doctrine within popular culture that age brings wisdom. This content analysis found that on prime-time programming, although women were still shown as significantly younger than men, the mean chronological age of females was 31, which was several years older than previous established (Gerbner & Signorielli, 1982), and consistent with more current research (Signorielli, 2004). The relationship between social age and intelligence showed that children and teens, those in the pinnacle of youth, were four times more likely to have an identifiable academic intelligence than the elderly.

More specific examination of the distribution of intelligence among characters found that the elderly were least likely to be portrayed as having high ratings of intelligence. The middle-aged, those in the prime of their career, were endowed with the highest ratings of academic- and practical-intelligence, sustaining the popular doctrine that age brings wisdom. Thus, there were advantages in the aging process, as these findings suggested that advancing age can bring dignity and wisdom, an important index of power in our culture.

When considering the relationship between intelligence and *social class*, one must acknowledge conventional wisdom that obtaining certain professional credentials and decrees can be costly; according to the Statistical Abstracts of the United States (2006), the average cost of tuition, excluding room and board, as well as financial assistance and/or scholarship, is $14,710 for public institutions and $21,131 for private institutions. When coupled with the average annual income being $37,765, one may expect there to be a correlation between academic intelligence and social class, and this finding was supported in this content analysis. This positive correlation is particularly salient when considering the small percentage of people that would be classified as upper class in the U.S. economic structure.

Another interesting finding was that comparatively smaller proportions of the lower class (9.5%) were coded as having high practical-intelligence; although Thomas and Callahan (1982) did not analyze street smarts, per se, they found that members of upper- and upper-middle-class were less frequently involved in central problems. In this study, more than one-third of the upper class characters (39.7%) were rated as having high practical-intelligence, as compared to less than one-tenth of those lower class characters (9.5%). Thus, in the fictional world, intelligence abounds with material wealth, despite what popular lore might suggest.

Extant scholarship and popular lore have traditionally made claims about characters' education, but this research study actually quantified and reliably coded for both the *education level* and for whether the education was revealed *explicitly* or *implicitly* within a program. It was found that level of education was most often revealed implicitly within the program, and that the majority of characters were working toward high school diplomas, bachelor degrees, and professional degrees. This is concomitant with U.S. Statistical Abstracts (2006) that report a twelve percent increase in the number of college graduates from 1970 to 2003, and in turn, provides support for the reciprocity between media representations of education and actual enrollment percentages.

Furthermore, there was a position correlation between intelligence and education; those characters assessed as having high academic-intelligence were more likely to have degrees in higher education. Yet contrary to expectations, intellectuals were not devoid of practical-intelligence or common sense; perhaps this reflects Hofstadter's (1964) claim that in our modern world, formal training has become a prerequisite for achievement; "in the practical world of affairs, then, trained intelligence has come to be recognized as a force of overwhelming importance" (p. 34). In other words, it is recognized that simple tasks, like making breakfast, require practical knowledge of mechanical devices. Thus, it would not be so unlikely that intellectuals would need, at most, average amounts of common sense to envision such devices. That is not to say, on the contrary, that intellectuals may not be resented for becoming too practical or important, however, because of their connection to power or leadership.

Prime-time television also provided important information about the world of work. Characters with high ratings of academic-intelligence were more likely to have degrees in higher education, and in turn, be employed as professionals or as students, aspiring for their undergraduate or graduate degrees. Conversely, the majority of characters with low academic-intelligence were portrayed as having high school diplomas and more than one-third (35.7%) were depicted in blue-collar jobs. This related to conventional wisdom insofar that credentials commensurate with employment opportunities.

In isolating specific occupations related to practical intelligence, findings were consistent with expectations based on convention; for instance, approximately 95% of law enforcement officers had average- to high practical-intelligence, as one might expect based on everyday tasks that involve rudimentary skills in logical deduction, reason, and organization.

What makes the confluence of intelligence and labor particularly salient is the finding that blue- and white collar workers were show to be more practically- and technically-intelligent than professionals. This finding sustained the work of Konrad and Szelenyi (1979), and Gramsci (1971), who considered the term intellectual in a broader occupational sense to refer to all those individuals involved in the production and dissemination of knowledge not limited to book smarts.

This content analysis also found that on television, intellectuals were most likely to lead, restrain, and organize the behavior of others (Lemon, 1978), whether in interpersonal or workplace contexts; in other words, characters with book, street and techno intelligence were consistently measured as having high ratings of *general life power*. Foucault (1977) and Coser (1965) considered power and, in turn, knowledge, to be crucial to the maintenance of ideological stability, with the former stating that "the intellectual is not intent on power but aims first at focusing the public mind upon a central issue and then brings to bear the force of public opinion upon the makers of policy" (p. 207). Although one might expect that the acquisition of knowledge would be concomitant with other important indices of vitality in our culture, like general life power, education, education and material prosperity, this study found, in contrast, an inverse relationship with those personality attributes that we may aspire to attain.

Of particular interest in this analysis was exploring how television programming upholds public opinion of intellectualism; previous scholars provided anecdotal evidence and historical analyses of intellectuals that described them as dull, awkward or pretentious (Gerbner, 1974; Hofstadter, 1964). When measuring and isolating *humor* in relation to specific types of intelligence, it was apparent that straight and sarcastic characters had high ratings of academic intelligence. Conversely, characters with wit, perhaps a more coveted personality attribute than sarcasm, were more likely to have practical intelligence. These distributions are consistent with social-order maintenance theory, insofar that characterizations on television reinforce the notion that intellectual knowledge is resented in our culture; within colloquial, public discourse, sarcasm is considered to be words of a sneering, jesting, or mocking nature, whereas wit is noteworthy as unusual, striking, or surprising remarks that conjure pleasure. Thus, these iterations within television programming might contribute importantly to the public's evaluation and desire for certain kinds of knowledge.

In general, the data for this study often conformed to expectations. Yet the question remains as to why, in this study, the distribution of intellectuality was more often coded as "average." Henry Adams claimed that "there is no such thing as an underestimate of average intelligence." But the concern here lies in its potential over-estimation and how this may impinge directly on the central research questions in this study and the possible methodological limitations that pertain to coding of intellectuality on prime-time television programming.

Limitations and Future Research

There were four limitations to this study. First, a larger sample size might have yielded more data that could have strengthened some of the findings in this study. Perhaps this study should have expanded the program sample to replace those variety, sports, news, and reality programs that were omitted for not being prime-time, fictional, television programs.

A second related limitation involves the generalizability of the findings; the frequencies in some of the cross-tabulations were too small to yield statistically significant data. To strengthen some chi-square values, certain variables had to be collapsed to more general categories during the data analysis. Additionally, multivariate cross-tabulations of theoretical interest were sometimes omitted for lack of statistical significance in part because of the small cell-frequency counts. Of course, to conduct a more substantial and perhaps generalizable study so that these statistical problems do not arise would have required resources that were beyond the scope of this research project.

A third limitation is related to differentiating types of plot-functional characters. As in many studies, the plot-functional character was the unit of analysis here. By the end of the coding process, it became apparent that major and supporting characters could vary importantly in terms of the same key variables, most notably, intelligence, in the same ways that "continuing" versus "guest" characters are invested with differing qualities. On television programming, more temporally enduring characters are often invested with the more positive attributes that writers and producers may not be as committed to embedding in one-time guest-featured performers (see Gitlin, 2000). Thus, it might be a good idea to distinguish between not only major and supporting characters, but those characters that endure and dominate the television landscape in continuing positions from those with only guest status. This may explain why this study found intelligence—particularly *academic intelligence*—to be so positively portrayed or, put another way, to be portrayed without more variance in its distribution.

A reexamination of this study's coding process reveals a final methodological limitation that may account for the surprising change in the representation of intelligence. More specifically, in our culture, the discreditation of intellectuality has a deep-long standing tradition. Yet the data collected here did not confirm to what previous scholarship would lead us to expect. When such divergence is found it can be explained by suggesting that social change has occurred, producing, in turn, changes in media representations. While this may explain why this content analysis uncovered data different from previous research, it does seem prudent to reexamine the operationalization of certain intelligence variables.

Through the coding process, coders did not have difficulty measuring the intelligence variables; as previously stated, Krippendorff's Alphas were robust, demonstrating that there was strong internal consistency of observations among coders. The operational definitions seemed to provide detail on exactly what was meant by the variable and what was included and excluded as evidence when measuring the variable. However, in hindsight, the operational definitions of *practical-intelligence* and *technical-intelligence* should be made more precise. Certainly, it is to be expected more nuanced definitions of intelligence would emerge from this content analysis inasmuch as there were almost no earlier quantitative empirical studies to reliably ground operational definitions of intelligence, particularly when distinguishing among the types of intelligence.

When measuring *technical-intelligence,* coders were not clearly made to differentiate between "manual" skills and "professional" skills. Thus, characters possessing technical-intelligence might have demonstrated skills in numerous types of scientific, mechanical, artistic, professional, or trade applications, but manual skills were not differentiated from other technical, but more cerebral endeavors. To this end, both the computer "geek" and the auto mechanic may have received the same high technical-intelligence rating. This, of course, becomes problematic insofar as the technically intelligent computer geeks might have been rated very differently in terms of labor or power, for instance, than the technically-intelligent auto mechanic.

Similarly, the operational definition for *practical-intelligence* may not have been sufficiently nuanced as well. In this case, it was not a question of differentiating between types of practical-intelligence (as with technical-intelligence above) but rather with the absence of coding the emotional or social results of a character's practical-intelligence. That is, based on the study's operational definition for practical-intelligence, a character that had, for instance, cleverly manipulated circumstances might have been rated high in practical-intelligence, even though she or he may also have been unkind or duplicitous and wound up with the benefit of the manipulation but wounded in another way. Thus, it might have been more helpful had there been separate measurements for different aspects of handling interpersonal interaction as well as an overall measurement of the character's emotional and social status at the end of the episode. With such adjustments, then, it might have been found that only certain types of practical-intelligence correlated with, for instance, education.

The limitations described above merely provide provisos and recommendations for future research, as this content analysis was innovative in its attempts to measure quantitatively varying types of intelligence. It can surely be argued that the data emanating from this content analysis are definitive and lead to a better understanding of how intelligence is presented on prime-time television.

Future research should seriously consider using the obtained content-analytic data as a basis for examining other important indices of vitality and power in our culture. For instance, the traditional discreditation of intelligence was hypothetically accomplished by making looks and sex the "price" that is paid for "brains" (Thomas & Krippendorff, 1988). It has always been assumed that the "dumb blonde" image would be prevalent in popular culture, for research in social psychology indicates that, "people depicted as outstanding in intellectual or accomplishment are not uniformly good-looking, nor are good looking people necessarily portrayed as smart and accomplished" (Eagly, Ashmore, Makhijiani and Longo, 1991, p. 112). It would also be expected that few intellectuals would be portrayed as engaging in sexuality activities, as intellectuals are rarely shown as successful in love and romance. A content analysis by Thomas and Krippendorff (1988) revealed that intellectual women were rarely coded as beautiful, having sexy figures and dress styles, and receiving male attention. Given the above methodological recommendation pertaining to intelligence, and in turn, to attractiveness and sexuality, future research should test this convergence of relationships. Additional research should also conduct a cultivation analysis, with scholars examining the relationship between viewers' assessment of the interrelationships among intelligence and the demographic attributes identified in this analysis, as related to their media consumption behavior.

References

Barbatsis, G. S., Wong, M. R., & Herek, G. M. (1983). A struggle for dominance: Relational communication patterns in television drama. *Communication Quarterly, 31*(2), 148-155.

Bell. J. (1996). Of fairy godmothers and witches: Network television and the teacher. In P. M. Lester (Ed.), *Images that injure: Pictorial stereotypes in the media* (pp. 167-171). Westport, CT: Praeger.

Benda, J. (1928). *The treason of the intellectuals.* (R. Aldington, Trans.) New York: W.W. Norton & Company.

Brown, B. E. (1980). *Intellectuals and other traitors.* New York: Ark House.

Butsch, R. (1992). Class and gender in four decades of television situation comedy: Plus ca Change . . . *Critical Studies in Mass Communication, 9*(4), 387-399.

Claussen, D. S. (2004). *Anti-intellectualism in American media: Magazines and higher education.* New York: Peter Lang.

Coser, L. A. (1965). *Men of ideas: A sociologist's view.* New York: Free Press.

DeFleur, M. L. (1964). Occupational roles as portrayed on television. *Public Opinion Quarterly, 28*(1), 57-74.

Dominick, J. R. (1979). The portrayal of women in prime-time, 1953-1977. *Sex Roles, 5,* 405-411.

Downs, C. (1981). Sex-role stereotyping on prime-time television. *The Journal of Genetic Psychology, 138,* 253-258.

Durkheim, E. (1951). *Suicide.* (G. Simpson, Ed. & Intro.) (J. A. Spaulding & G. Simpson, Trans.) New York: Free Press of Glencoe.

Eagly, A. H., Ashmore, R.D., Makhijani, M.G., & Longo, L.C. (1991). What is beautiful is good, but . . . : A meta-analytic review of research on the physical attractiveness stereotype. *Psychological Bulletin, 110,* 109-28.

Elasmar, M., Hasegawa, K., & Brain, M. (1999). The portrayal of women in U.S. prime time television. *Journal of Broadcasting and Electronic Media, 43*(1), 20-34.

Evans, W. A. (1995). The mundane and the arcane: Prestige media coverage of social and natural science. *Journalism and Mass Communication Quarterly, 72*(1), 168-177.

Foucault, M. (1977). *Power/knowledge: Selected interviews and other writings 1972-1977.* (C. Gordon, Ed.) (C. Gordon, L. Marshall, J. Mempham, & K. Soper, Trans.) New York: Pantheon Books.

Gerbner, G. (1958). On content analysis and critical research in mass communication. *AV Communication Review, 6*, 85-108.

Gerbner, G. (1974). Teacher image in mass culture: Symbolic functions of the "hidden curriculum." In P. Olson (Ed.), *Media and symbols: The forms of expression, communication, and education* (pp. 470-497). Chicago: The National Society for the Study of Education.

Gerbner, G. (1990). Epilogue: Advancing on the path of righteousness (maybe). In N. Signorielli and M. Morgan (Eds.), *Cultivation analysis: New directions in media effects research* (pp. 249-262). Newbury Park, CA: Sage.

Gerbner, G., Gross, L., Morgan, M., & Signorielli, N. (1981). Scientists on the TV screen. *Society, 18*(4), 41-44.

Gerbner, G., Gross, L.., Morgan, M., & Signorielli, N. (1994). Growing up with television: The cultivation perspective. In J. Bryant & D. Zillmann (Eds.), *Perspectives on media effects* (pp. 17-42). Hillsdale, NJ: Lawrence Erlbaum.

Gerbner, G., Gross, L., Morgan, M., Signorielli, N., & Shanahan, J. (2002). Growing up with television: Cultivation processes. In J. Bryant & D. Zillmann (Eds.), *Media effects: Advances in theory and research* (2nd ed., pp. 43-68). Mahwah, NJ: Lawrence Erlbaum.

Gerbner, G., & Signorielli, N. (1982, October). The world according to television. *American Demographics,* 15-17.

Gitlin, T. (1977). The televised professional. *Social Policy, 8*(3), 94-99.

Gitlin, T. (2000). Prime time ideology: The hegemonic process in television entertainment. In H. Newcomb (Ed.), *Television: The critical view* (6th ed., pp. 574-594). New York: Oxford University Press.

Glascock, J. (2001). Gender roles on prime-time network television: Demographics and behaviors. *Journal of Broadcasting and Electronic Media, 45*(4), 656-669.

Gouldner, A. W. (1979). *The future of intellectuals and the rise of the new class.* New York: Seabury Press.

Gramsci, A. (1971). *Selection from prison notebooks.* (Q. Hoare & G. N. Smith, Eds. & Trans.) London: Lawrence & Wishart.

Greenberg, B. S. (1982). Television and role socialization: An overview. In D. Pearl, L. Bouthilet, & J. Lazar (Eds.), *Television and behavior: Ten years of scientific progress and implications for the eighties, Volume II: Technical Reviews* (pp. 179-190). Rockville, MD: U.S. Department of Heath and Human Services.

Greenberg, B. S., Richards, M., & Henderson, L. (1980). Trends in sex-role portrayals on television. In B. S. Greenberg (Ed.), *Life on television: Content analysis of U.S. television drama* (pp. 65-87). Norwood, NJ: Ablex.

Habermas, J. (1989). *The structural transformation of the public sphere: An inquiry into a category of bourgeois society.* (T. Berger & F. Lawrence, Trans.) Cambridge, MA: The MIT Press (Original work published 1962).

Haskell, D. (1979). The depiction of women in leading roles in prime time television. *Journal of Broadcasting, 23*, 191-196.

Hegel, G. W. F. (1977). *Phenomenology of the spirit.* (A.V. Miller, Trans.) Oxford: Clarendon Press (Original work published 1807).

Henderson, L., Greenberg, B. S., & Atkin, C. K. (1980). Sex differences in giving orders, making plans, and needing support on television. In B. S. Greenberg (Ed.), *Life on television: Content analysis of U.S. TV drama* (pp. 49-63). Norwood, NJ: Ablex.

Hofstadter, R. (1964). *Anti-intellectualism in American life*. New York: Alfred A. Knopf.

Holderman, L. B. (2003). Media-constructed anti-intellectualism: The portrayal of experts in popular US television talk shows. *The New Jersey Journal of Communication, 11*(1), 45-62.

Janus, N. Z. (1977). Research on sex-roles in the mass media: Toward a critical approach. *The Insurgent Sociologist, 7*(3), 19-32.

Jeffery, L., & Durkin, K. (1989). Children's reactions to televised counter-stereotyped male sex role behavior as a function of age, sex and perceived power. *Social Behaviour, 4*, 285-310.

Jeffries-Fox, S., & Signorielli, N. (1979). Television and children's conceptions of occupations. In H. S. Dordick (Ed.), *Proceedings of the 6th annual telecommunications policy research conference* (pp. 21-38). Lexington, MA: Lexington.

Kadushin, C. (1982). Intellectuals and cultural power. *Media, Culture and Society, 4*, 255-262.

Kahlenberg, S. G. (1995). *Character portrayals on prime-time television: A content analysis*. Unpublished master's thesis, University of Delaware.

Kanigua, N., Scott, T., & Gade, E. (1974). Working women portrayed on evening television programs. *The Vocational Guidance Quarterly, 22*, 134-137.

Kant, E. (1992). *Theoretical philosophy: 1755-1770*. (D. Walford & R. Meerbote, Eds. & Trans.) New York: Cambridge University Press.

Konrad, G., & Szelenyi, I. (1979). *The intellectuals on the road to class power*. New York: Harcourt Brace Jovanovich.

Krippendorff, K. (1980). *Content analysis: An introduction to its methodology*. Beverly Hills, CA: Sage.

Lamont, M., & Fournier, M. (Eds.). (1992). *Cultivating differences: Symbolic boundaries and the making of inequity*. Chicago: University of Chicago Press.

Lemon, J. (1974). Dominant or dominated? Women on prime-time television. In G. Tuchman, A. Daniels, & J. Benet (Eds.), *Hearth and home: Images of women in the mass media* (pp. 51-68). New York: Oxford University Press.

Levine, L. W. (1988). *Hibrow / Lowbrow: The emergence of cultural hierarchy in America*. Cambridge, MA: Harvard University Press.

Marx, K., & Engels, F. (1970). *The German ideology*. (C. J. Arthur, Ed. & Intro.). New York: International Publishers.

McChesney, R. W. (1996). *Rich media, poor democracy: Communication politics in dubious times*. Urbana, IL: University of Illinois Press.

McNeil, J. C. (1975). Feminism, femininity, and the television series: A content analysis. *Journal of Broadcasting, 19*, 256-271.

Miller, M., & Reeves, B. (1976). Dramatic TV content and children's sex-role stereotypes. *Journal of Broadcasting, 20*, 35-50.

Mills, C. W. (1963). *Power, politics, and people: The collected essays of C. Wright Mills*. (I.L. Horowitz, Ed.). New York: Ballantine.

Morgan, M. (Ed.). (2002). *Against the mainstream: The selected works of George Gerbner*. New York: Peter Lang.

Mosco, V. (1996). *The political economy of communication: Rethinking and renewal*. London: Sage.

Neisser, U. (1976). General, academic, and artificial intelligence. In L. B. Resnick (Ed.), *The nature of intelligence* (pp. 135-144). Hillsdale, NJ: Lawrence Erlbaum.
Posner, R. A. (2001). *Public intellectuals: A study of decline.* Cambridge: Harvard University Press.
Rigney, D. (1991). Three kinds of anti-intellectualism: Rethinking Hofstadter. *Sociological Inquiry, 61*, 434-451.
Ross, A. (1989). *No respect: Intellectuals & popular culture.* New York: Routledge.
Said, E. W. (1994). *Representations of the intellectual.* New York: Vintage Books.
Schiller, H. I. (1973). *The mind managers.* Boston: Beacon Press.
Seggar, J. F. (1975). Imagery of women in television drama: 1974. *Journal of Broadcasting, 19*, 273-282.
Signorielli, N. (1993). Television and adolescents' perceptions about work. *Youth & Society, 24*(3), 314-341.
Signorielli, N. (2004). Aging on television: Messages relating to gender, race, and occupation in prime time. *Journal of Broadcasting & Electronic Media, 48*(2), 279-301.
Signorielli, N., & Bacue, A. (1999). Recognition and respect: A content analysis of prime time television. *Sex Roles, 40* (7/8), 527-544.
Signorielli, N., & Kahlenberg, S. G. (2001). Television's world of work in the 1990s. *Journal of Broadcasting and Electronic Media, 18*(2), 1-16.
Smythe, D. W. (1954). Reality as presented by television. *Public Opinion Quarterly, 18*, 143-156.
Squibb, P. G. (1973). The concept of intelligence—A sociological perspective. *Sociological Review, 21*, 57-75.
Statistical Abstracts of the United States (2006). Washington, D.C.: US Census Bureau. Retrieved December 10, 2005, from http://www.census.gov.
Sternberg, R. J. (1988). *The triarchic mind: A new theory of human intelligence.* New York: Viking.
Sternberg, R. J. (1990). *Metaphors of mind: Conceptions of the nature of intelligence.* New York: Cambridge University Press.
Sternberg, R. J., Wagner, R. K., Williams, W. M., & Horvath, J. A. (1995). Testing common sense. *American Psychologist, 50*(11), 912-926.
Sternglanz, S. H., & Serbin, L. A. (1974). Sex role stereotyping in children's television programs. *Developmental Psychology, 10*, 710-715.
Tedesco, N. (1974). Patterns in prime time. *Journal of Communication, 24*(2), 119-124.
Thomas, S., & Callahan, B. P. (1982). Allocating happiness: TV families and social class. *Journal of Communication, 32*(3), 184-190.
Thomas, S., & Krippendorff, M. (1988). *Beauty and brains: The sociopolitical concomitants of sexuality on television.* A paper presented at the Convention of the International Communication Association, New Orleans.
Tuchman, G. (1974). Introduction: The symbolic annihilation of women by the mass media. In G. Tuchman, A. Kaplan-Daniels, & J. Benet, *Hearth and home: Images of women in the mass media* (pp. 3-49). New York: Oxford.
Turow, J. (1974). Advising and ordering: Daytime, prime time. *Journal of Communication, 24*, 138-141.
VandeBerg, L. R., & Streckfuss, D. (1992). Prime-time television's portrayal of women and the world of work: A demographic profile. *Journal of Broadcasting and Electronic Media, 36*, 195-208.

Wober, M., & Gunter, B. (1988). *Television and social control.* New York: St. Martin's Press.

Part II

Social Class, Gender, and Youth Culture

Chapter Three

Better Keep the Egghead: Pragmatism in *The Simpsons*

Mary T. Conway

American culture loves to mock the intellectual. The cartoon television series *The Simpsons* adds a twist to this tradition. The series undermines the snobby, impractical and ego-driven intellectual, but it also mocks stupidity, champions practical knowledge, and demonizes capitalists. As such, the show only appears to be anti-intellectual: a closer reading shows that *The Simpsons* embraces Pragmatism, a distinctively home grown school of philosophy. Pragmatism enables the show to maintain these seemingly contradictory attitudes, and invites criticism of the American Dream. In its valuation of pragmatic intelligence, *The Simpsons* reveals an allegiance with the meritocracy, even if it is a show that does not show class. Finally, the series skewers Pragmatism with its own limits, revealing the weaknesses in the philosophy.

The Simpsons vs. Eggheads: Very Popular Culture's Assault on Intellectuals

The longest running television show, *The Simpsons* has won countless awards and accolades. Entire college courses and a handful of academic books, most notably *The Simpsons and Philosophy* (Irwin, Conard & Skoble, 2001), take the series as their subject. There are devout fans of all stripes, but one common trait is as American as apple pie: the disdain for "eggheads" (Hofstader, 1963). Egghead bashing is even more remarkable because of the series' reputation as smart: with Yale graduates writing inter-textual allusions to literature and film, the series is distinguished from most television. Not surprisingly then, the egghead-scripted series has a tense and dense relation to privilege, intelligence, and merit.

Lisa Simpson, the precocious, saxophone-playing, vegetarian 8-year-old is the most visible intellectual character. Lisa's ideas are usually based on reading, and this helps explain her many errors. Her youth faults intellectuals, who are lampooned for being book smart but not street smart, inexperienced but yet authoritative. However, her youth does not excuse her competitiveness, ego and unbridled drive to be recognized—these all align her with the worst sort of adult intellectual. In the "The PTA Disbands," Lisa implores her striking teacher for syllabi and testing (Crittenden, 1995). At home with Marge, Lisa explodes and demands to be evaluated and ranked, prompting Marge to slap a hastily scrawled "A" upon her daughter. Lisa's characterization also depicts intellectuals as benignly ineffectual. Lisa is likeable because of her insecurity, which illustrates her humanity, as she is dependent on the social world for her identity.

The rest of Springfield's intellectuals fare less well. In the series' sharpest egghead-bashing episode, "They Saved Lisa's Brain," the town's MENSA members take power after Mayor Quimby's corruption catches up to him (Selman, 1999). A fiasco results, driven by ego, hubris and a little bit of insanity. Although smart up to a point, Lisa, Professor Frink, Dr. Hibbert, Principal Skinner, and Comic Book Guy become enamored with their own individual ideas and chaos ensues. Their arrogance and lack of connection with reality are their undoing, but each possesses some likeable or pitiable quality. The exception is Comic Book Guy, who is uniformly contemptuous of the town, including other MENSA members. Socially and physically unfit, he has unhealthy fantasies about comic book heroines, falls in love with the caustic, octogenarian, Mrs. Skinner, attends Star Trek conventions alone, and spites children. Consoling himself in food makes him grossly overweight. Comic Book Guy might be the ultimate aggressive characterization of the intellectual: he is punished for his arrogance by the rest of his attributes.

The Simpsons' eggheads are lampooned, but they are also often depicted with sympathy. In "Homer Goes to College," Homer arrives at college expecting and reinforcing outdated dichotomies between jocks and nerds (O'Brien, 1993). The re-occurring nerd characters (who are simply, "the nerds") tutor Homer in Physics. While competent in computers and academics, they are so gullible that when Snake the criminal announces "Wallet Inspector," they comply. Long absent from their dorm room, they receive no phone messages. Nerds may possess information, but divorced from experiences and practices of the everyday, they are pathetic and often powerless in the social realm.

Eggheads vs. Eggheads: The Intellectual Assault on Intellectuals

Cartoon characters have as many lives as their animators dictate; they die one episode and live the next. Curiously, the same can be said of the American intellectual, who has been declared dead, or dying for the last fifty years. These declarations, however, are another form of egghead bashing. This death has been chronicled by Hofstader's *Anti-Intellectualism in American Life* (1963), Jacoby's *The Last Intellectuals* (1987), and most recently, Kramer's *Twilight of the Intellectuals* (an allusion to Nietzsche only intellectuals will get) (2000). But, rather than chronicling a death, Austin's speech act theory is at work: the death seems brought about by the reporting itself (1962).

Stranger still, the perpetrator is often the intellectual. From whom would you expect this statement?

> I think one must be careful in assuming that intellectuals have some kind of insight. In fact, if the track record of intellectuals is any indication, not only have intellectuals been wrong almost all of the time, but they have been wrong in corrosive and destructive ways (Eakin, 2003, p. 1).

The speaker is University of Illinois Professor Sander L. Gilman at the Theory's End Conference, hosted by the journal *Critical Inquiry*. Gilman is the downsizer's dream: imagine a worker in any other profession maintaining their job after such self-denigration. *The Simpsons* echoes Gilman, when Marge is accused of shoplifting and goes to trial (Oakley & Weinstein, 1993). The jury turns on their sweet, ethical neighbor, Marge, because of expert testimony. Dr. Frink shows the Zapruder film; in slow motion he zooms in on a spherical bush which he asserts is Marge, concluding that she was on the grassy knoll. Preposterous and illogical evidence, but Marge is sentenced to thirty days in prison.

The U.S. intellectual emerged in the 1920s, and sought experiences rather than ideas or theories to shape her opinions. They were not usually associated with a university since this link could mark one as too conservative and cautious to be a true intellectual. Today however, intellectuals are almost always associated with, and employed by universities. In *The New Criterion,* Kramer bluntly opines:

> By now, many of us have come to expect professors and indeed the cultural elite generally to act like spoiled adolescents. Often—not always, not everywhere, but frequently—there seems to be an inverse relationship between virtues like patriotism and common sense and the number of years spent at a university or similarly insulated cultural institution . . . (2002, p. 1).

In *Twilight of the Intellectuals,* Kramer insists that intellectuals no longer have the power to sway public opinion (2000). If the intellectual is dead, however, why ridicule? If the intellectual no longer has power, why attack with such overkill? The answer lies in the internal contradictions of the American Dream.

Paradoxes of the American Inheritance

Regardless the intellectual's sorry state, his power to make us feel dumb is not diminished. de Tocqueville observed that our two most cherished principles, freedom and equality, work at cross-purposes (1835/2001). While we say we have the freedom to pursue our loftiest goals, despite our background, we bristle at evaluation, which would mark us unequal. At its most benign, this tension between evaluation and equality is apparent in the children's game of Tee-Ball. To encourage participation, the unskilled and untalented are given endless opportunities to mask the discrepancy in skill. When carried into adulthood, the specter of the intellectual haunts such pap, reminding the free, but untalented, that she is, indeed, untalented.

The Simpsons reflects the fractured psyche of the American dreamer as well, as the editors of *The Simpsons and Philosophy* have argued (Irwin et al., 2001). The philosophers detect the series' ambivalence toward intellectuals, but because of their antipathy toward popular culture studies and pragmatic philosophy, they overlook the finer distinctions about intelligence. On one hand, they write, "Without question it is one of the most intelligent and literate comedies on television today. (We know that's not saying much, but still . . .)" (p. 2, parenthesis and ellipsis original). On the other hand, they worry that they've betrayed their colleagues, "Is it anti-intellectual for a PhD in philosophy to write an essay about a TV show" (p. 2)? They reinscribe the weakened divisions of intellectual labor (popular culture studies vs. traditional scholarship), and only begrudgingly praise *The Simpsons*. They reveal their limiting framework when they ask, "Do we admire or laugh at Lisa" (p. 2)? The pleasure most viewers get from the contradictory complexities of her character make us refuse this dichotomy: we both admire and laugh at Lisa because she is both an irrelevant, ego-driven egghead, and a practical, compassionately intelligent, girl-citizen. The ambivalence in *The Simpsons and Philosophy*, when considered in light of Pragmatism and intelligence (not just intellectuals), becomes more certain. *The Simpsons* lampoons pretentious, merely theoretical eggheads, while wit, practical knowledge, and modesty about smarts are championed. That is, if the intellectual appears as the pragmatist.

Capacious Philosophy: Pragmatism and the American Dream

The pragmatist is reborn almost as often as the intellectual dies. Modernism holds that reason leads only to the limits of reason, not to truth. Pragmatism advocates instrumentalism in response to this dilemma. Diggins writes in his comprehensive historical analysis:

> Pragmatism, America's one original contribution to the world of philosophy, had once promised to help people deliberately reflect about what to do when confronting "problematic situations." . . . Pragmatism advises us to try whatever promises to work and proves to be useful as the mind adjusts to the exigencies of events (1994, pp. 2-3).

With the quest for truth no longer the aim, practical questions with practical solutions take precedence. Understandings, influenced by context and contingency, replace universal truth.

Truth did not go quietly, however for early advocates of Pragmatism. Historian Henry Adams and philosopher John Dewey, forerunners of American Pragmatism, demonstrate the expansiveness of the philosophy. Dewey's faith in hard science, and Adams' faith in spirituality distinguish them from contemporary pragmatists. But, as each advanced instrumentalism in lieu of truth, Dewey was optimistic, and Adams pessimistic. This difference suggests that depending upon one's disposition, one is either freed or fraught with the openness of Pragmatism. Adams' pessimism is exceptional as Diggins explains:

> To the pragmatist the discrepancies and contingencies of the universe should not provoke meaninglessness and anxiety. On the contrary, life must be seen as a challenge and, an adventure, and a universe that is potential and "unfinished" presents abundant opportunities to create meaning through the exercise of intelligent control (1994, pp. 4-5).

One could argue that instead of being an exciting adventure, a pragmatic approach is saddled by relativism and nihilism—good reasons to be pessimistic. The most preeminent contemporary pragmatic philosopher, Rorty, has often been accused of "jumping from the frying pan of foundationalism into the fire of cultural relativism" (2004, p. 151). While embracing many of the philosophy's insights, *The Simpsons* flips the viewer from the frying pan to the fire, and back again, with a glee born of the limits of Pragmatism.

If there are no absolutes, and context determines truth, judgment becomes tenuous: how can one blame or praise if all judgments are qualified by shifting contexts? In the absence of absolute truth, we may scramble for standards and criteria with which to judge, and finding few, retreat, mute. Rorty argues that this retreat is unnecessary, perhaps even self-indulgent. He argues, "The fact that there aren't any absolutes of the kind Plato and Kant and orthodox theism have dreamt doesn't mean that every view is as good as every other. It doesn't mean that everything now is arbitrary" (2002, p. 375). Instead, understanding results from consensus, interpretive moves made by groups. Relativism and nihilism are only alternatives if one remains devoted to traditional problem-solving, and "Big T" truth. In *The Revival of Pragmatism,* Kloppenberg writes, "Pragmatism appeals to many American thinkers as a home-grown alternative to postmodernism that escapes the weaknesses of Enlightenment rationalism without surrendering our commitments to the values of autonomy and equality" (1998, p. 111). Pragmatism seems necessary and sufficient because the alternatives are untenable, at least to the pragmatist.

Pragmatism values experience and action over ideas; reflection, traditional philosophy's unifying method, too often breeds stagnation. With flux comes uncertainty, but action, "aims at adjusting and coping rather than knowing" (Diggins, 1994, p. 15). Pragmatism operates as the antithesis of the university intellectual, for whom deliberation trumps action. Teaching at Harvard in 1875, Henry Adams, irate at the university's reproduction of intellectuals, called the professorate and the students "prigs" who were "suffering under a surfeit of useless information." He asked, "Are we never to produce one man who will do something himself" (Diggins, 1994, p. 307).

Pragmatism does not privilege *a priori* principles, since context and contingencies make principles irrelevant. Dewey thought that principles covered up, rather than revealed, reality. This is a dramatic turn from traditional philosophy in which principles help discover truth. While we may laugh at Kant's principle that wigs are immoral, we daily subscribe to many of his other moral pronouncements, all of which are framed as principles.

Some academics are now turning to Pragmatism because of an increasing interest in the relations of knowledge and power. Pragmatism provides the theoretical framework to question many forms of elitism: when effective action trumps a pedigree, the arbitrary rules of elitism are visible. With Rorty, philosopher West is also responsible for the return to Pragmatism (2000). To West, social movements rekindled the interest in Pragmatism:

> Gone is the old humanist view that elevates the human agency of elite cultural creators and ignores social structural constraints that reinforce and reproduce hierarchies based on class, race, gender and sexual orientation (2000, p. 144).

Pragmatism both supports and undermines the American Dream. While Dewey hoped that free public education could solve class antagonisms, West reminds us of the continuing failures of the American dream, which sometime seem justified by Pragmatism.

Nonetheless, West is optimistic, if vague, about Pragmatism's role in ameliorating structural inequality. He writes that Pragmatism should be an, "attempt to reinvigorate our moribund academic life, our lethargic political life, our decadent cultural life and our chaotic personal lives for the flowering of many-sided personalities and the flourishing of more democracy and freedom" (2000, pp. 144-45). Pragmatism may not be *that* capacious, but as a school of philosophy that bears little resemblance to previous traditions, it is open to many readings. Repeating James' call for effective action, West writes that Pragmatism's, "basic impulse is a plebian radicalism that fields an anti-patrician rebelliousness for the moral aim of enriching individuals and expanding democracy" (2000, p. 145). West makes manifest what is latent in Pragmatism: class rebellion. However, there might be tenets of Pragmatism, especially those connected with utility and opportunity, that work at cross-purposes with West's idea.

To appreciate the contradictions that make West's proscription unlikely, we should return to Dewey's free, public education. Dewey argued that public schools create the opposite of what they intend: emphasizing memorization, tradition and authoritarianism, schools stifle intelligence. He argued for what he called progressive education, with students left to their own pursuits, rather than forced to learn impractical ideas. Unburdened, the most ambitious, creative and intelligent students would discover their talents, leading to better lives for all. Progressive education opposed indoctrination and a priori knowledge; education would be experience based. A century later, Dewey's philosophy has been adopted and adapted, but not everywhere.

Dewey's ideal classroom is most likely in wealthy, suburban, public school districts, or private schools. Better schools, while equipping students to succeed at Harvard (Dewey's alma mater), also allow the students' own pursuits. In its most absurd misinterpretation, his theory implied all ideas were equally valid (except for the teacher's whose could never be valid, since it was based on a priori knowledge). On the other hand, his nightmare classroom is in the vocational tracking, discipline-heavy schools in poorer areas. While poorer schools often teach vocational education, which might be beneficial, they offer no student choice, are authoritarian, and repetitious. Anyon's often-cited 1980 essay, "Social class and the hidden curriculum of work," chronicles the devastating discrepancy in teaching philosophies and practices between poorer and wealthier schools. Students in poorer districts are channeled into menial jobs that require little thought; classroom activities that are repetitious, and reward blind obedience reinscribe these limited choices. *The Simpsons* seems keenly aware of Dewey and the demise of public education. The series lauds Pragmatism, but reserves a full endorsement because of the unequal failures of free education. In this reservation, the series aligns with a meritocratic system which we've never had, but continue to mythologize.

Pragmatism in *The Simpsons*

Practical Knowledge

Even as Lisa's progressive pontifications are skewered, the series praises her practical knowledge. When a new friend is reassuring her, he says appreciatively that she stopped him from drinking saltwater. Because Lisa learned CPR while waiting for other less-smart kids to finish their math test, she is able to save Homer from a fatal poisoning by spider pesticide (Swartzwelder, 1994). She often rescues Springfield from catastrophes. Lisa is most likable when she uses her intelligence to help solve problems.

Homer is the buffoon ignorant of both theoretical and practical knowledge. When they get lost in the woods, Lisa uses the stars to navigate (Swartzwelder, 1990). But Homer dismisses her solution, and reminds her that they are in the woods, not a classroom. He believes he doesn't need help because he's an experienced woodsman for whom finding his way is second nature—he then walks off a cliff, fails to make a fire, or find food, and is attacked, by small forest animals. Meanwhile, abandoned and forgotten, baby Maggie pacifies a threatening grizzly, who then takes her in. At the end, Maggie saves Bart and Homer from a deadly mauling by communicating with the bears with pacifier-sucking. Maggie possesses superior practical intelligence, which is often interpreted as chance by the other characters.

"Homer Goes to College" prefers practical knowledge, but it also indicts ungrounded attacks against higher learning (O'Brien, 1993). Homer must return to college and pass a course in Nuclear Physics to keep his job at the nuclear power plant. While the three nerds who tutor Homer are naïve, foolish, and socially inept, they hack into the school's computer, change Homer's grade, and return him to work. When they come to live with the family over Marge's objections, Homer describes them as the solution to all of their problems, and gleefully imagines putting their genius to work to make them millionaires.

Deadly Deliberation Versus Action

Common stupidity is targeted at least as often as the overly-deliberative intellectual. In a cockamamie scheme to get the three nerds readmitted to college, Homer will pretend to run over the Dean, allowing the nerds to rescue him, prompting the Dean to readmit them. As the nerds debate the theories of timekeeping, they miss their chance, and Homer crushes the Dean, who now needs a hip replacement. In this case, excessive deliberation precludes action, with nearly fatal consequences.

Theory does not always lose, however, especially in contrast to foolish action. When the Professor of Nuclear Physics introduces a Proton Accelerator, Homer rudely interrupts saying that because he has worked at a nuclear power plant, he in fact is the real expert (O'Brien, 1993). When the professor asks Homer to demonstrate, he explodes their ivy-covered building. In "Bart Gets an Elephant," Marge wisely protests an African elephant living with the family. Homer, parroting intellectual banter, says that he agrees with Marge in theory, but reminds her that communism worked in theory, too—while Stampy the elephant destroys the house (Swartzwelder, 1994). Theoretical knowledge trumps practical misinformation. The tensions between action and deliberation are settled by clear criteria; context and contingency determine the value of deliberation or action. The a priori principle that action and experience outsmart ideas and theory, is undermined by another of Pragmatism's tenets.

A priori Principles

The Simpsons loves to upend a priori principles. Even as eggheads are bashed, egghead bashing is itself challenged when Homer misapprehends the Dean in "Homer Goes to College." At a party, students object to Homer's spiked punch; Homer is shocked that college kids don't want to get drunk, and blames the mean old dean (O'Brien, 1993). But, this Dean is supremely cool: he plays guitar for the Pretenders, hacky sack with the students, and wears a ponytail. When hospitalized after Homer's hare-brained car accident, the Dean is absurdly understanding. Still Homer remains governed by a priori ideas when he growls that he's anxious to take revenge on the stuffy dean.

Lisa's a priori ideas, particularly those progressive ones, are dramatically skewered, to the horror of many left-leaning critics. In a scene reminiscent of *Free Willy*, she frees a Marine Park dolphin (Lazebnick, 2000). Instead of a grateful goodbye, the dolphin faces the camera with a devilish grimace. He is the evil Dolphin King who will flood Springfield and subjugate humans. Lisa's commitment to a priori principles, freedom for all animals, results in the catastrophic destruction of the town. With no land in sight, the Simpsons family floats on their sofa, reaching for food in the filthy currents. The tyranny of a priori principles leads to child abuse, neglect, and even starvation, in "A Streetcar Named Marge" (Martin, 1992). Maggie goes to the Ayn Rand Nursery School, where the fascistic headmistress withholds bottles from the toddlers because bottles announce dependency. Instead, kids should develop "the bottle within." But regardless how rational the theory, real-life experience tells us that babies die of malnutrition if food is withheld.

Anti-elitism, Anti-display

Pragmatism's anti-elitism derides the display of intellect, and the power those displays create and abuse. Henry Adams, though writing in 1875, echoes one effect *The Simpsons* writers may aim for, "I am preaching a crusade against Culture with a big C . . . I mean to irritate everyone about me to a frenzy by ridiculing all the idols of the University and declaring a University education a swindle" (Diggins, 1994, p. 306). Adams' words show how a branch of philosophy can seem anti-intellectual on *The Simpsons*. The series upbraids those who seek the limelight for their intelligence. Genuinely smart characters shun attention, and their actions speak louder than words: in the case of Maggie, the pre-verbal toddler, her actions speak volumes.

Maggie's pacifier, comforting and communicative, is withheld by the Ayn Rand Day Care headmistress. But, with instrumental ingenuity, Maggie assembles the necessary tools, enacts and perfectly times a sophisticated abstract plan, and liberates the kids' pacifiers. When Homer, Lisa and Bart arrive to collect Maggie, they carefully step through a room of silent, sucking, helpless toddlers. However, they only appear helpless; the scene evokes Hitchcock's *The Birds* and hints at an eerie, hidden power of the weak. Maggie is clearly and practically intelligent; she is also a hero to the other toddlers. Most importantly, she is silently intelligent, known only to the silent kids. Maggie and Lisa both demonstrate that pragmatic intelligence is valued, but its obvious display is not.

Maggie's intelligence has leverage, paradoxically, because it is nearly invisible. Lisa is smart, but when she merely performs smartness, she is ridiculous. Without experience, her ideas are often mere mimicry of those she has read. Her panicked demand to be evaluated and ranked mentioned earlier reveals her identity's dependence on the display of intelligence. In "Marge and the Monorail," Lisa's ego stops her from derailing a monorail swindle. Initially suspicious of the con man, she is disarmed when he compliments her intelligence. Taking advantage of her swooning under his false flattery, he says that if he did answer, only Lisa and he could understand the answer (Martin, 1992).

Measures of intelligence are also often questioned. Lisa is rankled when others think Maggie may have superior intelligence, in "Smart and smarter" (Omine, 2004). Her identity is threatened and her panic is palpable. However, investigation reveals that Maggie learned language like a dog, through classical conditioning—pairing gesture with sound, not word with intent. Later, Maggie breaks the fourth wall of the stage when she winks at the camera: she is smart—too smart for their measurements and their investigations of their measurements. In "Bart the Genius," the school psychiatrist asks Bart questions, which conclude he is gifted, rather than a juvenile delinquent (Vitti, 1990). The psychiatrist asks if he is bored at school. Bart emphatically agrees, but he is not bored because he is a genius, as the psychiatrist and Dewey might assume. In this case Skinner (both the principal and the behaviorist) is right, but if measures of intelligence are suspect, so are those deemed intelligent.

Pragmatism's Blindspots: Separate and Unequal

One need not be a devotee of Pragmatism to acknowledge its value, nor a critic to see its limitations. These limits are especially keen regarding structural inequality, education and class mobility. In *The Simpsons* the limitations take form in depictions of education, meritocracy's failure, evil capitalists, and mob rule.

Free, Public, and Poor: Springfield's Education Opportunities

Springfield's public school failures resonate in the series' depictions of the meritocracy and the American Dream. Dewey believed free public education to be the fulcrum upon which the American Dream is launched or alternately, falls flat. He believed that, "Schools in a democracy should give students the opportunity to grow by allowing them the freedom to question; the students should also be able to acquire the knowledge and techniques that enable them to control their physical and social environment" (Diggins, 1994, p. 310). Dewey's faith in the transformative power of free public education was so strong that he objected to Marxism on the grounds of its lack of necessity. Class antagonisms become irrelevant if the principle of equality creates the environment for the freedom to succeed.

Dewey's dream classroom sits empty at the heart of Pragmatism's limits. Ruling with authoritarianism, control, blind obedience, and repetition of meaningless tasks, Springfield Elementary defies Dewey who believed that, "Education, like democracy, must resist authoritative principles that promise the false security of ready-made truths" (Diggins, 1994, p. 310). In *The Simpsons,* the faculty, administration, and staff are unenthusiastic, when not obstructionist, about the children's futures. Principal Skinner and Mrs. Krabappel have given up on their own pathetic lives, are jealous when bright kids shine, and employ insane techniques for disciplining children. In "The Boy Who Knew Too Much," the school acquires hard, uncomfortable, new chairs designed to stop slouching (Swartzwelder, 1994). This episode depicts Principal Skinner in pursuit of Bart, who, inspired by daydreams of Huck Finn, Abe Lincoln and Mark Twain, plays hooky on a beautiful day. Skinner is inhuman: he walks through roaring rivers and scales sheer cliffs. At the episode's end, when Bart has done the right thing, even though it is unpopular, Skinner confirms that he is petty. In a grade school principal, this is chilling, even if he is a yellow cartoon figure.

Creativity is squashed in working-class schools like Springfield Elementary. When Lisa is left to her own devices during the teachers' strike, she blossoms, defies the laws of physics and builds a perpetual motion machine. Homer chastises and reminds her that in his house the laws of thermodynamics are to be obeyed (Crittenden, 1995). Free public education hinders class mobility; Lisa is dependent upon, and suffocated by the inept system.

Springfield's gifted schools show Pragmatism's irrelevance in education and class mobility, and make Dewey a wide-eyed utopist. Bart spends time in a school for gifted children, when he is mistaken for a genius (Vitti, 1990). There, students pursue their own interests, learning is fun and funny, and fourth graders are naturally inclined to work on calculus. When Bart sits mutely during a discussion on paradox, his teacher probes him for an example, stating that often the smartest student is the quietest one. His clichéd reply—damned if you do, damned if you don't—wows the class. He only temporarily fools them, though, as he repeatedly blows up the chemistry lab. While the school sees this as the trials of a creative genius, the viewer knows that Bart's public schooling has failed him. Bart loses his lunch, day after day, in contests of intelligence with others (the smart school's bullying). The scene forecasts a future of swindles (peacefully and legally) between Bart and smarter mates. When Bart wants to research Springfield's ordinary kids, the psychiatrist likens him to Jane Goodall studying chimps. Ordinary children are a different and lesser species than the gifted, and no amount of training can bridge this divide. In *The Simpsons,* education *maintains* class divisions; Dewey's dream classroom is only available to, and productive for, those born privileged.

Lisa's uncertain future, despite her intelligence and ambition, confirms the effects of the early divisions. The series has multiple, contradictory accounts of her future. In one episode, Lisa's ambition, hard work and intelligence lead her to be the first female President; in another episode, she becomes a grossly overweight, immobile hillbilly, who watches soap operas, as their kids (descendants of Bart and Lisa, together) help her shift her huge frame (Greaney, 2000). She often dreams of life at a Seven Sister college, and dreads the state school alternative. But her intelligence and ambition are not guarantees of success. In the many references to the Ivy League, and in Lisa's anxiety about failing, the series indicates an awareness of the betrayal of the American Dream. In its oblique nods to class immobility and the deceptions of the meritocratic system, there may be an allegiance, however oblique, to the working person.

Privilege Trumps Merit

There are few direct references to class, either in passing, or as episodic themes. *The Simpsons* Archive webpage categorizes references to nearly every concept imaginable—except class. Yet, just as content analysis can deceive, this absence is misleading. The series registers the effects of inequity, and markers of class, rather than the direct concept of class, a move that reflects contemporary American discourse. *The Simpsons* have money woes occasionally, but unlike most impoverished families, these troubles are not chronic, or painful. In contrast, the bully Nelson and the hillbilly Cletus are dirt poor, and they suffer. Their poverty is explained culturally, however, and not in terms of an unequal, arbitrary system. Nelson is poor because his Dad left (he was kidnapped by the circus), and his Mom is a loose sauce-hound. Cletus' stupidity explains his poverty: he sees Lisa as smart because she understands letters and how they make up words. Cletus and his wife, who is also his sister, have more than a dozen children, who share the family's one tooth.

The Simpsons may be television's most subtle and accurate measure of our changing notions of class. The recent *New York Times* series chronicles, "class as Americans encounter it: indistinct, ambiguous, the half-seen ghost that upon closer inspection holds some Americans down while giving others a boost" (Scott & Leonhardt, 2005, p. 2). Class may be less obvious but more divisive, and even though class is more determinative than ever, most people view themselves and others as middle class. Americans do not believe that class marks differences, because of mobility myths. Scott and Leonhardt write that, "The contours of class have blurred; some say they have disappeared" (2005, p. 1). But they argue, the lines are only less apparent, and more complexly bound up with other factors. While the poverty level is the highest in years, personal spending is up, and not by the rich. Consumption, conspicuous and otherwise, is in part responsible for the blurring of class markers.

Strangely, as real mobility declines, people's belief in mobility increases. The overloaded engine of mobility is a problematic new meritocracy:

> A paradox lies at the heart of this new American meritocracy. Merit has replaced the old system of inherited privilege, in which parents to the manner born handed down the manor to their children. But merit, it turns out, is at least partly class-based. Parents with money, education, and connections cultivate in their children the habits that the meritocracy rewards. When their children then succeed, their success is seen as earned. (Scott & Leonhardt, 2005, p. 2)

In Lisa, we see the perils of such a system. Regardless how hard she works, and how high her scores, she still may lack those necessary intangibles which Homer, Marge and her public school are not even aware of. If she makes it to any college, let alone a Seven Sister, she may find that "colleges have come to reinforce many of the advantages of birth. On campuses that enroll poorer students, graduation rates are often low. And at institutions where nearly everyone graduates . . . more students today come from the top of the nation's income ladder than they did two decades ago" (Leonhardt, 2005, p. 1).

More than ever, then, the American dream is a false hope, but the causes are obscured. This is because of the complex nesting of privilege in education, and of a good education in mobility:

> Grades and test scores, rather than privilege, determine success today, but that success is largely being passed down from one generation to the next. A nation that believes that everyone should have a fair shake finds itself with a kind of inherited meritocracy (Leonhardt, 2005, p. 2).

Old systems of advantage, those Dewey could object to, attract too much attention, and don't appear fair. In place of obvious elitist barriers, there are "new ways of transmitting advantage that are beginning to assert themselves" (Scott & Leonhardt, 2005, p. 2). *The Simpsons* writers, all Ivy League educated, might be more aware than those for whom the system does not work. The Gentleman's C and the legacy freshman blow the Dream's cover, but are red herring: real privilege transmission is transparent and often appears fair.

The Simpsons is aware of the complex structures of class inequities, and never offers a utopic revolution. This disappoints Wallace, whose limited ambitions for comedy include policy solutions or accurate depictions (2001). In *The Simpsons*, he sees only "mean-spirited humor that includes no hope for progress" (2001, p. 249). While the series is often mean-spirited, and sometimes at those moments, exceptionally comic, it is not hopeless. The hope I see emanates from a pragmatic, anti-foundational notion of authority and knowledge. Wallace, as a Marxist, has little leeway in the standards he sets for the series:

> A Marxist might contend, that if we truly recognized the violence done to workers, the human costs of stereotyping and scapegoating, the devastation sanctioned in pursuit of profit, we couldn't possibly find *The Simpsons* comical. In fact, *The Simpsons* would have to be considered the worst kind of bourgeois satire since it not only fails to suggest the possibility of a better world, but teases us away from

> serious reflection on or criticism of prevailing practices, and finally, encourages us to believe that the current system, flawed and comical as it sometimes is, is the best one possible (2001, pp. 250-251).

Wallace labors under the impression that didactic, overly political entertainment can be interesting. Because such entertainment follows a political proscription, and is always already known, it isn't likely to interest. Didactic entertainment fails because there is no suspense, and no mystery (Conway, 2004). Governed by a priori principles, his position admits of no contexts, or contingencies. And *The Simpsons* doesn't address this anachronistic Marxist reading. Like Dewey's description of philosophical questions before Pragmatism, the Marxist question is solved by disappearing.

Class does not disappear completely, however. In the series and in our lives, the Marxist notion of class is transformed because of a more complex relation with other factors, which sometimes produce paradoxical results. A Bourdieuan notion of class is more helpful here than a simplified Marxist one (1984). The capacious, messy category of culture determines one's place in the structured system, what Bourdieu calls the habitus. Income or education alone cannot account for one's success or failure, and for lack of mobility. Instead, there are infinite attitudes, behaviors, preferences and habits of mind that we learn from our earliest days; these in turn help determine our aptitudes, appetites, and skills. Lisa is a stunning example of the power of the habitus. She wants to escape the habitus by mimicking the traits of the educated, but she's encumbered by her original preferences. We see this when the family goes to the opera. They know little about opera and less about opera house behavior, and therefore make a scene. Protesting and embarrassed at first, Lisa eventually capitulates to the low-brow shenanigans the family engages in, which she enjoys much more than the opera. Bourdieu's idea of forced choice illustrates Lisa's dilemma: she likes what she has, because she has what she likes. That is, we come to prefer what we have available, but what is available is determined by all the factors of the habitus, and her interaction with it.

Mocking the Evil Wealthy

Nothing as simple as a meritocracy could intervene in the complex structuring of the habitus; in contrast, the American Dream appears naïve. *The Simpsons* seems aware of the contradictions of adopting a system that can't absolutely judge values and truth. As often as the series mocks intellectuals, it criticizes unbridled greed, most often in the characterization of sadistic billionaire Monty Burns. On a date with Mr. Burns, a charmed Mrs. Bouvier giggles and calls Monty a devil, prompting a found-out Burns to ask how she knew (Oakley, 1994). He then recovers, realizing she was complimenting him, rather than detecting his demonic connections. At their wedding, the only guest on the groom's side is a Kaiser-like character, evoking the means-ends corroboration between the Nazis and capitalists in WWII.

However, Burns need not align with the Nazis in order to be despicable. He is an old fashioned American tycoon who engages in any business practice as long as it's lucrative. Memory can humanize by invoking mortality, but Burns stands above (or below) such communal traits. In "$pringfield," Burns seeks to tighten his grip on the town by building a casino on the boardwalk (Oakley, 1993). A flashback recalls his childhood: a wee, but evil Burns repeatedly slams his bumper car into their housepainter, who begs him to stop, crying out for his abandoned, hungry wee ones. The adult Burns chuckles to himself about the crippled Irishman. At the end of this episode, Burns becomes Howard Hughes: tissue box slippers, long fingernails, and delusions of power and paranoia. Inconceivably rich with the casino, Burns proclaims that he can't be stopped except perhaps by microscopic germs. Burns also evokes the worst consequences of colonialism. When he takes a ride on Stampy the elephant, he cries out to Wayland Smithers that the experience reminds him of a fat man he used to ride (Swartzwelder, 1994). In "Last Exit to Springfield" Burns laughs as a window washer plunges to his near-certain death (Kogen, 1993). Later, when he shuts off power to the town, from Hell he cackles a goodbye to Springfield. A demon, indeed.

When a theme directly addresses class inequality, it takes the form of cultural differences. In "Scenes From the Class Struggle in Springfield," Marge is mistaken for one of the elite because she wears a *Chanel* suit she got at a discount (Crittenden, 1996). The family is invited to join the country club, and Marge, nervous that they will reveal their true colors, admonishes them to be on their best behavior. The club hosts events and activities the family would choose, especially Lisa, who longs to ride horses. The club members are not depicted as exceptionally callous; instead, it is Marge's desire to keep up with them that creates conflict within, and hostility toward the family. While money is one element that divides the family from the rest of the club, they have important similarities with the others as well. Eventually Marge decides that the stress of the cultural masquerade is too much, as the family decides they don't want to belong anyway. Attitudes and behaviors long-ingrained, not simple economic differences, drive the Simpsons back to their suburban lower-middle class life.

Consensus of the Mob

Like the capitalist, "the mob" has only an ugly side. Such depictions underscore the perils of consensus. *The Simpsons* often depicts an irrational, torch-carrying mob: typically the furor builds, the mob calls for extreme action, and reasoned truth seeking seems much better in comparison. In "Marge vs. Monorail," the town misspends three million dollars because of mob rule (O'Brien, 1993). Springfield is swindled by a con artist who provides a costly and deadly monorail, and their potholes remain unfilled. As the episode ends, we see the other foolish projects the mob has created: the Popsicle stick skyscraper, the fifty-foot magnifying glass trained upon it, and the escalator to nowhere. Mob rule, the ugly flip side of consensus, conflicts with the town's best interests.

The Linguistic Turn

Although it often rankles traditional academics, *The Simpsons* merits their approval for its creative approaches to tired themes. In a similar vein, analytic philosophers oppose Rorty's position, since he sees, "Pragmatism as primarily therapeutic philosophy—therapy conducted on certain mindsets created by previous philosophers. [. . .] It makes you think, 'Gee, I never knew you could look at it that way before' " (Rorty 2002, p. 373). His "linguistic turn" in philosophy argues for truth-justification, not truth-detection. Instead of verifying reality and truth, he wants us to ask, "Which of all the available languages would be best for describing this particular thing?" (Rorty, 2002, p. 388) We can see this underway in the series where, although visually sophisticated, plot movement and most jokes are linguistic. Characters are inconsistent, and truths shift with context and contingency. The series' focus on useful, modest displays of intelligence indicates that while it does indeed mock intellectuals, it values intelligence: Pragmatism seems like the clear winner in these contests of wit and wisdom.

But, Pragmatism is far from perfect. Its anti-elitist thread recalls the meritocratic impulse of U.S. democracy. The series betrays its class-consciousness when it loosens the meritocratic linchpin—free and equal, public education. While *The Simpsons* values Pragmatism, the series is not itself pragmatically useful. Rather, the series depicts a snapshot of a pragmatist in practice. And this snapshot, however limited, may be its true value. While one may be tempted to glean knowledge or messages from the show, that wouldn't be very pragmatic, since understanding is context-based and contingent.

The Simpsons resists being a guide for ethical behavior, but it does model truth-justification and challenges to long-held beliefs. Matheson is especially concerned about this, when he writes that today's comedies, "are *hyper-ironic*: the flavor of humor offered by today's comedies is colder, based less on a shared sense of humanity than on a sense of world-weary cleverer than thouness" (2001, p. 109). He argues that in *The Simpsons*, knowledge is replaced by knowingness, and "oh-so-clever intellectual one-upmanship" (2001, p. 123) replaces authority. The displacement of authority and knowledge are the aims of Pragmatism as well. Rorty writes, "Pragmatism should pride itself on being a form of low cunning rather than being exciting. Insofar as Pragmatism privileges the imagination over argumentation, it's on the side of the Romantics. Insofar as it prizes intersubjective agreement, it's on the side of plain ordinary democratic politics" (Rorty, 2002, p. 383).

References

Anyon, J. (1980). Social class and the hidden curriculum of work. *Journal of Education*, 162, 1, 67-92.

Austin, J.L. (1962). *How to do things with words.* Cambridge, MA: Harvard University Press.

Bourdieu, P. (1984). *Distinction: A social critique of the judgment of taste.* (R. Nice, Trans.) Cambridge, MA: Harvard University Press.

Conway, M. T. (2004). A becoming queer aesthetic. *Discourse* 26.3, 166-189.

Crittenden, J. (Writer), & Dietter, S. (Director). (February 4, 1996). Scenes from the class struggle in Springfield [Television series episode]. In *The Simpsons,* Beverly Hills, CA: Twentieth Century Fox Studios.

Crittenden, J. (Writer), & Scott, S.O. (Director). (April 16, 1995). The PTA disbands [Television series episode]. In *The Simpsons,* Beverly Hills, CA: Twentieth Century Fox Studios.

de Tocqueville, A. (2001) *Democracy in America.* New York: Penguin. (Original work published 1835).

Diggins, J. (1994). *The promise of Pragmatism: Modernism and the crisis of knowledge and authority.* Chicago: University of Chicago Press.

Eakin, E. (2003, April 19). The latest theory is that theory doesn't matter. *New York Times.* Retrieved June 19, 2005 from http://www.nytimes.com/2003/04/19/arts/19CRIT.html

Greaney, D. (Writer) & Mercantel, M. (Director). (March 19, 2000). Bart to the future [Television series episode]. In *The Simpsons*, Beverly Hills, CA: Twentieth Century Fox Studios.

Hofstader, R. (1963). *Anti-intellectualism in American life.* New York: Vintage.

Irwin, W., Conard, M.T., & Skoble, A. J. (Eds.) (2001). *The Simpsons and philosophy: The d'oh! of Homer.* Chicago: Open Court.

Jacoby, R. (1987). *The last intellectuals.* New York: Basic Books.

Kloppenberg, J. (1998). Pragmatism: An old name for some new ways of thinking? In M. Dickstein (Ed.), *The revival of pragmatism.* Chapel Hill, NC: Duke University Press. pp. 83-127.

Kogen, J. & Wolodarsky, W. (Writers), & Kirkland, M. (Director). (March 11, 1993). Last exit to Springfield [Television series episode]. In *The Simpsons,* Beverly Hills, CA: Twentieth Century Fox Studios.

Kramer, H. (2000). *Twilight of the intellectuals.* Chicago: Dee Publishers.

Kramer, H. (2002, December 4). Notes and comments. *The New Criterion,* 21, 4, Retrieved August 12, 2005 from http://www.newcriterion.com/archive/21/dec02/notes.htm

Lazebnick, R. (Writer), & Nastuk, M. (Director). (November 1, 2000). Tree house of horror XI [Television series episode]. In *The Simpsons,* Beverly Hills, CA: Twentieth Century Fox Studios.

Leonhardt, D. (2005, May 24). The college dropout boom. *New York Times.* Retrieved May 24, 2005 from http://www.nytimes.com

Martin, J. (Writer), & Moore, R. (Director). (October 1, 1992). A streetcar named Marge [Television series episode]. In *The Simpsons,* Beverly Hills, CA: Twentieth Century Fox Studios.

Matheson, C. (2001). *The Simpsons*, hyper-irony, and the meaning of life. In W. Irwin, M.T.Conard, and A. Skoble (Eds.), *The Simpsons and philosophy: The d'oh of Homer*. (pp. 108-125). Chicago: Open Court.

Meyer, G., Simon, S., Swartzwelder, J., & Vitti, J. (Writers), & Archer, W. & Gray, M. (Directors). (April 15, 1990). Crepes of wrath [Television series episode]. In *The Simpsons,* Beverly Hills, CA: Twentieth Century Fox Studios.

Oakley, B. & Weinstein, J. (Writers), & Archer, W. (Director). (December 16, 1993). $pringfield [Television series episode]. In *The Simpsons,* Beverly Hills, CA: Twentieth Century Fox Studios.

Oakley, B. & Weinstein, J. (Writers), & Archer, W. (Director). (May 12, 1994). Lady Bouvier's lover [Television series episode]. In *The Simpsons,* Beverly Hills, CA: Twentieth Century Fox Studios.

Oakley, B. & Weinstein, J. (Writers), & Reardon, J. (Director). (May 6, 1993). Marge in chains [Television series episode]. In *The Simpsons,* Beverly Hills, CA: Twentieth Century Fox Studios.

O'Brien, C. (Writer), & Reardon, J. (Director). (October 14, 1993). Homer goes to college [Television series episode]. In *The Simpsons,* Beverly Hills, CA: Twentieth Century Fox Studios.

O'Brien, C. (Writer), & Moore, R. (Director). (January 14, 1993).Marge vs. the monorail [Television series episode]. In *The Simpsons,* Beverly Hills, CA: Twentieth Century Fox Studios.

Omine, C. (Writer), & Moore, S.D. (Director). (February 22, 2004). Smart and smarter [Television series episode]. In *The Simpsons,* Beverly Hills, CA: Twentieth Century Fox Studios.

Rorty, R. (2002). Worlds or words apart? The consequences of Pragmatism for literary studies. [Interview with E.P. Ragg]. *Philosophy and Literature,* 26.2, 369-396.

Rorty, R. (2004). Review of H. Putnam, *The Collapse of the Fact-Value Distinction and Other Essays. Common Knowledge,* 10.1, 151.

Scott, J. & Leonhardt, D. (2005, May 19). Shadowy lines that still divide. *New York Times*. Retrieved May 19, 2005, from http://www.nytimes.com

Selman, M. (Writer), & Michels, P. (Director). (May 9, 1999). They saved Lisa's brain [Television series episode]. In *The Simpsons,* Beverly Hills, CA: Twentieth Century Fox Studios.

The Simpsons Archive. (n.d.). Retrieved September 16, 2005, from http://www.snpp.com/

Skoble, A.J. (2001). Lisa and American anti-intellectualism. In W. Irwin, M.T., Conard, and A. Skoble (Eds.), *The Simpsons and philosophy: The d'oh of Homer*. (pp. 24-34). Chicago: Open Court.

Swartzwelder, J. (Writer) & Archer, W. (Director). (February 18, 1990). Call of the Simpsons [Television series episode]. In *The Simpsons,* Beverly Hills, CA: Twentieth Century Fox Studios.

Swartzwelder, J. (Writer), & Lynch, J. (Director). (May 5, 1994).The boy who knew too much [Television series episode]. In *The Simpsons,* Beverly Hills, CA: Twentieth Century Fox Studios.

Swartzwelder, J. (Writer) & Reardon, J. (Director). (March 31, 1994). Bart gets an elephant [Television series episode]. In *The Simpsons,* Beverly Hills, CA: Twentieth Century Fox Studios.

Swartzwelder, S.J., Tombkins, S., & Cohen, D.S. (Writers), & Anderson, B. (Director). (October 30, 1995). Tree house of horror VI [Television series episode]. In *The Simpsons,* Beverly Hills, CA: Twentieth Century Fox Studios.

Vitti, J. (Writer) & Silverman, D. (Director). (January 14, 1990). Bart the genius [Television series episode]. In *The Simpsons,* Beverly Hills, CA: Twentieth Century Fox Studios.

Wallace, J. M. (2001). A (Karl, not Groucho) Marxist in Springfield. In W. Irwin, M.T. Conard, and A. Skoble (Eds.), *The Simpsons and philosophy: The d'oh of Homer.* (pp. 235-251). Chicago: Open Court.

West, C. (2000). *The Cornel West Reader.* New York: Basic Books.

Chapter Four

Keeping the Intelligent Woman "In Her Place" within the Patriarchal Social Order: Containing the Unruliness of Genius Brenda Chenowith on *Six Feet Under*

Kylo-Patrick R. Hart

When the HBO television series *Six Feet Under* debuted in June 2001, it stood out for several reasons: because it was intelligently written and beautifully filmed; it confronted the television taboo of death on a weekly basis in shockingly graphic and realistic ways; it regularly portrayed funeral-home workers as responsible, caring, non-ghoulish individuals; it consistently featured a dark tone, dramatic intensity, and characters who refused to let the fact that they are dead prevent them from talking to the living; and it offered special sorts of psychological realism and philosophical rigor that are rarely encountered on popular television. As *San Francisco Chronicle* television critic Tim Goodman (2004) has explained:

> The greatness that welled up in *Six Feet Under* relied almost entirely on viewers buying into an audacious, dizzying mix of deadpan drama from a laconic, disturbed family that ran a funeral home. Theirs was a macabre, often hilarious juxtaposition of fantasy and dreamlike storytelling, which felt a little bit as if Salvador Dali co-authored each weekly script with Ernest Hemingway (p. E1).

Over the course of its five seasons, the show tackled a range of important issues, including drug abuse, familial relationships, gay relationships, grieving, heartbreak, infidelity, religion, romantic and sexual desire, widowhood, and the acceptance of one's own mortality. Essayist and real-life funeral director Thomas Lynch (2004) concluded that "many who found the holy blood and gore of Mel

Gibson's *Passion [of the Christ]* quite acceptable will find the all-too-human flesh and blood of *[Six Feet Under's]* cast of characters unacceptably disturbing in its aching, uncertain, struggling humanity, weeping and giggling at the awkward facts of life and death" (p. 22).

The premiere episode of *Six Feet Under* began with a bang—literally—as Nathaniel Fisher (played by Richard Jenkins), father and owner of the Los Angeles-based Fisher & Sons Funeral Home, was killed on Christmas Eve in a collision with an oncoming bus while he attempted to light a cigarette—and sang "I'll Be Home for Christmas"—in the new hearse he was driving. He was on his way to the airport to pick up his prodigal son and namesake, Nate (played by Peter Krause), who was returning from Seattle to Southern California to celebrate the holidays with the man and their additional dysfunctional family members: repressed mother Ruth (played by Frances Conroy), who found herself pregnant, married Nathaniel at the age of 19, and has spent the better part of the past three decades cleaning their house and raising their offspring; younger brother David (played by Michael C. Hall), a gay man who has been working alongside his father at the family's funeral home for more than a decade; and younger sister Claire (played by Lauren Ambrose), the baby of the family, who found herself virtually invisible to her busy loved ones in the years preceding her father's untimely demise.

While his father was being crushed by the bus, Nate was enjoying steamy casual sex in an airport janitor's closet with a young woman he had just met on his flight; she demanded that he stop talking and start having sex with her before adding, unconvincingly, that she never really does this sort of thing. The viewer soon learns that this woman is Brenda Chenowith (played by Rachel Griffiths), one of the most highly intelligent women ever to appear on popular television. Over the course of the first two seasons of *Six Feet Under*, it is revealed that Brenda is, in fact, a genius. While a young girl, she was found to possess an IQ of 185 and was studied extensively by a team of psychiatrists, who attempted to understand the complexity of her thinking and her manipulative social interactions. Their findings ultimately served as the basis of a best-selling book, titled *Charlotte Light and Dark*, written by psychologist Gareth Feinberg, Ph.D.

It is perhaps no surprise, therefore, that the unpredictable, outspoken Brenda's introduction into the extended Fisher family, as son Nate's adult girlfriend, posed a serious threat to the patriarchal social order of the story world. The Fisher family has clearly always been dominated by men: recently deceased father Nathaniel, eldest son Nate, and middle child David; by comparison, mother Ruth and daughter Claire have remained relatively disempowered and assumed comparatively insignificant familial and social roles in the shadows of their surrounding males. Brenda's extreme intelligence, coupled with her confident outspokenness and sexually aggressive ways, contributed substantially to her status as an "unruly woman" in the story world, one who emerged as a strong contrast to the Fisher women and as a perceived threat to the traditionally privileged patriarchal status of the Fisher men.

A common strategy in patriarchal societies, past and present, has been to contain the threat posed by female unruliness by socially constructing women as beings that are inherently inferior to men. Whether intentionally or not, that is precisely what the producers and writers of *Six Feet Under* did to the character of Brenda Chenowith during the years this highly acclaimed television series aired. As such, this chapter demonstrates how *Six Feet Under* ultimately represented one of the most intelligent women ever to appear on popular television—by constructing her as unruly and then narratively containing her unruliness—as an individual who could not pose any serious threat at all to the patriarchal social order of her surrounding story world.

Women on Top: A Brief History of Unruliness in Relation to Women

As Natalie Zemon Davis (1975) and Kathleen Rowe Karlyn (2003) have demonstrated in great depth, concern over the unruly woman in social and literary history—or the concept of the female out of her place—focuses primarily on the phenomenon of "women on top" who, through female disorderliness and sexual inversion, threaten the social and political order of a patriarchal society by having what has ideologically been constructed as belonging "below" (i.e., women) usurp the power, position, and privilege of what belongs "above" (i.e., men). For example, as Davis (1975) explains:

> With the woman, the [unruliness] was founded in physiology. As every physician knew in the sixteenth century, the female was composed of cold and wet humors (the male was hot and dry), and coldness and wetness meant a changeable, deceptive, and tricky temperament. Her womb was like a hungry animal; when not amply fed by sexual intercourse or reproduction, it was likely to wander about her body, overpowering her speech and senses. . . . The male might suffer from retained sexual juices, too, but he had the wit and will to control his fiery urges by work, wine, or study. The female just became hysterical (pp. 124-125).

Thus, as Davis points out, women have historically been viewed as being disorderly and unruly, in contrast to their surrounding men, because they have been socially constructed as being ruled by their lower regions and readily falling victim to the desires of the uterus, which in turn forces them to want to rule over those "above" them or leads to even more extreme outcomes, such as husband-beating or practicing witchcraft, in order for them to become more powerful than men. She also notes that doctors in the late seventeenth century and beyond maintained that, even though men might occasionally suffer from moments of hysteria, women were much more prone to experience uncontrollable outbursts and exhibit unruly behaviors because their temperaments were inherently more fragile and unsteady than those of men, which readily suggested that women were thereby inherently 'inferior' in nature to men. Similar sorts of social constructions have persisted into the late twentieth and early twenty-first centuries, such as when individuals state that women who are experiencing PMS are entirely "out of control" and "need to be avoided."

Davis reveals that, historically speaking, the proposed remedies for female unruliness have included religious training that taught women modesty, humility, and their expected social roles in a patriarchal society; selective education that instilled expected gendered social roles and values but did not unnecessarily fuel a woman's imagination or her desire to speak in public; and laws and constraints that encouraged marriage and subjected the woman to the desires of her husband. Clearly, the social construction of female unruliness, as well as the proposed remedies for that condition, served primarily as a ready means by which to ensure the subjection of women to men. Davis maintains that men who perpetuated the belief that women were unruly and therefore threatening by nature were vindictive, envious, or dissolute misogynists. She concludes that such sexual symbolism pertaining to women versus men was remarkably effective at maintaining the perceived status of women as subordinates to the men in their society, who were presumed to be naturally abler and stronger. Accordingly, Davis notes, girls in a patriarchal society are raised to believe that they must obey their husbands, and boys are raised to believe that they have the right to exert power of correction over their wives.

Karlyn (2003) elaborates on Davis' rich historical insights by revealing that modern-day women are regarded as being unruly whenever they are considered to be excessive, such as by being too sexual or too mouthy, when compared to the norms of conventional gender expectations. She writes:

> Through body and speech, the unruly woman violates the unspoken feminine sanction against 'making a spectacle' of herself. I see the unruly woman as prototype of woman as subject—transgressive above all when she lays claim to her own desire. The unruly woman is multivalent, her social power unclear. She has reinforced traditional structures, as Natalie Davis acknowledges. But she has also helped sanction political disobedience for men and women alike by making such disobedience thinkable. She can signify the radical utopianism of undoing all hierarchy. She can also signify pollution (dirt or "matter out of place," as Mary Douglas might explain). As such she becomes a source of danger for threatening the conceptual categories that organize our lives (p. 253).

According to Karlyn, examples of unruly women throughout history have included Sarah of the Bible's Old Testament (who laughed at God), Mrs. Noah from the medieval Miracle Plays (an obstinate individual who refused to board the Ark until she was entirely ready), the typical heroine of 1930s screwball comedies, Miss Piggy, Tammy Faye Bakker, Leona Helmsley, and Zsa Zsa Gabor. She proceeds to note that the genre of melodrama regularly punishes the unruly woman when she asserts her desire, making her the target of the audience member's laughter while denying her pleasure and power. As such, Karlyn's list might also have included Brenda Chenowith of *Six Feet Under*, yet another unruly woman in contemporary media offerings who is constructed as a gendered subject "in the language of spectacle and the visual" (Karlyn, 2003, p. 254).

Constructing and Containing Brenda's Unruliness on *Six Feet Under*

Without a doubt, *Six Feet Under* substantially defied television conventions by killing off a major character, father Nathaniel Fisher, within the first four minutes of the show's premiere episode. From a narrative standpoint, this unexpected development appears to represent the symbolic death of patriarchal order, or the law of the father, within the series from its opening moments, suggesting, in the words of television scholar Janet McCabe (2005), "a unique opportunity for producing a female subject who is beyond patriarchal constraints" (p. 121). It is perhaps no coincidence, therefore, that the first female character the viewer encounters immediately following Nathaniel's death sequence is Brenda Chenowith. As she appears to use Nate in the broom closet simply to fulfill her own sexual desires and needs, Brenda initially comes across as a feminist force to be reckoned with. However, as the series progresses, Brenda is ultimately presented not as a powerful feminist figure but rather a stereotypically unruly woman who is unable to pose any significant threat to the patriarchal social order of the Fisher family's lives.

An excerpt from *Charlotte Light and Dark*, published in the book *Six Feet Under: Better Living Through Death* (a companion book to the television series featuring a range of items from within the story world written by, or about, the show's various characters), reveals that, from an early age, Brenda has been a psychologically complex, stubborn, self-aware individual who rejects traditional gender roles and expectations and objects to limits of any kind (Ball & Poul, 2005). As such, she held the potential—as a result of her extreme intelligence and at-times-shockingly outspoken manner—to emerge as a 'woman on top' both in her relationship with Nate Fisher and in the world of the Fisher family more generally. The viewer soon learns, however, that Nathaniel Fisher refuses to go quietly into the afterlife; instead, he appears frcm time to time over the show's five seasons to his various family members, most commonly to his two sons, to offer advice and help guide them in leading what he believes will be successful, fulfilling lives. Whenever the ghost of the father materializes to talk to his family members, therefore, he represents both the patriarchal social order generally as well as their internalized versions of its social expectations. This means that the ghost of Nathaniel Fisher functions in conjunction with the show's narrative developments to contain Brenda's unruly threat to the patriarchal social order of the story world once it has been constructed and perceived.

In an earlier analysis of the character Brenda Chenowith on *Six Feet Under*, popular culture scholar Erin MacLeod (2005) argues that Brenda, as a result of having been studied so extensively while a child, tends as an adult to challenge and rebel against gendered boundaries and patriarchal borders of all kinds. While I agree that this characterization applies well to the Brenda that viewers encounter during the earliest episodes of *Six Feet Under*, it becomes far less applicable as the first season continues and as the show's remaining four seasons progress. Although Brenda initially appears to find the socially expected roles of wife and mother for women to be a bit distasteful, it becomes apparent soon thereafter that she is looking for the love of a good man to make her feel complete. As her relationship with Nate continues to develop in season two, and she asks Nate to marry her, Brenda becomes fascinated with the idea of writing a novel based on her lived experiences, in part to counter the patriarchal picture that was painted of her years earlier in *Charlotte Light and Dark*. Despite her extreme intelligence and creative mind, however, Brenda—a character who has been described alternately as "fascinating," "complex," "intimidatingly frank," and "slightly psycho" all within the same feature article (Deziel, 2004, p. 60)—soon finds that she suffers from writer's block, which leaves her unable to create any interesting content whatsoever.

In an attempt to get her creative juices flowing, Brenda begins spending increasing amounts of time with Melissa (played by Kellie Waymire), a local prostitute who inspires her to begin acting out her most secret and extreme sexual fantasies. Brenda, who earns her living as a massage therapist, feels that prostitution is a more empowering profession. Early in their friendship, she accompanies Melissa to a sexual encounter with a john where she serves as the "watcher," observing every moment as Melissa gives her client a lengthy blowjob. Immediately thereafter, Brenda rushes home and records all of the details, as experienced by the fictional character "Christina," in the pages of the novel she is writing. In the days and weeks that follow, Brenda proceeds to give one of her message clients a handjob, to allow a random male stranger she meets in a department store to put his hand inside of her on the sales floor, to have casual sex with a different male stranger on the floor of a bookstore bathroom and with yet another at the farmer's market, to enhance the experiences of a married couple at a sex party, and to sleep with two surfer dudes simultaneously in her own home, in the bed she shares regularly with Nate. About the latter incident, Brenda writes in the pages of her novel:

> Christina feels herself getting wet as Asswipe #14 and Big Prick #5 exchange glances. . . , where surfer boy and surfer boy will kneel above me and sneak glances at each other's surfer cocks. . . . "Fuck me. Fuck me harder." The rest, in her head only: Fuck me harder, surfer boy, with your fat little crooked cock and your please-tell-me-you're-not-serious shaved balls. . . . The day went on until the sun rose up and fell into the beach waves, crashing, and everyone came. EVERYONE in the entire world came, except, of course, for Christina. She'll come later, after they leave, safe alone with her bedspread, remembering the cock

> show as a slide show only. . . . She is proud that she did something brave, she never has to tell anyone. This was her afternoon (Ball & Poul, 2005, pp. 160-161).

Before ending her friendship with Melissa, Brenda confesses that these various instances of sex with random strangers make her feel hallucinatory, euphoric, and truly alive. Shortly thereafter, Brenda becomes indignant and self-righteous when Nate confesses that he slept with, and impregnated, his friend Lisa (played by Lili Taylor) on a recent visit to Seattle; however, she conceals her own, far more extreme sexual infidelities entirely. It is only after Nate catches on that Brenda has actually experienced all of the things she is writing about in her novel that she confesses she is powerless, out of control, and uncertain why she's been doing such things.

By the end of season two, it is clear that Brenda, as a sex addict, has become the stereotypical unruly woman: a hysterical, inferior individual ruled by her lower regions who possesses a deceptive and tricky temperament in addition to her insatiably hungry womb. It is quite telling in this regard that HBO's official summary of the concluding episode of season two reads, in part, "Brenda attempts to deal with *her true nature*, making her future with Nate even more of a question mark [emphasis added]" (HBO, n.d.). Because her sexual needs have effectively overpowered her mind and senses, Brenda no longer poses a serious threat to her surrounding men in the patriarchal story world. Instead, her unruliness must now simply be readily contained, whether through the historically proposed remedies of religious training, education, or marriage that will instill her with greater knowledge of expected gendered social roles and subject her to the control of a husband. It is perhaps entirely unsurprising, therefore, that Brenda, in the remaining three seasons of *Six Feet Under*, completes a twelve-step recovery program for her sexual addition, which enables her to successfully resist the sexual advances of her manic-depressive brother and inspires her to attend graduate school to become a licensed therapist. It also helps to explain why Brenda reunites with—and ultimately marries—Nate after he has become a widower and gives birth to Nate's (second) child, in the months after she had entered into a romantic (and sexually kinky) relationship with another man.

During a key moment in *Six Feet Under*'s second season, matriarch Ruth tells Brenda that she has grown to love and admire her, primarily because Brenda is independent, spirited, and does not apologize for being a woman who, on the surface at least, appears to be quite different from the Fisher women. By the time all is said and done, however, the construction and subsequent containment of Brenda's unruliness in the show leaves her as powerless and unimpressive as Ruth—who has been described by more than one critic as a mere "doormat for the show's producers to step on" (Lesser, 2001, p. 28)—and Claire, who has played a relatively meaningless and marginalized role within the patriarchal story world from the moment of her birth to the moment of her father's death and beyond. Over the course of the series' five-season run, Ruth endured a string of unfulfilling romantic entanglements and remarried badly, while Claire similarly took a series of short-term lovers and ended up terminating a pregnancy. In the end, it is evident the substantial degree to which all three of these women have "imbibed dominant cultural models of female accomplishment, whereby marriage and children equal happiness while singledom results in nothing but loneliness or death—despite what feminism tells [them]" (McCabe, 2005, p. 127).

Growing up, Brenda identified strongly with the female title character of the (fictitious) children's book *Nathaniel and Isabel*, an excerpt of which appears in *Six Feet Under: Better Living Through Death* (Ball & Poul, 2005). It is clear from that excerpt, as well as from the way this children's book is discussed in several episodes of the show, that Isabel is an intelligent, resourceful individual who is not only able to hold her own against the various individuals who surround her, but also one who consistently outsmarts and emerges superior to them at every turn. In other words, it becomes readily evident that Isabel is a prime example of a 'woman (in training) on top.' It is all the more surprising, therefore, that Brenda—an individual whom Claire describes as being so much smarter than the professionals who were analyzing her as a child that she was regularly able to confound and torment them—turns out to be anything but. Brenda confirms Claire's assessment when she explains that, after hearing the term 'borderline personality,' she visited a library, researched the disorder's symptoms, and intentionally began exhibiting them in order to frustrate and confuse the doctors who were studying her.

Actress Rachel Griffiths, who played Brenda, was not entirely comfortable with the direction her character was taken during *Six Feet Under*'s second season. "I don't think you could have taken her much further without her catching something very nasty," she explained in a 2003 interview, adding that, as an actress, she took the onscreen sexual developments about as far as she had the capacity to go ("Why Rachel," 2003, p. B1). "I was [also] nervous as to whether [the] character [was] redeemable from that point," Griffiths stated ("Why Rachel," 2003, p. B1). As the interview proceeded, Griffiths pointed out that it is about time we "stop judging people on what they choose to do with like-minded people who are choosing to do it with them" ("Why Rachel," 2003, p. B1). She continued: "Girl has sex with a few people: who cares? I don't know whether the writers of *Six Feet Under* are consciously putting that across but I think we, as a society, really need that message. Let's take sex out of the ethical and moral way we appraise people" ("Why Rachel, 2003, p. B1). In a patriarchal social order, however, that is not really a viable nor likely possibility.

On a related note, about *Six Feet Under*, gender studies scholar Robert Deam Tobin has argued that the series "attempts to provide a positive answer to the question of how society should develop without patriarchal guidance" (p. 87). With regard to genius Brenda Chenowith, however, the one female character who seemed to stand the greatest chance of successfully molding a fulfilling post-patriarchal social identity, the series has not lived up to that potential. As it has continuously throughout history, the sexual symbolism associated with female unruliness functioned in this series to reinforce traditional notions of patriarchal order and subordination, rather than to effectively challenge them (Davis, 1975). Brenda's unruliness, as represented in the early episodes of *Six Feet Under*'s first season, appeared initially to be aimed at redefining the limits of acceptable female behavior in a potentially post-patriarchal story world. However, when all was said and done over the show's entire five-season run, this unruliness served ultimately and primarily to keep Brenda securely 'in her place' within the patriarchal social order.

As television critic Tim Goodman (2004) has so eloquently stated, "At its most basic, *Six Feet Under* is about death, and life, and the complications that come from living it—mostly the complications from sex, but also from being unable to find your true path in life, while beating your head against people you meet going in the wrong direction" (p. E1). In this regard, although the narrative developments involving Brenda Chenowith held the potential to be quite profeminist in nature, the resulting representation of Brenda-as-intelligent-woman ultimately turned out to be remarkably retrograde, instead. This reality is especially surprising given the fact that series creator Alan Ball, best known for writing and winning an Oscar for the film *American Beauty*, was given the green light by HBO executives to make the program's contents smart, distinctive, and quite unlike the sorts of representations presented regularly on broadcast network television (McCollum, 2001).

By portraying Brenda as a deceptive individual with a 'secret life' who was unable to control her wildest sexual desires, *Six Feet Under* ultimately represented one of the most intelligent women ever to appear on popular television as an individual who could not pose any serious threat at all to the patriarchal social order of her surrounding story world. Instead, the series trivialized her potential power and status by representing Brenda as just another woman who fell victim to her hungry, overpowering womb and simply became hysterical, rather than as a powerful, intelligent force to be reckoned with.

References

Ball, A., & Poul, A. (Eds.). (2005). *Six Feet Under: Better living through death.* China: Melcher Media.

Davis, N. Z. (1975). *Society and culture in early modern France*. Stanford: Stanford University Press.

Deziel, S. (2004, June 14). Crazy Brenda and a baby. *Maclean's, 117*, 60.

Goodman, T. (2004, June 11). 'Six Feet Under' gives up the ghost. *San Francisco Chronicle*, p. E1.

HBO. (n.d.). *Six Feet Under episode guide: Episode 26.* Retrieved November 12, 2005, from http://www.hbo.com/sixfeetunder/episode/season2/episode26.shtml

Karlyn, K. R. (2003). Roseanne: Unruly woman as domestic goddess. In J. Morreale (Ed.), *Critiquing the sitcom: A reader* (pp. 251-261). Syracuse: Syracuse University Press.

Lesser, W. (2001, July 22). Here lies Hollywood: Falling for *Six Feet Under*. *The New York Times*, pp. S2, S28.

Lynch, T. (2004, November 2). Grave affairs. *Christian Century, 121*, 18-22.

MacLeod, E. (2005). Desperately seeking Brenda: Writing the self in *Six Feet Under*. In K. Akass and J. McCable (Eds.), *Reading Six Feet Under: TV to die for* (pp. 135-145). London: I.B. Tauris.

McCabe, J. (2005). "Like, whatever": Claire, female identity and growing up dysfunctional. In K. Akass and J. McCable (Eds.), *Reading Six Feet Under: TV to die for* (pp. 121-134). London: I.B. Tauris.

McCollum, C. (2001, May 31). HBO continues to redefine the TV series with *Six Feet Under*. *San Jose Mercury News*, p. D1.

Tobin, R. D. (2002). *Six Feet Under* and post-patriarchal society. *Film and History, 32*(1), 87- 88.

Why Rachel Griffiths is going all the way. (2003, June 14). *The Australian*, p. B1.

Chapter Five

Being a Nerd and Negotiating Intelligence in *Buffy the Vampire Slayer*

Holly Randell-Moon

In the fourth season episode of *Buffy the Vampire Slayer* (1997-2003) called "Doomed" (4.11[1]), the character Willow Rosenberg (Alyson Hannigan) protests that a college student whom she knew in high school still thinks she is a "nerd." Willow explains that attending college and having dated a musician should prove sufficiently socially transgressive enough to warrant a delineation with the status she was located within in high school. Because of these social changes, Willow protests that she has not "been" a nerd for some time. Her frustration centers on the category of "nerd" and the frightening idea that, like the episode title suggests, Willow's nerd status was not a transitory identity but one that is "doomed" to become the basis for her whole subjectivity. For although Willow may have, or feels she has, "changed," she is still struggling with the socially marginalizing category of "nerd" that she was interpellated into in high school due to her intellectual ability. Willow's ambivalent relationship with her nerd status reveals the ways in which the representation of intelligence is contingent upon subjectivity and identity.

Notions of intelligence and how they are embodied are articulated through discursive cultural practices that implicate knowledge in the formation of subjectivity. Anthony Giddens writes in *Modernity and Self-Identity* (1991) that in late-modernity "each of us not only 'has', but *lives* a biography reflexively organized in terms of flows of social and psychological information about possible ways of life" (p. 14). Social constructions of intelligence as the concentration of knowledge and information as a disembodied mental ability, work to externalize the reception of knowledge from the self-reflexive articulation of subjectivity that Giddens points to. Further, conceptions of intelligence and intellectuality cannot be unbound from social and cultural contingencies such as race, gender, class, and sexuality. Popular television programs such as *Buffy the Vampire Slayer*, where the use of knowledge is a central part of the narrative structure, contribute to the circulation of meanings attached to intelligence. Buffy's (Sarah Michelle Gellar) sidekick Willow, characterized as a "nerd" due to her scholarly aptitude, demonstrates how knowledge interpellates subjects into certain positions so that intelligence is ambiguously related to power. Willow's transformation to a villain utilizing her intellectual knowledge of witchcraft underlines how notions of intelligence are informed by gender, race, and sexuality in the formation of subjectivity.

Buffy focuses on a young vampire slayer named Buffy and her friends' attempts to deal with the demonic forces that reside in fictional Sunnydale, California. The show attempts to subvert conventional gender expectations of superficial and overtly feminine characters (like Buffy) as helpless or vulnerable in the horror genre by recoding them as empowered. Similarly *Buffy's* central premise that Sunnydale High sits atop a Hellmouth (a subaltern portal to a demon dimension), allows horror and fantasy metaphors to represent dramatic conflicts; ignored girls turn invisible and abusive boyfriends are literally demons. Fundamental to the process of overcoming these demons is the use of knowledge and research by Buffy's friends Giles (Anthony Stewart Head), Xander (Nicholas Brendon), and Willow, as a method of aiding Buffy's physical slaying power, where libraries or archives often act as settings for the show. Wall and Zryd (2001) write that this "intellectual labour" endorses "the violence meted out to forces of evil" (p. 54), identifying such knowledge as being either "hard" or "soft" (p. 54). "Hard" knowledge comprises the scholarly expertise of ancient demonic texts, prophecies, or magic books, whereas "soft" knowledge is connected to contemporary technology such as computers. The elder character Giles' expertise in "hard" forms of knowledge frequently situates him as a source of exposition and guidance for the younger Buffy and friends. As a computer hacker and witch, however, Willow's proficiency in both forms of knowledge renders her uniquely powerful within the Buffyverse.

Intellectuality and Knowledge

The role of research and knowledge gathering in *Buffy* is undertaken by all of its main characters in one form or another, yet it is possible to designate some characters as more intelligent or cleverer than others. Intelligence is usually differentiated from knowledge as a specialized form of expertise, most commonly academic. As such Willow is characterized as intelligent mainly due to her scholarly aptitude in high school and college; for example, during her senior year, Willow's high school principal reminds her she has been accepted by almost all known colleges and universities ("Doppelgangland," 3.16). Further, Willow's proficiency with computer technology, useful for accessing encrypted city council plans such as sewer systems—a common means of travel for vampires and those who prefer darkness, can be similarly described as intelligent. Therefore although Buffy and her friends, to different degrees, may research and acquire knowledge within the Buffyverse, Willow is distinguished as a "nerd" within social settings of the show because of her intellectuality. The popular Cordelia (Charisma Carpenter) refers to both Xander and Willow as "losers" in the first of episode of *Buffy* ("Welcome to the Hellmouth"), yet Willow is further differentiated as a "nerd" due to her talents in computer science and academic achievements in school. It is the relationship between Willow's identity and research skills that locates her as smarter or more "nerd"-like than the similarly socially marginalized Xander, suggesting intelligence is formed by specific notions of knowledge acquisition.

Intellectuality and Knowledge as Cultural and Contingent

In the context of criticizing American society's "love-hate relationship with the notion of the intellectual" in *The Simpsons*, Aeon J. Skoble (2001) argues "those who champion the common man ought not to do so in ways that belittle the achievements of the learned" (p. 24, 34). Intelligence connects knowledge to subjectivity in particular ways, which is evidenced in Skoble's comments. Intelligence, in its corporeal articulations, is related to notions of gender, race, class and sexuality. Skoble's treatment of the representation of intellectuality focuses on the character of Lisa Simpson, an eight-year old whose intelligence and intellectuality are out of proportion to her age as well as the other mainly obtuse inhabitants of Springfield. But, Skoble does not attempt to contextualize Lisa's intelligence within identity and instead relates Lisa's characterization to an "ambivalent" cultural hostility towards expertise. Skoble writes "we seem to be on the verge of a new 'dark ages,' where not only the notion of expertise, but all standards of rationality are being challenged" (p. 25). Intelligence is here posited as a disembodied mental ability supported by a transhistorical conception of knowledge with its own neutral value system. Skoble is critical of conceptions of knowledge as socially and culturally contingent saying, "even physical science is said to be value-laden and non-objective" (p. 28).

This is an inadequate way of conceptualizing what Skoble terms "anti-intellectualism" as it suggests that such antagonism derives from a misunderstanding of the inherent values of expertise rather than a critique that such expertise has specific cultural and social values that work to exclude certain subjects from acquiring it. As Elizabeth Grosz (1995) writes, this type of "objective" knowledge is "governed by a singular, exclusive, and privileged access to true representations and valid methods of knowing reality" (p. 30). These "standards of rationality" are neither "objective" nor neutral; for example Skoble uses the phrase "common man" despite analyzing a female television character. The separation of knowledge from an embodied, and therefore gendered, subjectivity derives from "the historical privileging of the purely conceptual or mental over the corporeal . . . the inability of Western knowledges to conceive their own materiality and the conditions of their (material) production" (Grosz, 1995, p. 26).

We can draw on Grosz's arguments concerning knowledge to say that the model of intelligence cited by Skoble is one that disguises the social and cultural formation of knowledge as a meaning-making practice through a separation of "mind" from subjectivity and identity. *The Simpsons*' representation of knowledge is more complicated than simply "anti-intellectualism" towards those with expertise since the gendering and age of Lisa Simpson is ostensibly humorous because it parodies cultural understandings of intellectuality as aged or masculine. My argument is not to suggest that there is a proper or otherwise way of analyzing representations of intelligence, but however knowledge is historically and culturally conceived of is reflective of how intelligence is represented.

Knowledge and Subjectivity

Before looking at the ways in which Willow is represented as intelligent in *Buffy*, the idea that the relationship between subjectivity and knowledge underwrites how the representation of intelligence is rendered culturally meaningful needs to be further unpacked, in order to provide a background for the way that notions of subjectivities and how they come to be articulated or performed is an important theme in *Buffy*.

Let us return to the normative conception of intellect as a mental ability disarticulated from embodiment and subjectivity. Such an understanding derives from a philosophical tradition known as dualism inherited from the seventeenth-century philosopher René Descartes. Dualism posits a naturalized differentiation between mind and body where the two are seen to function independently of one another with their own "incompatible characteristics." Grosz (1994) explains that such a differentiation "places the mind in a position of hierarchical superiority over and above nature, including the nature of the body" (p. 6). For example Descartes writes that "I now know that even bodies are not really perceived by the senses or the imaginative faculty, but only by intellect; that they are perceived, not by being touched or seen, but by being understood" (1962, p. 75). "Mind" can be differentiated from "body" through an understanding of knowledge acquisition as "objective" and extraneous to the body. In this way, dualism seeks to deny an embodied subjectivity and instead presumes an ostensibly "rational" and knowing subject.

Cartesian dualism is translated into representations of knowledge in television through analyses of *Buffy* that unproblematically reproduce the normative coherent and "knowing" subject. The use of research and knowledge in *Buffy* works as a narrative device to portray what demon it is that Buffy and her friends must face, but also how, or whether, they need to defeat it. For some critics this extends to an allegory where self-knowledge forms the basis of character progression. DeCandido (1999) argues that the *Buffy* characters are motivated by "the thirst to know . . . to know the forces of darkness, to name them, and hence to defang them; to know themselves, as they dance on the edge of maturity; to search out the specifics of how to overmaster a particular demon along with the principles of how knowledge can lead to larger truths" (p. 2). This posits knowledge as a transhistorical mode of accessing "truth" in order to form a coherent subjectivity, so that, as Wandless describes, "the pursuit of knowledge . . . turns inward for each of them" (2001, para. 3). Following this, the demons often represent fears that this self-knowledge will be thwarted. Hence they render visible the internal or repressed desires of the characters that they must externalize and defeat as a form of character development. Foucault (1988) writes that this confessional mode of subjectivity, typical of liberal humanism, "implies that there is something hidden in ourselves and that we are always in a self-illusion which hides the secret" (p. 46). This humanist notion of the coherent subject (finding her or his inherent destiny) through an internal struggle with their demons posits a mind/body opposition whereby it is a character's mental "will" to disembody these fears through knowledge.

The conception of knowledge as enabling a character progression by highlighting internal "truths" fails to adequately describe the complex ways in which subjectivity is shown to be a contested process in *Buffy*. *Buffy's* portrayal of knowledge and the representation of Willow as intelligent are related to notions of subjectivity and identity as unstable owing to their cultural and social contingencies. One of the ways in which the model of intelligence as a disembodied mental ability is subverted in *Buffy* is the underlining of how it is discursively constructed through gender. For Grosz (1995) the binary opposition between mind and body:

> Function(s) in lateral alignments, which are cross-correlated with other pairs—particularly the distinction between male and female—the body has been and still is closely associated with women and the feminine, whereas the mind remains associatively and implicitly connected to men and the masculine (p. 32).

The representation of the "mind" through academic intelligence is traditionally gendered male and raced white, demonstrated in *Buffy* by the way Giles' characterization as an erudite Englishmen is contrasted with Buffy and Willow's more reflexive and adaptive use of ancient sources. As Playden (2001) notes *Buffy* and Buffy's mission is to subvert the "monumental cemetery full of undead white males, the grand narrative of Western thought from Freud back to Plato" that legitimizes the marginalization of those groups unable to access such information (p. 121), and similarly, we can say this extends to androcentric and ethnocentric computer technologies. Willow's intellectuality then, positions her as capable of acquiring access to and understanding of Giles' demon library and utilize this information with computer skills to aid Buffy, dramatizing a feminist subversion of knowledge acquisition.

Representing Intellectuality

Given that we have shown how knowledge can be situated as a contingent and value-laden practice, it is possible to de-code the ways in which the representation of intelligence is rendered culturally meaningful. It is important to look at how Willow is discursively portrayed as intelligent and as a "nerd" in the first episode of *Buffy*, "Welcome to the Hellmouth," since this forms a reference point for her later character changes and demonstrates the ways in which *Buffy* shows how subjectivity is continuously contested and re-articulated. Stuart Hall (1997) defines a "system of representation" as consisting "not of individual concepts, but of different ways of organizing, clustering, arranging and classifying concepts, and of establishing complex relations between them" (p. 17). These systems of representation enable meanings to be produced by "constructing a set of correspondences or a chain of equivalences between things" (p. 19). Through various visual and aural techniques Willow is represented as a "nerd" through a common set of learned cultural meanings that signify intelligence.

As the sidekick to Buffy's hero, Willow's characterization is less dependent on her sexuality and an essentialized appearance of femininity. Although Willow's portrayer, Alyson Hannigan, is thin, "feminine" looking and white, she is not as conventionally attractive as Buffy's portrayer Sarah Michelle Gellar. Willow's red hair color also serves to physically differentiate her, at least until the fourth season, from the sexualization of other female characters such as Buffy and Cordelia.[2] Willow is introduced in "Welcome to the Hellmouth" wearing conservative clothing, a long sleeved tartan dress, compared to Buffy and Cordelia, who wear short skirts and trendy tops. In a scene culminating with Cordelia humorously testing Buffy's "cool factor," the girls pass by Willow and Cordelia explains that if Buffy wishes to "fit in" at Sunnydale High she ought to avoid "losers" like Willow. Once you can identify them all by sight, they're a lot easier to avoid."[4] Buffy's expression of sympathy for Willow means she eventually rejects Cordelia's friendship in favor of socializing with Willow, and later Xander. This plays an important role in establishing the friendship of Buffy, Willow, and Xander as the core relationship in the show. Dechert (2002) contends that for Buffy and her friends "it is a liberating discovery as they realize that together they are, indeed, quite powerful, even though they are marginalized" (p. 218). Willow's costuming then acts to semiotically convey a cultural conception of "bookish" or "nerdish" girls as unfashionable in order to represent her as socially marginalized.

Buffy's negation of further social contact with Cordelia in favor of Willow marks a recognition of the performativity of identities where Willow's social standing is an arbitrary one based on cultural modes of social behavior. Indeed, like Buffy, Willow's appearance masks the powerful role she plays in defeating the demons that attack her fellow students if read according to the misleading social standards in Sunnydale High that Cordelia invests in. But if Willow becomes interpellated into the socially marginalizing category of "nerd" by a reading of her appearance by others, such an interpellation also requires an identification with the "nerd" discourse by Willow. As Hall (1997) writes, "all discourses . . . construct subject-positions, from which alone they make sense . . . Individuals . . . will not be able to take meaning until they have identified with those positions which the discourse constructs" (p. 56). Language is one of ways the notion of "nerd" is performed and constructed as the underlying interior of subject hood for intelligent characters. Rob Cover (2004) observes that language in *Buffy* is not only related to the verbal use and translation of spells and ancient texts but becomes discursively associated with certain kinds of performative subjectivities (para. 9). For example, Willow's speech patterns alternate between short sentences and pauses but gain in confidence when she speaks of computer technology and books; when the latter occurs, Willow's dialogue is often spoken by Hannigan without punctuation or periods.[3] Overbey and Preston-Matto (2002) note that for Willow, "the words sometimes get away from her; unlike Xander, she is awkward in talk, blushing, stammering" (p. 78). The representation of Willow's communication skills draws on notions of intelligence as a mental ability that is disembodied, but differs with Cartesian dualism in that "nerd"-like characteristics are portrayed as out of proportion to bodily control—the idea is that Willow is so smart she cannot not express the titles of all the books she has read or computer programs she owns.

Additionally, "the voices of the actors" contributes to "the sonic meaning potential" of representational media such as film or television and work to semiotically locate their character's roles in the narrative structure (Cranny-Francis, 2007, p. 96). Willow's voice is high-pitched, with a low timbre, and sonically differentiates her from Buffy's more low and sonorous voice, characteristically associated with sexually alluring female characters (p. 99). This works to aurally heighten Willow's "stammering" and disjointed way of speaking. We may further note the difference between Willow's and Giles' intelligence through their accents. Giles' clipped British dialect and measured tone of voice conveys a learned knowledge associated with the conventional idea of white masculine scholarly expertise noted above, in comparison to Willow's, albeit nervous, but more adaptive use of language.

Finally, Willow's surname "Rosenberg" is the means by which the writers convey her Jewishness, but also functions alongside the representation of Willow as intelligent. Dowling (2003) explains that "with the disclosure of her surname, Internet discussions began in earnest about Willow's ethnic/religious identity . . . Subsequent confirmation that Willow was Jewish thus came as no surprise, since in the American construct of difference, 'being Jewish is the equivalent to being smart'" (By contrast, Willow section, para. 2, 3). In this way, Willow's surname signifies a cultivated access to knowledge through representational stereotypes of middle/upper class Jewish characters as intelligent.

The representation of knowledge aptitude as "nerd"-like works as a discursive confirmation of Willow's underlying intelligence. Willow's costuming, language—her linguistic excitement with books and computers and the way this is aurally interpreted by Hannigan, and her name, are the tools through which a causality between ineptitude in sartorial skills and social communication renders Willow's nerd status and intelligence culturally meaningful. Further, this episode points to the tension between intelligence as an essential aspect of "nerd" identity and a culturally learned mode of behavior that Willow struggles with throughout *Buffy's* seven seasons.

Intelligence and Subjectivity in the Buffyverse

Willow's transformation from shy school "nerd," to witch, then lesbian, and finally a villain who attempts to destroy the world is represented as an ambivalent struggle to articulate a coherent subjectivity. This characterization is mobilized through an interpellation and identification of and by Willow as intelligent. Cover (2005) argues that "Willow's fear of the return to marginalization" (p. 94) leads her to become empowered "through utilizing the tool which gave her the nerd status—her expertise with books and knowledge as it leads to magic and thereby power" (p. 96). Willow is able to wield enormous power, but her "empowerment is located in a struggle with the identity into which she was interpellated at school" (p. 93). The representation of Willow as intelligent speaks to the ways in which cultural conceptions of intellectuality are embodied differently according to gender, sexuality, or demon-slaying power, in the Buffyverse.

Willow is a complex character since her narrative arcs not only signify character development but also position her differently from season to season. In the first two seasons of *Buffy* she is situated as an intelligently shy, computer nerd, and best friend to Buffy. Her growing interest in witchcraft renders Willow a witch from season three onwards. She becomes romantically involved with another witch Tara (Amber Benson), and identifies as a lesbian after season five. During season six, Willow's social insecurities lead her to rely increasingly on magic use as a form of relief, a practice that becomes allegorically referred to by the characters within the show as an "addiction" (see "Lessons," 7.1). When Tara is accidentally murdered by Warren Mears (Adam Busch), a nemesis of Buffy, in "Seeing Red" (6.19), Willow's use of "dark" magic culminates in a grief-fuelled "bender" (Forster, 2003, p. 9) and she tries to destroy the world. Willow's character arcs throughout the seven seasons of *Buffy* necessitate an understanding of representation as linked to discursive notions of subjectivity.

In "From Butler to Buffy" Cover (2004) proposes "a strategy for identity analysis" that would "look at the concept of discourse not only within *Buffy* or as constituting *Buffy* (or the character Buffy), but *as Buffy*" (para. 34). It is possible to analyze discursive formations in and as constituting *Buffy* because "under the guise of having to 'accept' unwanted or frightening facts about themselves and their lives" the character's "performativity is altered as they struggle to maintain coherence and individuality" (para. 24). Performativity here refers to Judith Butler's thesis that gender is constituted by a series of social and cultural behaviors learned through repetition rather than an inherently "natural" gender identification (1990, p. 145). Cover suggests television shows such as *Buffy* translate a notion of subjectivity as performative by portraying characterization through "discursive co-ordinates, categories and signifiers which are then cited and come to 'make' the character in an ongoing process of performance that aims for coherence over a longer period of time" (2004, para. 26). Cover's thesis is useful for analyzing the ways in which Willow attempts to recapitulate her subjectivity as deriving from something other than the socially marginalizing category of "nerd" through another more ostensibly empowering identity.

Willow negotiates her increasing skill in witchcraft as a mode of empowerment and transformation, which continues in her relationship with another witch, Tara, who substantially augments and enhances Willow's powers. Extra-textually, censorship from *Buffy's* then network, the Warner Brothers Television Network (the WB), necessitated Willow and Tara's relationship play out as a witchcraft metaphor until the actors were allowed more physical expressions.[4] This conflation of a lesbian identity with witchcraft within the diegesis serves to heighten Willow's ambivalent self-identity. The literal power Willow wields from her witchcraft skills is frequently cited by Willow as a means of social differentiation from her former "nerd" self. Whilst searching for Buffy's sister Dawn (Michelle Trachtenberg), who has spent a night out on the town with her friends and given Buffy false information concerning her whereabouts, Willow and Tara check the local music club the Bronze as a possible haunt for the errant teen. Willow justifies to Tara a search at the Bronze by saying she would be there if she were fifteen but then corrects herself and notes at fifteen she probably would not be at the Bronze due to her previous "nerd" status ("All the Way," 6.6). Cover (2005) writes that "Willow's confidence, through knowledge and the mastery of 'appropriate' discourse, allows an initial (surface) identity transformation from her past as 'geek'—an identity conferred through both harassment and the network of social relations in which she was recognized and recognizes herself" (p. 93). Willow's relationship with Tara confers on Willow a lesbian identity and combined with witchcraft allows Willow to revise a subjectivity whereby her intelligence becomes a way of obscuring the socially marginalizing category of "nerd." This points to the ways in which for Willow, her identities as a witch and in her relationships with others, define her sense of self, and poses an interrogation of subjectivity as being comprised of an essentialized or coherent inner self.

The fourth season "Restless" (4.22), an episode where the narrative is comprised of a dream from Willow, Xander, Giles, and Buffy, portrays Willow's ambivalence about her sense of self. Escaping a drama class requiring Willow to appear onstage, Buffy and Willow enter an empty classroom. Buffy stops, and referring to the drama class sequence, wonders why Willow has not yet taken off her "costume."[7] Willow, dressed in a designer yellow top with flared pants, replies impatiently that she is not wearing a "costume" just her "outfit." Buffy again asks Willow to remove her "costume" and reaches for her clothes. A ripping noise is heard on the soundtrack and we see a shot of Buffy holding Willow's clothes saying triumphantly that Willow now looks "realistic." In the next shot Willow is standing in front of a now full class, including Xander, her former boyfriend Oz (Seth Green), and current girlfriend Tara, wearing the same outfit she wore in "Welcome to the Hellmouth"—the first episode of *Buffy*. As noted above, Willow is initially less sexualized than Buffy or other female characters in the show due to her representation as bookish and shy. But her increasing confidence is equated with her looking more conventionally attractive so that Willow's season one "nerd" outfits give way to fashionable clothing. But Buffy's reference to Willow's ensemble as a "costume" in "Restless" extratextually deconstructs this attractiveness by showing the audience that the same actor can be rendered more attractive through their character's appearance. This also works to textually signify Willow's ambivalence concerning her performativity of intelligence. At this stage in the series, Willow's attendance at college stimulates her intellectual interests. She also develops a relationship with Tara, and expands her witch skills. Buffy's actions reveal Willow's ambivalence about her newfound confidence. Willow fears that it is her witchy-ness and her friends that constitute her identity; that underneath Willow's character changes is the Willow of the first episode of *Buffy*. But even this Willow is not the "real" Willow; as her outfit indicates, this is merely another costume Willow used to wear.

Cover notes that the idea of the coherent humanist subject has been undermined by postmodern conceptions of identity as unstable and fragmented, as theorized by Fredric Jameson for example. However, this shift does not erase the discursive power of the coherent subject, rather there is a "push-and-pull relationship" between "a continuing, residual, humanist imperative for coherent identities and . . . an ongoing cultural assertion that identities are not authentic but can be made and re-made" (2004, para. 5). Conventional television programs:

> Rely on episodic closure through identity resolution and depend on an articulation of a foundation or "inner identity core" that is variable or denied only as an "obstacle" within an episodic narrative and which must be rearticulated or restored in the closure of an episode, whereas more recent television which has embraced the multi-episodic and multi-seasonal narrative arc more readily espouse an identity that is represented as *process* (Cover, 2004, para. 7)

In this way Willow's ambivalency regarding her intelligence can be read as a complex process of performativity and identification through the citation and destabilization of the coherent subject. Fifarek (2001) describes "Restless" as "a reminder that, with no external forces, no demons to fight, the only battles left would be with ourselves" (para. 38). Yet Willow's "battle" to know herself provides no assurance that there is a "self," showing the Cartesian model of subjectivity to be just one of the many ways in which subjectivity is articulated. As Cover argues, one of *Buffy's* themes is "that selfhood does not merely reflect a coherent, recurring and ongoing identity but is shown up to have no roots or grounding to which to return in the process of performative articulation" (2004, para. 17). That is, if the demons are representative of inner conflicts that must be materialized and "known" in order to produce a coherent self, then Willow cannot externalize them—since she does not have a stable being to displace deeds and fears onto, and so her "battles" are not simply representations of an internal struggle.

Willow's transformation to villain at the end of season six is the culmination of her ambivalence about her self-definition through intelligence. In "Seeing Red" (6.19) Tara is murdered and Willow's corresponding grief pushes her to the "dark" side of magic. Tara's death leaves Willow literally selfless, she uses magic to become something other than "Willow"— for example, Willow describes her history as a "loser" in third person ("Two to Go," 6.21). Distancing herself from the "Willow" that used to exist enables her to confront Tara's killer, Warren, torture him before flaying him alive, and then attempt to destroy the world. If the only way intelligence is marked is as a socially marginalizing category of "nerd," then this marginalization is taken to its excessive logical conclusion. Willow is represented as "evil" because she literally becomes "the magic"—the knowledge—that produced her marginalization. This is visually represented by Willow draining the text from numerous "dark" magic books—the ink crawling over her arms towards her head, turning her hair and eyes black.

Willow's transformation can be read as a disruption to the male oriented mind/body hierarchy of disembodied knowledge. As noted above, Grosz argues that Cartesian dualism functions to align "mind" with masculine and "body" with feminine. Read this way Willow's intellectual knowledge of witchcraft is a form of "excess" knowledge tied to Willow's emotions as feminine instability—as if the "dark" magic text inside Willow's head can no longer exert control over her body. Willow is capable of conquering minds through her intelligence, casting memory loss spells on Tara and herself ("Tabula Rasa," 6.8), whereas the reverse occurs here and she becomes intelligence embodied. But the defeat of "evil" Willow does not arrive through a reordering of mind over body. It occurs through a blurring of the coherent and essentialized notion of subjectivity.

In "Grave" (6.22) Willow attempts to destroy the world by raising the effigy of a she-demon named Proserpexa to funnel her anger into, burning the world from the inside out. Xander appears, expressing sympathy for her pain, and relays a story about how Willow broke her yellow crayon in kindergarten. He says he loves Willow, including the Willow-who-broke-her-crayon and the Willow-who-turned-evil. This allows Willow to express her grief for Tara and resist the "dark" magic—Willow's hair returns to its red color. Xander's statement that he loves Willow may be construed as thwarting her despair over a lack of self by revealing that there is an essential "Willow" behind her actions as evil or crayon breaking. Willow's ambivalence about self-definition leads her to constitute her identity through others. When this is no longer possible she seemingly has no self at all. But we see in this scene with Xander that Willow's "battle" in relation to subjectivity is necessarily contradictory. Willow may want to be just "Willow" but she is also the Willow-who-broke-her-crayon and the Willow-who-turned-evil and the tool through which she articulates this struggle, her intelligence, is not easily locatable within a stable binary of mind and body.

Willow's power through knowledge reflects an ambiguity surrounding the marginalization of intelligence through the social category of nerd. Willow's attempts to reconfigure an alternative underlying essence for her subjectivity results in an extreme mind/body separation. Through the recognition of multiple and sometimes contradictory identities in "Grave" we see that Willow's characterization is less about the control of her power, or the social power of interpellation into culturally meaningful modes of knowledge, but locating it within an ongoing process of subjectification.

Conclusion

Buffy's portrayal of subjectivity as a process located within both Cartesian dualism and postmodern representations of subjectivity, as well as its feminist critique of patriarchal conceptions of knowledge, are both effected through the representation of Willow as intelligent. The understanding of knowledge as culturally contingent and socially situational, by feminist writers such as Grosz, has influenced the representation of intelligence in contemporary television shows such as *Buffy*. In this way, notions of intellectuality as a disembodied mental ability fail to account for the means through which intelligence is embodied and contingent upon race, gender, class, and sexuality. Despite not being a "nerd" at various points in her character arcs, Willow's intellectuality is not simply an isolated subject position articulated through "objective" knowledge acquisition. Being a "nerd" structures and is informed by all aspects of her subjectivity, represented in *Buffy* as a shifting and self-reflexive process.

Notes

Thank you Elaine Kelly and Jessica Cadwallader for your intelligent suggestions and comments for this paper.

1. Denoting season 4, episode 11.

2. I am aware that Hannigan's hair color in the first season of *Buffy* is more brunette than red, and becomes visibly reddened by the second season. This change in hair color though, from Willow's initial introduction, does not substantially alter my point.

3. Due to copyright issues around transcribing dialogue from DVD versions of *Buffy*, I have had to rely on paraphrasing character's quotes rather than reproducing them in full. This raises an interesting question in that visual scenes can be transcribed without infringing copyright whereas the transcription of dialogue into text (which arguably does not reproduce the "authenticity" of the original work as dialogue is written to be performed rather than read) becomes problematic due to the privileging of language in its written form. These copyright issues reinforce the concerns of this chapter, which argues that knowledge is privileged in language as disembodied and therefore excludes certain kinds of representations such as the visual or aural from counting as valuable knowledges.

4. *Buffy's* final two seasons were aired on another network, United Paramount Network (UPN), who did not censor the relationship.

References

Butler, Judith. (1990). *Gender Trouble: Feminism and the Subversion of Identity*. London and New York: Routledge.

Cover, Rob. (2004, Spring). From Butler to Buffy: Notes Towards a Strategy for Identity Analysis in Contemporary Television Narrative. *Reconstruction: Studies in Contemporary Culture, 4* (2). Retrieved August 17, 2005, from http://www.reconstruction.ws/042/cover.htm

Cover, Rob. (2005, March). "Not to Be Toyed With": Drug Addiction, Bullying and Self-empowerment in *Buffy the Vampire Slayer*. *Continuum: Journal of Media & Cultural Studies, 19* (1), 85-101.

Cranny-Francis, Anne. (2007). Mapping Cultural Auracy: The Sonic Politics of *The Day the Earth Stood Still*. *Social Semiotics, 17* (1), 87-110.

DeCandido, GraceAnne A. (1999, September). Bibliographic Good vs. Evil in Buffy the Vampire Slayer. *American Libraries, 30* (8), 44-47.

Dechert, S. Renee. (2002). "My Boyfriend's in the Band!" *Buffy* and the Rhetoric of Music. In Rhonda V. Wilcox & David Lavery (Eds.), *Fighting the Forces: What's at Stake in Buffy the Vampire Slayer* (pp. 218-226). New York: Rowman & Littlefield.

Descartes, René. (1962). Second Meditation: The Nature of the Human Mind: It is Better Known than the Body. In Elizabeth Anscombe & Peter Thomas Geach (Trans. and Ed.), *Philosophical writings* (pp. 66-75). Edinburgh: Nelson. (Original work published 1641)

Dowling, Jennifer. (2003, March). "We Are Not Demons": Homogenizing the Heroes in *Buffy the Vampire Slayer* and *Angel*. *Refractory: A Journal of Entertainment Media, 2*. Retrieved September 11, 2003, from http://www.sfca.unimelb.edu.au/refractory/journalissues/index.htm

Fifarek, Aimee. (2001, June). "Mind and Heart with Spirit Joined": The Buffyverse as an Information System. *Slayage: The Online International Journal of Buffy Studies, 1* (3). Retrieved October 22, 2005, from http://www.slayage.tv/

Forster, Greg. (2003). Faith and Plato: "You're Nothing! Disgusting, Murderous Bitch!" In James B. South (Ed.), *Buffy the Vampire Slayer and Philosophy: Fear and Trembling in Sunnydale* (pp. 7-19). Chicago: Open Court.

Foucault, Michel. (1988). Technologies of the Self. In Luther H. Martin, Huck Gutman, & Patrick H. Hutton (Eds.), *Technologies of the Self: A Seminar With Michel Foucault* (pp. 16-49). Amherst: University of Massachusetts Press.

Giddens, Anthony. (1991). *Modernity and Self-Identity: Self and Society in the Late Modern Age*. Cambridge: Polity Press.

Grosz, Elizabeth. (1994). *Volatile Bodies: Toward a Corporeal Feminism*. Sydney: Allen & Unwin.

Grosz, Elizabeth. (1995). *Space, Time and Perversion: Essays on the Politics of Bodies*. St Leonards: Allen & Unwin.

Hall, Stuart. (1997). The Work of Representation. In Stuart Hall (Ed.), *Representation: Cultural Representations and Signifying Practices* (pp. 15-64). London: Sage, in association with The Open University.

Overbey, Karen Eileen, & Preston-Matto, Lahney. (2002). Staking in Tongues: Speech Act as Weapon in *Buffy*. In Rhonda V. Wilcox & David Lavery (Eds.), *Fighting the Forces: What's at Stake in Buffy the Vampire Slayer* (pp. 85-97). New York: Rowman & Littlefield.

Playden, Zoe-Jane. (2001). "What you Are, What's to Come": Feminisms, Citizenship and the Divine. In Roz Kaveney (Ed.), *Reading the Vampire Slayer: An Unofficial Critical Companion to Buffy and Angel* (pp. 120-147). New York: Tauris Parke.

Skoble, Aeon J. (2001). Lisa and American Anti-Intellectualism. In William Irwin, Mark T. Conrad, & Aeon J. Skoble (Eds.), *The Simpsons and Philosophy: The D'oh of Homer* (pp. 25-34). Chicago: Open Court.

Wall, Brian, & Zryd, Michael. (2001). Vampire Dialectics: Knowledge, Institutions and Labour. In Roz Kaveney (Ed.), *Reading the Vampire Slayer: An Unofficial Critical Companion to Buffy and Angel* (pp. 53-77). New York: Tauris Parke.

Wandless, William. (2001, January). Undead Letters: Searches and Researches in *Buffy the Vampire Slayer*. *Slayage: The Online International Journal of Buffy Studies, 1* (1). Retrieved October 22, 2005, from http://www.slayage.tv/

Filmography

Whedon, Joss (Creator). (1997-2003). *Buffy the Vampire Slayer* [Television series]. 20th Century Fox Television, Mutant Enemy Inc., Kuzui Enterprises, Sandollar Television.

Chapter Six

Is School Cool? Representations of Academics and Intelligence on Teen Television

Amy Richards Franzini

Who is cooler: Ryan from *The O.C.*, Dylan from *Beverly Hills, 90201*, Vinny Barbarino from *Welcome Back Kotter*, or The Fonz from *Happy Days*? It's a pretty hard call. You could argue for any of these characters, and cite numerous episodic anecdotes to support your position. But if asked who was *smarter*, could you do the same? If you were a teenager, what crowd would you want to hang out with—the underachievers from *That 70's Show* or *Kotter*, or the over-achievers of *Saved by the Bell* or *Head of the Class*? Would you prefer to date The Fonz or Richie Cunningham of *Happy Days*? Stereotypical portrayals of television characters as being either "cool" or "smart," but *rarely* both, have always been popular on television. It seems that "cool" and "smart" tend to be mutually exclusive on television programming. These conflicting representations can present a dilemma for adolescents as to which trait they should value more in themselves and in others.

Adolescence is a critical time in a person's development, academically and socially. A well-rounded individual is typically considered both academically and socially intelligent. Primary emphasis on the value of only one area of development may lead an adolescent to ignore the other, rather than striking the preferred balance. Socialization supplies adolescents with necessary information on society's values and norms relative to both. Theories suggest that television is a primary socialization tool in which adolescents learn what is valued in society.

Theories of Socialization Through Television

Bandura's (1986) Social Cognitive Theory suggests that people may identify with characters on television and develop expectations about what it means to be someone who is "like" that character. Therefore, teens develop expectations about what it means to be a teen through the teen characters they watch on television.

Gerbner's Cultivation Theory (Gerbner & Gross, 1976) takes a more long-term, cumulative approach to socialization. Cultivation Theory attends to the common themes that intersect all the television programs a person grows up watching. Consequently, if a person sees certain values and norms over-and-over again (like you can't be both "cool" and smart), that person is more likely to "cultivate" similar values.

Both Parsons (1951) and Geertz (1973) recognized the power of socialization and enculturation to show members of a culture not only codes of behavior, but also the norms and values. Television offers a lifelong inculcation of a culture's values and norms. Therefore, it is important to look at television's symbolic content to discover what is valued, rather than simply compare that symbolic content with reality to see if it is accurate. In other words, while it may be interesting to compare education on television with education in the "real world" to ascertain if the portrayal is truthful, it is arguably more critical to uncover enduring themes relating to the portrayal of education, as that enlightens us as to the *value* of education in our culture. Given that, and given that fact that adolescence is a critical time in a person's value-formation, this chapter investigates such enduring themes relating to academics, intelligence and "coolness" on programs featuring adolescent characters.

Previous studies regarding the representation of academia on television tend to focus on shows specifically written in an educational setting, usually from the teacher-perspective. Mayerle and Rarick (1989) analyzed forty years of such programming and found that such series were rare and did not remain on the networks' schedules for very long. They did find that on such programming, teachers were portrayed as dedicated and interested in helping students. Still, there is a body of literature that emphasizes the cultural importance of how education and intelligence is portrayed on television (see for example Gerbner, 1974). The emphasis of much of this literature posits a cultural anti-intellectualism (Gerbner, Gross, Morgan & Signorielli, 1981; Hofstadter, 1964; Holderman, 2003; Ross, 1989; and Thomas & Krippendorf, 1988) that suggests that such representations contribute to the cultural-power status quo, by portraying intelligence in a non-appealing way that is therefore less likely to be modeled.

Rather than focus on academically set shows, the present study examines the representation of academics and intelligence on "typical" adolescent programming. Arguably, if a person watches a program "about" school, she or he may be *more* conscious of the portrayals, and therefore *less* likely to integrate what she or he has learned. In contrast, Thomas (1978) argues that if a person is watching a program purely for "entertainment," they are actually more likely to learn cultural values and norms, precisely because they don't *think* they are learning anything.

Adolescents spend more time watching television every day (3:04 hours) than they do interacting with their parents (2:17 hours) or friends (2:17 hours) (Rideout, Roberts & Foehr, 2005). There is little argument that television is popular among adolescents, and popular rhetoric would suggest that there is much public concern about the content adolescents view on television. The present study does not attempt to evaluate the "appropriateness" of such content, but rather seeks to identify the common themes relating to two integral areas of adolescent socialization—academics and intelligence on the one hand, and "coolness" and popularity on the other. The study asks, "Is school cool?" More specifically, how are academia and intelligence represented on television shows popular among adolescents and how are academics and intelligence related to popularity and romance?

Methods

The first step in operationalizing "cool" was choosing a sample of programs that were popular with adolescents. The initial sample included all episodes (excluding reruns) of an entire season of prime-time, network shows most viewed (according to Nielsen) by adolescents (twelve to seventeen year olds). The sample involved a total of ten shows from the Fall 1998 to Spring 1999 season: *Beverly Hills, 90210*; *Boy Meets World*; *Dawson's Creek*; *ER*; *Friends*; *Home Improvement*; *King of the Hill*; *Party of Five*; *Sabrina, The Teenage Witch*; and *The Simpsons*. The total sample included 231 episodes. Since this sample involved a variety of program genres (drama, sitcom, and cartoon) from one specific season, a second, follow-up sample was then studied that focused on the program genre that was found to focus more on long-term enduring plotlines—the drama—and that spanned a multitude of years. Three "quintessential" teen dramas, which all enjoyed huge adolescent popularity and were often featured in the popular media were chosen for this secondary sample. The first program, *Beverly Hills, 90210* first aired on FOX in 1990; the second program, *Dawson's Creek* premiered on the WB in 1998; and the final program, *The O.C.* debuted on FOX in 2003, providing a thirteen year span of adolescent programming. The first three episodes of the premiere season of each program were coded to provide a good indication of each show's thematic-direction and focus.

The unit of analysis was the individual character. More specifically, the study analyzed all plot-functional characters that were adolescent or older. A plot-functional character was one playing an integral role in the narrative of the show. This character-as-unit-of-analysis provided the clearest route to determine how intelligence correlated with popularity and romance.

The coding instrument for the initial sample of ten programs individually coded each character in each episode of his or her appearance. The coding instrument included four variables coding characters intelligence and three additional variables that were intended to measure character "coolness." When coding the follow-up sample of "quintessential" teen shows, the same unit of analysis was used—the adolescent plot-functional character. However, storyline was also coded, to investigate overall storyline-focus on academics. More specifically, the amount of time characters were seen in the classroom (or engaging in academic activities) was monitored.

The intelligence variables analyzed characters intellect and, if a character was in school, his or her *academic success* was also coded. Additionally, if a character received any *specific praise or criticism* related to academic performance it was also coded. Character's overall *intelligence* was assessed in terms of a conventional distinction: first, characters were evaluated for their erudition (i.e., how learned or educated they were), and second, characters were assessed in terms of "street smarts" (i.e., mundane, interpersonal cleverness). For the follow-up sample, characters' intelligence was ranked with more streamlined variable values. Intelligence was rated as *intelligent* (evident through characterization that character is erudite), *average* (character's intelligence is neither obviously intelligent nor noticeably unintelligent), or *unintelligent* (evident through characterization that character is dimwitted). As with the initial sample, if a character was specifically praised or criticized for their academic performance, that was coded as well.

The "cool" variables rated characters' popularity and romantic lives. The authorial intent of characters' popularity was coded in simple terms—either yes, the character was popular or no, she or he was not. Authorial intent considers how it was inferred that the character was "written," or how she or he was "supposed" to be seen by other characters. The follow-up sample added a third value—rather than simply yes or no, characters were rated as being a member of the "in" crowd, a social outcast, or just considered "regular."

Finally, any sexual activity experienced by a character was coded, from kissing to sexual intercourse.

Results

For the initial sample, an entire season of ten programs, resulting in 231 episodes were analyzed, and a given character was coded only once per show. Collectively, the ten programs featured 144 adolescent-or-older, plot-functional characters that made a total of 1656 appearances in the 231 episodes. For the follow-up sample, eighteen regularly-featured adolescent characters were coded for the first three episodes of the premiere seasons of these archetypical teen programs. General findings relating to intelligence will be discussed first, followed by the relationship between intelligence and "coolness."

Academic Success

When applicable, the *academic success* of characters (in school) was indexed. Table 6.1 provides the frequency of these appearances.

Table 6.1 Frequency of *Academic Success*

	N	%
Highly Successful	41	2.5
Above-Average	35	2.1
Nondescript/Average	142	8.6
Unsuccessful	32	1.9
Not indicated / inapplicable	1406	84.9
Total	1656	100.0

Most of the time (85%) characters were either not students (inapplicable) or were not shown in the classroom setting (not indicated). When characters were shown in the classroom during an episode, their academic accomplishments were mostly *average* or *nondescript.* In approximately 2% of appearances, characters were coded *highly successful* in academics; and another 2% were *above-average.* Characters were coded as academically *unsuccessful* fewer than 2% of the time.

Academic Praise and Criticism

Tables 6.2 and 6.3, respectively, provide frequencies for any instances of characters being *criticized* or *praised* with respect to academic success. *Academic praise* or *criticism* was present in less than 1% of total appearances.

Table 6.2. Frequency of *Academic Criticism*

	N	%
Yes	4	0.2
No	1652	99.8
Total	1656	100.0

Table 6.3. Frequency of *Academic Praise*

	N	%
Yes	10	0.6
No	1646	99.4
Total	1656	100.0

Intelligence

Table 6.4 shows assessment of characters' overall *intelligence.*

Table 6.4. Frequency of *Intelligence*

	N	%
Above-Average (both)	374	22.6
Intellectually Erudite	22	1.4
Street smart	94	5.7
Average	1053	63.6
Dumb	113	6.3
Total	1656	100

Characters were generally assessed as possessing *average* intelligence (about two-thirds of the sample). The *average* value accounted for characters that were neither particularly smart nor substantially dumb, as well as characters for which an intelligence rating was not available.

Perhaps it might be argued that "average" and "not available" are not the same thing and should appear as separate values. However, in this case the "average" value is used to code *both* characters who are assessed as not having extreme (or remarkable) attributes or behavior in either direction *or* for characters whose screen-time/performance does not provide an opportunity for such assessments to be made (i.e. normally, "n/a" values). The logic behind combining these two values—although analytically separate—is as follows: when considering narrative structure, authorial intent, or any other term that may be used to identify what the story is seemingly representing, the "average" and "n/a" values indicate that the story is *not* defining or drawing attention to a character in terms of some non-ordinary quality. In this sense, then—especially if the subsequent analysis concentrates on representations of the more extreme qualities—the "average" and "n/a" values *are* functionally quite similar. Clearly, in terms of variables where either "average" and/or the particular *absence* of an event *has* theoretical significance, in and of itself, collapsing the two *would not* be appropriate.

Keeping that coding caveat in mind, then, again, in about two-thirds of the sample, characters' intelligence was typically measured as *average*. About 20% of characters were coded as *above-average*—moderately learned and socially capable. Fewer characters were coded as *street smart, intellectually erudite*, or *dumb*, with the fewest being *erudite*. Generally, the characters seen on the television shows most viewed by teenagers were neither remarkably bright nor noticeably dimwitted; in fact, their intelligence was rather unremarkably and unnoticeably *average*.

Time and Focus on Academics

During the first three episodes of the premiere season of *Beverly Hills, 90210*, characters were shown in the classroom 37% of the time (a total of 50 minutes of the total 135 minutes of programming). During the first three episodes of the premiere season of *Dawson's Creek*, characters were shown in the classroom 20% of the time (a total of 27 minutes of the total 135 minutes of programming). *The O.C.* did not show characters in the classroom at all during the first three episodes of their premiere season. In fact, after subsequent viewing, it was found that *The O.C.* did not show characters in the classroom at all until the ninth episode. In total, then, characters on these three programs were shown in the classroom close to twenty percent of the time.

However, while characters were sometimes shown in the classroom, the only storylines or thematic focus relating to academics involved adjusting to a new school. On *Beverly Hills, 90210*, the two main characters, Brandon and Brenda Walsh, were starting their first day of school in Beverly Hills, after moving from Minnesota. While school was the focus, per se, it really only served as a backdrop for the real storyline focus on the emerging friendships of the characters and the struggles of being the "new kid in town." *Dawson's Creek* also featured a "new kid"—Jen Lindley. Jen upset the balance of the friendship between the shows two main characters, Dawson and Joey. Again, while the characters were seen in school, the storyline really focused on the relationships between the characters, not *how* or *what* they were doing in school.

Intelligence and "Coolness" Correlates

The intelligence variables were crosstabulated with the "coolness" variables to discover any relationships between the two. The relationships found between intelligence and romantic activity will be discussed first, followed by popularity.

Romance/Sex Correlates

Table 6.5 show that appearances by intellectually erudite characters featured the least amount of sexual activity. In fact, among appearances by the *intellectually erudite* characters there was virtually no sexual activity to be had: in the entire season of all episodes, only one intellectual engaged in romantic activity, and that did not go beyond kissing.

Table 6.5 Intelligence Correlated with Sexual Intimacy

	Intelligence					
Sexual Intimacy	Smart	Erudite	Street smart	Average	Dumb	Total
No Sex						
N	355	20	89	858	109	1431
% within Intimacy	24.8	1.4	6.2	60.0	7.6	100.0
% within Intelligence	94.9	95.2	94.7	81.5	96.5	86.4
Low Sex						
N	12	1	3	117	2	135
% within Intimacy	8.9	0.7	2.2	86.7	1.5	100.0
% within Intelligence	3.2	4.8	3.2	11.1	1.8	8.2
Sex						
N	7	0	2	79	2	90
% within Intimacy	7.8	0	2.2	87.8	2.2	100.0
% within Intelligence	1.9	0	2.1	7.4	1.8	5.4
Total						
N	374	21	94	1054	113	1656
% within Intimacy	22.6	1.3	5.7	63.6	6.8	100.0
% within Intelligence	100.0	100.0	100.0	100.0	100.0	100.0

χ^2 =79.151, df=8, p=.000

Besides overall intelligence, characters' actual academic performance in the classroom was coded and correlated with characters' sexual intimacy. Table 6.6 shows that of the 1656 total character appearances, 482 involved characters in school. In 232 of those 482 appearances, those characters performance in school was not indicated (48%). When performance was indicated, most of those characters that were highly successful (95%) or successful (83%) were not engaging in any level of sexual activity. Five percent of highly successful and 11% of successful characters kissed/necked, while none of the highly successful characters and six percent of the successful characters had sex. These relationships reinforce a long-standing "discreditation" of intellectuality (Gerbner, 1974), which, it is argued, in turn serves to sustain social structure (see Thomas, 1986; Holderman, 2003).

Table 6.6. Academic Performance Correlated with Sexual Intimacy

	Academic Performance					
Sexual Intimacy	Highly Successful	Successful	Passing	Unsuccessful	Not Indicated	Total
No Sex						
N	39	29	93	26	213	400
% within Ac Per	95.1%	82.9%	65.5%	81.3%	91.8%	83.0%
% within Sex Int	2.9%	7.3%	23.3%	6.5%	53.3%	100.0%
Low Sex						
N	2	4	45	5	13	69
% within Ac Per	4.9%	11.4%	31.7%	15.6%	5.6%	14.3%
% within Sex Int	2.9%	5.8%	65.2%	7.2%	18.8%	100.0%
Sex						
N	0	2	4	1	6	13
% within Ac Per	0.0%	5.7%	2.8%	3.1%	2..6%	2.7%
% within Sex Int	0.0%	15.4%	30.8%	7.7%	46.2%	100.0%
Total						
N	41	35	142	32	232	482
% within Ac Per	100.0%	100.0%	100.0%	100.0%	100.0%	100.0%
% within Sex Int	8.5%	7.3%	29.5%	6.6%	48.1%	100.0%

χ^2=55.571, df=8, p=.000

Character Intelligence and Popularity

As seen in Table 6.7, almost two-thirds of all appearances were made by popular characters. A large majority of those characters were characterized as having average intelligence (73%), with a little more than 20% being smart (both intellectually and street-smart). The singular character considered to be only intellectually "book" smart was not considered popular.

In the follow-up sample of archetypical teen shows, three of the eighteen regularly-featured characters on these programs were rated as intelligent (17%). The remaining fifteen regularly-featured characters were characterized as having average intelligence. Throughout the first three episodes of each program, there were no instances in which a character was praised for their academic performance, and four in which they were criticized (three from *90210* and one from *Dawson's Creek*).

Table 6.7. Intelligence Correlated with Popularity

	Intelligence					
Popularity	Smart	Intellectually Erudite	Street smart	Average	Dumb	Total
Yes Popular						
N	16	0	0	52	3	71
% within Int	88.9%	0.0%	0.0%	55.9%	100.0%	61.7%
%within Pop	22.5%	0.0%	0.0%	73.2%	4.2%	100.0%
No Not						
Popular	1	1	0	41	0	44
N	5.6%	100.0%	0.0%	44.1%	0.0%	38.3%
% within Int	2.3%	2.3%	0.0%	95.3%	0.0%	100.0%
%within Pop						
Total						
N	18	1	93	93	3	1656
% within Int	100.0%	100.0%	100.0%	100.0%	100.0%	100.0
%within Pop	15.7%	0.9%	80.9%	80.9%	2.6%	100.0

χ^2 =17.542, df=6, p=.007

None of the characters who were characterized as intelligent were considered to be a member of the "in" crowd. Two of the three intelligent characters were considered "normal" while the third intelligent character was characterized as an outcast. All six characters that were characterized as being part of the "in" crowd were characterized as having average intelligence.

Discussion

This study sought to identify the common themes relating to two integral areas of adolescent socialization—academics and intelligence on the one hand, and "coolness" and popularity on the other. The study asks, "Is school cool?" More specifically, how are academia and intelligence represented on television shows popular among adolescents? How are academics and intelligence related to popularity and romance? Overall, the findings from both the original and follow-up samples suggest that, in the main, school is not generally considered to be cool.

The most critical evidence supporting this statement centers on the lack of time characters are shown in an academic setting actually performing academic tasks. In the initial season-long sample of the first study, character appearances in the classroom hovered around fifteen percent. Similarly, in the follow-up sample, characters were shown in the classroom 7 percent of time. When characters were seen in the classroom, the academic performance of characters was infrequently indicated in any way. On these programs popular with adolescents, school seems to serve merely as a backdrop for the social relationships between characters. The uses and gratifications approach would suggest that this highlights the escapist nature of these programs—student's spend much time in the classroom, so they want to escape that in the programming they choose . . . or, it may just be, perhaps, that this belief dominates the mindset of the producers of such shows. In this light, it would be interesting to collect the Nielsen demographics for shows whose major focus *was* high school, from *Room 222* through *Fame* to *Boston Public*.

From a socialization perspective, however, it does not provide much in the realm of social modeling. If these programs serve as a "super-peer," then these programs' emphasis on relationships, as well as the scant storyline-focus surrounding academics could arguably skew adolescents' ever-shaping value system. This study's findings support the anecdotal evidence of the mutual exclusivity of intelligence and "coolness." The Fonz's and Dylan McKay's of the television world who are depicted as the "cool" guys are rarely if ever seen as being erudite and involved in academics. The opposite seems true as well—the stereotypical "nerds" like Screech or Urkel that are portrayed as being smart are almost always socially inept and unpopular. Considering such portrayals, social modeling theories would suggest that teens would need to make a decision (conscious or not) as to which trait they would strive to emulate—intelligence or "coolness"—because the cultural lesson they are constantly being presented with is that you can't be both cool and smart. In fact, the complete lack of any real focus or emphasis on academics—not only through limited classroom storylines but also through the plethora of adolescent characters depicted as "average" students—can instill in adolescents the value of simply being *average*.

Further study is needed to ascertain more precise data regarding the valuation of intelligence found in this type of programming. Perhaps more detailed character analyses and long-term studies could provide more information. For instance, *Beverly Hills, 90210* ran for ten seasons. It might be interesting to track the long-term success of the characters relating to intelligence. For example, do those character's characterized as being "smart" in the "book-sense," make smart life choices? Given that most "loyal" adolescent audiences watch these programs for multiple seasons, and that values are embedded over time, this type of longitudinal study could provide some enlightening information relative to the representation of intelligence on programming popular among adolescents.

Another trend that could provide interesting data involves reality television. For instance, reality programs on the cable channel MTV, such as *The Real*

World and *Road Rules* have long been popular with adolescents. A "character" study of how intelligence is characterized by the real people who are cast in these types of programs and the actions they engage in could shed light onto how these programs value intelligence and academics.

No matter what the genre, looking at the relationship between intelligence and adolescent values on television is important. Adolescence is a critical time in the formation of a person's value system, particularly relating to academics, as they are regularly attending school and evaluating how important education is to them on a personal level. The potential role that television content contributes to these values should not be ignored. And, finally, on a larger scale, this lack of emphasis on academics, together with its "uncoolness" underscores our society's devaluation of intelligence, thus keeping intelligence unattractive to people from an early age.

References

Bandura, A. (1986). *Social foundations of thought and action: A social cognitive theory.* Englewood Cliffs, NJ: Prentice-Hall.

Geertz, C. (1973). *The Interpretation of Cultures: Selected Essays.* New York: Basic Books.

Gerbner, G. (1974). Teacher image in mass culture. In P. Olson (ed.), *Media and Symbols* (470-497). Chicago: National Society for the Study of Education.

Gerbner, G., & Gross, L. (1976). Living with television: The violence profile. *Journal of Communication*, 26(2), 173-196.

Gerbner, G., Gross, L., Morgan, M., & Signorielli, N. (1980). The 'mainstreaming" of America: Violence profile No. 11. *Journal of Communication*, 30(3), 10-29.

Hofstadter, R. (1964). *Anti-intellectualism in American Life.* New York: Alfred A. Knopf.

Holderman, L. (2003). Media-Constructed Anti-Intellectualism: The Portrayal of Experts In Popular US Television Talk Shows. *The New Jersey Journal of Communication.* 11(1), 45-62.

Mayerle, J. & Rarick, D. (1989). Image of Education in Primetime Network Television Series 1948-1988. *Journal of Broadcasting and Electronic Media*, 33 (2), 139-157.

Parsons, T. (1951). *The Social System.* Glencoe, IL: Free Press.

Rideout, V., Roberts, D. and Foehr, U. (2005). *Generation M: Media in the lives of 8-18 Year-olds.* Menlo Park, CA: Kaiser Family Foundation.

Ross, A. (1989). *No respect: Intellectuals and Popular Culture.* Routledge, NY: Chapman and Hall.

Thomas, S. (1978). Reality, Fiction and Television. *Journal of the University Film Association*, Spring, 1978, pp. 29-33.

Thomas, S. (1986). Mass media and the social order. In G. Gumpert & R. Cathcart (Eds.), *Inter/Media: Interpersonal communication in a media world* (third edition) (pp. 611-627). New York: Oxford University Press.

Thomas, S. & Krippendorf, M. (1988). *Beauty and Brains: The sociopolitical concomitants of sexuality on television.* A paper presented at the Convention of International Communication Association, New Orleans.

Part III

Scientists and Science Fiction

Chapter Seven

Sexy Nerds: Illya Kuryakin, Mr. Spock, and the Image of the Cerebral Hero in Television Drama

Cynthia W. Walker and Amy H. Sturgis

Nerds are in fashion these days (Williams, 2003), no doubt about it, and they're being celebrated not only for their intelligence, but for their sex appeal as well (Taormino, 2002; Lomrantz, 2005). Several currently popular television series like *Numb3rs, House*, and *The O.C.* even feature brainy characters in leading roles.

Most articles in the popular press that discuss this current "revenge of the nerds" treat the phenomenon as if it were a new one. But, it's not—far from it. In her profile of actor David Duchovny, Anita Gates (2000) noted that *X-Files* fans think Duchovny's character, Fox Mulder, "comes across as . . . the [small] screen's first sexy nerd." While Mulder definitely belongs to that rare breed, the cerebral television hero (i.e., "the sexy nerd"), he is neither the first, nor most certainly the last. Indeed, the history of the sexy nerd, at least on the small screen, dates back more than forty years.

Although the slang term "nerd" apparently has been in wide use since the mid-1960s, both the *Merriam-Webster Online Dictionary* (2005) and *Wikipedia* (2005) trace the origin back to Dr. Seuss's *If I Ran the Zoo* (1950). In that children's book, a "nerd" is pictured as a short, hairy, disheveled fellow with a sour, grumpy expression. Currently, *Merriam-Webster* defines a nerd as "an unstylish, unattractive or socially inept person especially one slavishly devoted to intellectual or academic pursuits" (2005).

The word "nerd" has a number of synonyms, among them dweeb, dork, geek, brain, brainiac, egghead, wonk, grind, member of the pocket protector brigade, computer jock, and Dilbert (after the popular Scott Adams comic strip). After half a century, the stereotype has become fairly well defined. Nerds are introverted, shy, emotionally repressed, and socially awkward. Alienated loners, they are either boring conversationalists or nearly mute. They smile weirdly or seldom at all, enjoy little romantic success, and possess no social life to speak of. Either through choice or necessity, nerds generally abstain from alcohol, sex, and physical sports. Except for typical nerd pastimes such as chess, stamp collecting, and more recently, computer programming and video gaming, they don't seem to know how to have much fun (Hollandsworth, 1994).

Even with wider opportunities available to women, the classic nerd remains essentially male (Didio, 1996). Women who are timid, wall-flowerish, and unusually smart have always managed to occupy a more acceptable place in the social order (Hollandsworth, 1994). In fact, the mousy, unattractive woman (usually a librarian) who takes off her spectacles and lets down her hair to reveal a stunning beauty is a common cliché in popular culture. As a common social stereotype, the male nerd is more rigidly defined. Physically, the nerd is unfashionably or carelessly dressed with uncombed hair or a bad haircut. He is generally weak—either too thin or too fat—and usually wears thick, ugly glasses with black frames, often repaired with a piece of masking tape.

All of these negatives notwithstanding, the nerd has one thing going for him: his intelligence. Nerds are brainy, although their prowess is often limited to the sciences and technology (usually computer-based), and they have an obsessive and exhausting penchant for facts, figures, and details that no one else really cares about. The nerd's impressive intelligence and unimpressive physique make him the exact antithesis of another popular male stereotype, the (often dumb) "jock" (Jones, 1996).

In television comedy, the nerd is the object of jokes and derision. Indeed, television comedy has even offered a subspecies of the nerd: the "idiot nerd." For all the brainy *Quiz Kids* (1949-1956), *Mr. Peepers* (1952-1955), and Steve Urkels (*Family Matters*, 1989-1998), there are also enough goofy Walter Dentons (*Our Miss Brooks*, 1952-1956), Barney Fifes (*The Andy Griffith Show*, 1960-1968) and *Beavis and Buttheads* (1993-1997) to balance them out.

In drama, however, nerds must be more pro-active characters. As villains, they are likely to be brilliant, boorish, impatient, self-righteous, irritating, socially alienated, or just plain crazy. As heroes, on the other hand, some of their stereotypical weaknesses are re-interpreted as strengths. Eggheadedness becomes reasoning, rationality, and a source of important arcane knowledge. Resentment is expressed though a sarcastic, deadpan wit. Alienation is recoded as detachment. Introversion and shyness become mystery.

Cerebral heroes are most likely to appear in genres in which information is particularly valued—that is, crime and detective stories, espionage stories, and science fiction. The most famous cerebral hero who immediately comes to mind is, of course, Sherlock Holmes. Arthur Conan Doyle gave Holmes many of the qualities we have come to associate with the classic nerd long before the term and the concept were coined. Indeed, Holmes is probably the first techno-hero, the flip-side of such Victorian characters as Dr. Frankenstein and Dr. Jekyll—

villainous nerds who epitomize science and rationality gone horribly wrong (Maertens, 1996).

It is important to note, however, that Holmes is the hero, and quite a charismatic one at that (Watson, after all, idolizes him.) Holmes is the dragonslayer while Watson, the regular, married, gregarious everyman, functions in the Sancho Panza sidekick role (Skene-Melvin, 1996). This sort of partnership, with the nerd in the alpha position, can function effectively only in a genre that emphasizes inquiry over action. In the classic detective story, which favors reasoning, minimizes violence, and is physically confined mostly to interior settings, the nerd has the advantage.

Holmes and his successors, characters like Nero Wolfe and Ellery Queen, have appeared on television from time to time.[1] But for a variety of reasons, the majority of television series adopt formulas closer to the action/adventure story. Of course, the demands of adventure, as opposed to mystery, are quite different (Cawelti, 1976). The obstacles are physical rather than intellectual, the playing field is wide and unrestricted, and the emphasis is on movement, action, and novel encounters (Newcomb, 1974).

In such an environment, too much thinking is a definite weakness. Indeed, in Western culture, the activity of thinking is associated with feelings of powerlessness. Writing in *The Humanist*, Todd Jones observed, "We think and scheme when we feel we lack the knowledge, authority, or sheer brawn to solve the problems with which we find ourselves confronted" (1996, p. 44).

To integrate the nerd hero into an action/adventure story, one of two strategies is generally employed. In the first, the "superhero strategy," the nerd and the person of action (i.e., the jock) are combined in the same schizophrenic character, à la Clark Kent/Superman. In the second, a variation of the very popular "Odd Couple" partner strategy, the nerd becomes the friend, the sidekick, the outsider whose peculiar (and even opposing) attitudes and talents aid the good, moral, intuitive (but decidedly less cerebral) jock hero in accomplishing the ultimate task or mission. The best partnerships are between complementary characters who create a symbiotic whole greater than its parts (Skene-Melvin, 1996).

In an action/adventure formula, the positions of Holmes and Watson are reversed. This relationship is more reminiscent of Ralph, the athletic natural-born leader, with Piggy, the fat, bespectacled, intellectual outsider in William Golding's *Lord of the Flies* (1954). However, because action/adventure on screen nearly always involves romance, the nerd hero, though not conventionally attractive, will be found appealing—that is, sexy—by at least some of the women characters he encounters, though he will either deny that appeal or be unaware of it. As a result, he will also function as an identity figure and/or object of erotic interest for many in the audience. The ideal nerd hero, then, combines the social position of Piggy with the personal charisma of Holmes.

Illya Kuryakin: "Unlikely" Sex Symbol

Arguably, television drama's first "sexy nerd" was Illya Kuryakin in the wildly popular *The Man From U.N.C.L.E.* (1964-1968). Indeed, Stephanie Schorow (1997) regards Illya Kuryakin's popularity as such an important pop culture phenomenon that she dubs any instance in which a co-star surpasses the star in popularity "The Illya Factor."

Norman Felton, whose Arena Productions created *U.N.C.L.E.* along with such other hits as *Dr. Kildare* (1961-1966) and *The Eleventh Hour* (1962-1964), has often told a story to interviewers of being challenged by a female comptroller at the BBC in the early 1960's. "Why must the leads in your American series always be big, tall and muscular?" she asked the producer (Felton, 1995; Heitland, 1987). Her question, Felton recalled, made him take a hard look at the television series his children were then watching. He found himself resenting what he saw as a distorted view of heroism and thinking about the possibility of an action/adventure hero who could be "intelligent, not massive in size, witty, and interesting" (Felton, 1968). A year or so later, when Felton was approached to create an escapist spy series for television, he finally had the opportunity to test his theory.

Originally, *U.N.C.L.E.*'s lead, Napoleon Solo, was supposed to be a smart, witty, but also an obviously regular guy (Felton, 1962). Felton even cast Robert Vaughn, an actor of average height who was charming and wry but self-admittedly not an imposing, larger-than-life figure like the ex-Mr. Universe, Sean Connery (Vaughn, 1997). With the success of Ian Fleming's 007 character in film, however, industry and audience expectations that *The Man From U.N.C.L.E.* would be a sort of "James Bond for television" were too great to resist (Walker, 2004). So Vaughn's Solo became a suave, sophisticated small-screen version of Bond and Felton's hope for an alternative hero shifted to Solo's sidekick.

In *The Man From U.N.C.L.E.*'s promotional booklet for advertisers, Illya Kuryakin, played by David McCallum, is described as something of an enigma: clever, physically adept but also an introverted loner whose only passion is a love of jazz and a collection of records secretly stashed away ("The Man from U.N.C.L.E.: Coming on NBC-TV," 1964). The aired episodes reinforced and expanded on Kuryakin's character. He is highly intelligent, having received his Ph.D. in quantum mechanics from Cambridge, followed by postgraduate work at the Sorbonne. He has a wide range of knowledge, ranging from chemistry to arcane gypsy lore, and he exhibits interest and expertise in computers, espionage technology, and mechanical devices. In the partnership, he is the one who can fix an engine, reprogram a computer, or defuse an incendiary device.

Kuryakin is taciturn and emotionally distant. He doesn't smile much and almost never broadly. During the course of one first season episode, "The Quadripartite Affair," he repeatedly advises a female companion to think of him as part of the background—a rock or a tree, one of the walls—until she exclaims in exasperation that she wishes he would try, just once, to act like a human being. Although Kuryakin is presented as physically adept, particularly in swimming, gymnastics, and the martial arts, he is short and slender and wears black-framed glasses when he reads. In the course of a mission, he is also more likely to be captured, beaten, and tortured than his partner. The fact that he is a Russian functioning within Western society in the mid-1960s also lends him an air of alienation. Interacting with female characters, Kuryakin is particularly shy. He doesn't enjoy compliments, he has little patience for small talk or time for flirting, and women were more likely to pursue him rather than the reverse. As Felton put it, "Illya doesn't kiss girls; girls kiss him" (Heitland, 1987).

When *The Man From U.N.C.L.E.* became a surprise hit late in its first season, due in no small part to the popularity of the Illya character,[2] media commentators were baffled. After all, the introverted, shaggy, eggheaded Kuryakin had few of the qualities usually associated with a traditional heartthrob. *TV Guide* called him an "unlikely" sex symbol ("The Greatest Thing," 1965). *Look* compared him to the "maniacal mediocrities, cutesy panelists, and pea-brained cowboys" and dubbed him a "new kind of TV idol" (Brossard, 1965). *The New York Times* theorized that because Illya looked like a "cherubic esthete" and was more vulnerable than the usual "hulking hero," children and teenagers could identify with him (Thompson, 1965).

There is no doubt that *The Man From U.N.C.L.E.*'s prime audience was young and a good portion of them were college-educated, or brainy teens headed for college ("Inside U.N.C.L.E.," 1965; Raddatz, 1966; Borie, 1965). *Newsweek* attributed the sudden upturn in ratings to the teenagers and college students who discovered the series over Christmas vacation ("Inside UNCLE," 1965). *TV Guide* reprinted a fan letter from the Bronx High School of Science to represent the series' following (Raddatz, 1966). And finally, when Wellesley co-eds petitioned the producers not to allow Illya to romance any female characters on the series, so their own, personal fantasies would have no competition, they received wide publicity (Borie, 1965; "The Greatest Thing," 1965). A guide for *U.N.C.L.E.* scriptwriters acknowledged that the series "has evolved to the point where we are the darlings . . . of the super-sophisticates who populate Cal Tech and the space-age laboratories" ("The Man from U.N.C.L.E. Information for Writers," circa 1967). This same audience, incidentally, would later prove to be loyal fans of *Star Trek.*

U.N.C.L.E. fans today remember that they were attracted to Kuryakin's physical appearance, to his "mystery," and most of all, to the fact that he more often used brains than brawn. They also relate their fascination with Kuryakin to a similar contemporary character, *Star Trek*'s Mr. Spock. "What struck me about Illya, first and foremost," recalls one longtime fan, "was that he was *smart* . . . [he] was a revelation. He read. He knew things. He was Spock without the pointed ears." Observes another, "I would say Spock was Illya without the charm, but I agree [that Illya's] intelligence was a major draw" (Channel_D and Channel_M, responses from fans, 1999).

Mr. Spock: Cerebral Heroes Take to the Stars

Star Trek, created and produced by Gene Roddenberry, premiered in September, 1966, at the beginning of *The Man From U.N.C.L.E.*'s third season. On the heels of Illya Kuryakin came Mr. Spock, the ever-logical First Officer of the *U.S.S. Enterprise*, played by Leonard Nimoy. Roddenberry first designed the part for a female character, the stoic genius called Number One. However, the network could not accept a woman who was one step away from command of a starship, and so the character of Spock appeared, inheriting her emotionless intellect and high rank before the program hit the airwaves (Okuda, Okuda, & Mirek, 1994).

Spock immediately made a splash with viewers, particularly feminine ones (Gerrold, 1984; Bacon-Smith, 1986). Not surprisingly, the Vulcan takes Kuryakin's mysterious detachment to another level. Whereas Illya is politically alienated, Spock is literally alien. His Vulcan heritage never leaves the viewers' mind, especially since characters such as Dr. Leonard ("Bones") McCoy insist on pointing it out at every turn, frequently commenting on his pointed ears and green blood, highlighting his distinctive Otherness (Tullock, 1990). His uncanny ability to mark time and figure probabilities underscores his unique, computer-like mentality. And while Kuryakin's cool logic is a question of personality, Spock's is akin to a religion. In *Star Trek V: The Final Frontier* (1988) we see that the culture of the planet Vulcan, Spock's home, breeds mathematicians and scientists and disinherits those, like Spock's half-brother Sybok, who embrace emotion. At times during the series and subsequent feature films, Spock seeks to eradicate his Human half altogether, and in *Star Trek: The Motion Picture* (1979) even (unsuccessfully) attempts the mystic discipline of *Kohlinar* to seek pure logic. Spock's internal struggle remains one of the centerpieces of the original *Star Trek* series, the heroic man confronting himself against the background of space, the final frontier.

Star Trek's Science Officer also revises the image of the nerd created by Illya Kuryakin. Though slender and unassuming like the Russian agent, Spock is far stronger and more physically powerful. His Vulcan heritage gifts him with superhuman strength, endurance, and athletic ability. As he demonstrates in several episodes, he can even best his muscled Captain with little effort. And unlike the androgynous Kuryakin, Spock is far from asexual. Thanks to pon farr, the seven-year mating cycle of Vulcans, Spock's drive can be intense, aggressive, and all-consuming. While Captain Kirk oozes virility in a playful, sensual manner, Spock is animal-like, violent, and frankly erotic when aroused ("Amok Time," 1966; *Star Trek III*, 1984).

Star Trek in general, and Spock in particular, appealed from the very beginning to an upwardly-mobile, educated, literary-minded audience. Fans adapted Robert Heinlein's term from the science fiction classic *Stranger in a Strange Land* to proclaim "I Grok Spock" on bumper stickers (Helmik, 1994). Fan fiction stories such as Terry Endre's "The Adventure of the Vulcan Detective" (1980) and James Van Hise's "Sherlock Spock" (1984) led to speculative essays such as "Was Sherlock Holmes a Vulcan?" (Dunn 1990). Dedicated fan authors and artists began underground circulation of Spock erotica that thrives to this day, portraying the Vulcan alternately as devoted mate, lustful animal, and even collared love slave (Bacon-Smith, 1992).

While *Star Trek*'s original series set the standard for the sexy science fiction nerd with Spock, but there were more to come. *Star Trek: The Next Generation* (1987-1994) introduced the human-like android Data, a character who drew a legion of fans and inspired feminist critic Camille Paglia to name him the thinking-woman' s sex symbol when she wrote her confessional essay, "Dear Mr. Data, You Made Me Love You" (1995). Often cast as the Holmes of the *Enterprise* (Erisman & Erisman, 1996), Data proved he is completely functional in the bedroom in the episode "The Naked Now." *Star Trek: Deep Space Nine* (1992-1999) continued the *Trek* tradition of sexy nerds with Dr. Julian Bashir and Constable Odo. Both are alien, the former due to genetic engineering and the latter as a unique example of a mysterious species that, when at last found, proves to be on the opposite side from Odo of a violent war. Both characters also echo earlier archetypes: Bashir becomes a secret agent of sorts for the mysterious Section 31, a futuristic Illya Kuryakin, a scientist-spy, while Constable Odo serves as the space station's Sherlock Holmes, reading fictional mysteries when he is not investigating and solving real ones. *Star Trek: Voyager* (1995-2001) and *Enterprise* (2001-2005) altered the pattern by offering female sexy nerds in the forms of part-Borg Seven of Nine and Vulcan T'Pol, both women who are isolated from their respective cultures and machinelike in either body or mind.

Although the cerebral heroes of the various *Star Trek* series have worn many faces, all of them share the alienated heroism originally exemplified by Mr. Spock. Twelve years after Spock appeared to fans, a British science fiction series yielded a new sexy nerd: the *Blake's 7* cynic in black leather, Kerr Avon, played by Paul Darrow. Avon, however, would illustrate what happens when nerds go bad.

Kerr Avon: Mad, Bad Cerebral Hero

If *Star Trek* was, as Gene Roddenberry first described it, "Wagon Train to the Stars," then the British series Blake's 7 (1978-1981) was, as creator Terry Nation termed it, "The Dirty Dozen in Space" (Muir, 2000). The hero, Blake, an action-oriented idealist much like Captain James T. Kirk, is the only one of the crew of the *Liberator* who is not a criminal. Among the thieves, smugglers, and murderers who make up his band of rebels is Kerr Avon, a brilliant computer programmer. Like Spock, Avon is often compared to the computers with which he works. Unlike Spock, who routinely faces, even volunteers for, dangerous derring-do, Avon's mantra is that he's not stupid enough to risk his life. The other crewmembers distrust, even fear Avon, because they know he would sacrifice them if need be for his own ends. In the episode "Cygnus Alpha," for example, he tries to leave in *The Liberator*, stranding Blake and the others planetside, so he can keep the vessel and its riches for himself.

His ambition does not stem as much from greed as it does from self-protection, however. Over and over again, Avon has been betrayed. He tries to end the pattern of pain by anticipating the lesson of *The X-Files*: he trusts no one.

This nerd's social alienation, then, is not a matter of natural personality, but rather an artificial device to save him from emotional and psychological hurt. Avon's subtle undercurrent of sorrow and anger adds to the mystery so many viewers have found attractive. As one fan admits, "I like him because he's such a compelling bastard" ("Siubhan," 1999).

Attractive, also, is Avon's usual attire. Unlike his crewmates, who much of the time wear either space-age leisure suits or bloused shirts reminiscent of Sherwood Forest, Avon's choice of apparel is severe at best. He often, in fact, sports silver-studded black leather outfits that suggest the barely-contained eroticism of a bondage fetishist (Nazarro & Wells, 1997). His dark clothes accent the fact that, like Illya Kuryakin, he is not only a loner, but also a slightly-built man, certainly unable to match many of his comrades physically. Yet his arrogance as an Alpha-grade (a space-age aristocrat) and his cleverness with concealed weapons and dirty fighting make up for his size. Friend and foe alike find him intimidating. And though he may be bad, and even mad, Kerr Avon remains an appealing character to fans nonetheless. As another fan explains, "He is allowed by story conventions to be many things that I as a brainy woman would like to have permission to be. He can be cold, arrogant, driven, 'difficult', curious, eccentric, ill-tempered. And yet he is still loveable. I'm jealous" ("Alison," 1999).

Although *Blake's 7* was popular with the British audience during its original run, the series did not travel to the United States until 1986. It struck a responsive chord with American fans hungry for a more dark and cynical fictional universe. To this day, fan writers and artists on both sides of the Atlantic maintain a faithful and enthusiastic following for the character. Despite the fact the series introduced more than a dozen main characters, one fan clearinghouse noted in 2005 that more Blake's 7 fan stories had been published about Avon than about all other of the series' characters combined (Hermit.org, 2005). Meanwhile, members of the Avon Club, the Liber Avonis, Disgusting Slavering Paul Darrow Groupies, and A.N.G.E.L.S. (Avon's New Generation of Eager Love Slaves) continue to carry the torch for their sexy nerd (Hermit.org, 2005).

Fox Mulder: Sexy Nerd as Leading Man

The cerebral hero continued to make progress through the late 1980s and into the 1990s. It was certainly no coincidence that Illya Kuryakin, television's first sexy nerd, appeared when college enrollment began to increase sharply, particularly among women (National Center of Education Statistics, 2005). Just as Sherlock Holmes, the first techno-hero, emerged during the Industrial Revolution, the various appearances of the cerebral hero on television seem to correspond roughly to developments in computer technology: the second generation supercomputers in the mid 1960s, Apple's introduction of the Macintosh personal computer in 1984, and the development of the World Wide Web.

It wasn't really until 1993, however, that U.S. television provided an arc-based paranoid drama to rival *Blake's 7*. Called *The X-Files*, this new series spoke directly to the disenchanted members of Generation X.

A sidekick no more, the cerebral hero finally arrived at star status with FBI Special Agent Fox Mulder, played by David Duchovny. A recognizable nerd, Mulder is a wealth of trivial information and ludicrous expertise, and is exiled to a basement office as punishment for his unorthodoxy. Though called "Spooky" by his superiors and colleagues, he is befriended by a fellow nerd—this time, a female—named Dana Scully. Like Dr. Bashir, she internalizes the scientific aspects of the stereotype by being an obsessively dedicated and well-trained medical doctor. Wilcox and Wilcox take the comparison further, naming Scully the Watson to Mulder's Holmes (1996).

When his back is against the wall, his weapon is not his sidearm, but his mind. Like Kuryakin and Spock before him, Mulder appears slim, almost delicate-featured, and somewhat androgynous. His slight though somewhat athletic frame is no match for that of his superior, the well-muscled Assistant Director Walter Skinner. And though Mulder is a star, he relies on Scully and Skinner to save him from a variety of dangers. Another fan-produced music video, "As Girls Go," highlights the wussiness in Mulder, showing scenes of the agent crying, screaming, and facing general peril throughout the episodes, not unlike, as the song lyrics describe, "a damsel in distress" (Gwyneth, 1998).

Nonetheless, female fans nonetheless extol his good looks and obvious intelligence. One fan comments that Mulder is "sexy. He's smart. There is serious angst. His personality (and ditto with Spock, by the way) grabs you by the brain and lurches you into the fictional world" ("JL," personal correspondence, 1999) Another fan goes further and links her own college experience with her appreciation for the FBI agent: ". . . we were shown a bespectacled, studious-looking, academic type who was totally devoted to his arcane obsessions, who worked in a cramped, crowded office in a shabby part of the university—oopsie, I mean FBI building—and who was constantly under fire from his departmental colleagues . . . Hey, who can relate to that?" ("Kathy," 1999).

Bree Sharp's 1999 hit song, "David Duchovny" praises Mulder, "so smooth and so smart" as "The American Heathcliff/brooding and comely" (1999). His web-savvy followers continue to keep the series alive in memory, meeting as the Duchovniks, the Red Speedo Appreciation Society, Hunkus Intellectico, the Mulder Droolers, the David Duchovny Estrogen Brigade, the Church of Our

Guy David Duchovny, and MulderTorture Anonymous ("because we always hurt the ones we love") (DuchovnyNet, 2005).

Mulder's Successors: Nerds in Leading Roles

With all the media attention and celebrity status of "entreprenerds" like Bill Gates, Steve Jobs, and others in the early 1990s (Gordon, 1996), one might have expected to see other cerebral heroes besides Mulder on television in the lead roles and, indeed, this was the case. *MacGyver* (1985-1992), for example, was a kind of cerebral hero, a special agent who lived by his wits and technical expertise. Nevertheless, Richard Dean Anderson who played him was too conventionally handsome and MacGyver, himself, had none of the physical limitations or emotional baggage usually found in nerd characters.

Five years later, however, in *Stargate SG-1* (1997-present), Anderson was cast as a more traditionally macho male lead, and teamed with a true sexy nerd, a bespectacled, shaggy haired archaeologist played by Michael Shanks. In discussing the appeal of Shanks' enormously popular character, McNamara (2002) noted that Dr. Daniel Jackson "occasionally launches into tedious Spock-like discourses on obscure academic subjects." In the words of one fan, "Who wouldn't want to spend an hour in an Egyptian tomb listening to that soft, sensitive voice explaining the technicalities of hieroglyphs? There is something uniquely attractive about a man so absorbed in discovery. Plus I suppose also there is the thought of what it might be like to be the focus of that intensity." When the producers, pursuing a younger, male demographic, killed off the Illya and Spock-like Jackson in 2001, *Stargate SG-1* fans were infuriated and waged a successful campaign to have Shanks returned to the show (McNamara, 2002; SaveDanielJackson.com, 2005).

VR.5 (1995) featured a female computer whiz named Sydney Bloom who did exhibit many authentic nerd qualities, including computer expertise and social ineptitude, despite the fact that she was played by a lovely young actress, Lorie Singer. The series lasted barely two months, hardly enough to even register with the general audience, although cult fans remember it well. It's noteworthy that the cast included the former Illya Kuryakin, David McCallum, as Sydney's researcher father, and Anthony Stewart Head as a mysterious spy. Head, of course, would become famous as a father-figure himself, Rupert Giles, who was Sunnydale High's librarian and official "Watcher" on *Buffy the Vampire Slayer* (1997-2002).

Joss Whedon and the Evolved Nerd

In creating Buffy, its companion series, *Angel* (1999-2003), and the science fiction series *Firefly* (2002) (and its related 2005 film *Serenity*), writer/producer Joss Whedon became the best, most reliable source of sexy nerd heroes since the *Star Trek* franchise. It is not surprising that Whedon and *Buffy*'s supervising producer, Marti Noxon, are self-confessed high school nerds themselves.

"I was one of those kids who no one pays attention to so he makes a lot of noise and is wacky," Whedon admitted to *Rolling Stone* (Udovitch, 2000). I was an egghead," remembers Noxon, "and I didn't date until college. I was totally antisocial and I was very, very, shy. I couldn't talk to boys . . . there's no one as geeky as me on this show. There's nobody as awkward and introverted and creepy as I was" (Golden and Holder, 1998, p. 259).

Buffy the Vampire Slayer follows the adventures of Buffy Sommers (Sarah Michelle Gellar), the one young woman of this generation chosen to defend mankind against vampires, demons and other creatures of darkness. Although she is very much a typical teenage American girl, Buffy also has a good deal in common with other conventional jock heroes. She is physically attractive, extroverted, athletic, more shrewd than brainy—a good fighter, a fashionable dresser, and a natural-born leader. Throughout the series, Buffy is supported and assisted, by several nerd heroes, most notably her best friend and sidekick, Willow Rosenberg (Alyson Hannigan), a shy, smart, computer whiz (for an extended analysis of Willow's nerd qualities, see Randell-Moon in this volume). Aware of her uncool status, in the episode "Prophesy Girl," Willow observes that she's not ashamed to be a nerd because it's the age of computers and nerds are now a good thing to be.

The most important cerebral hero in Buffy's life, however, is the aforementioned Rupert Giles played by Head. Cool, reserved, and bookish with more than a hint of British stuffiness, Giles would be at home at 221 Baker Street, drinking tea and talking shop with Holmes and Watson. For the first three seasons of the series, Giles is notably the school librarian, (the stereotypical profession of the female nerd) and considered a potent sex symbol by his fans, librarians (DeCandido, 2001) and non-librarians alike (The Giles Appreciation Society Panters or "G.A.S.P.", 1997). He is nearly always found among his beloved books filled with occult and arcane knowledge. As Buffy's counselor, advisor, and erstwhile father figure, Giles also shares the usual befuddlement and cluelessness of sitcom fathers trying to understand the fads and fashions of the younger generation. Although he is the adult of the group, Giles is the most physically vulnerable, repeatedly being beaten, kidnapped, and tortured by various demons. Indeed, the only character more victimized is another Watcher, Wesley Wyndam-Pryce (Alexis Denisof), introduced in Buffy's second season episode, "Bad Girl" who later joined the cast mid-season during *Angel*'s first year.

Introducing even nerdier nerds in order to shore up the status of cerebral hero is, incidentally, a common ploy. Outside of dramatic television, the famous match-up of Charles Van Doren with the far nerdier-looking and less appealing Herb Stempel on the quiz show, *Twenty-One* (1956-1958) comes immediately to mind. Wesley began as an even nerdier-nerd, a fact immediately recognized by the sharp-tongued Cordelia when, in "Consequences" she notes sarcastically that Wesley is just a newer edition of Giles. Although Denisof bears a resemblance to Pierce Brosnan (Cordy notes in "The Prom" that Wesley in a tuxedo would look like Bond), his character is painfully naive and physically ineffectual. Giles has little use for him: he calls Wesley a twerp in "Bad Girls" and remarks on Wes' lack of maturity in Gingerbread." At the end of the battle of the third season finale, "Graduation, Part 2," Giles notes that he must attend to Wesley, who was knocked out by a single, early blow, to see if he's still whimpering.

Nevertheless, from his first appearance on the *Angel* spin-off series as a leather-jacketed, motorcycle riding self-described "rogue demon hunter" to his poignant death in the series' final episode, Wesley slowly evolves from a joke to a dark, ruthless and psychologically complex hero. As a result, he earned the sympathy and appreciation of female fans (Berner, 2004).

Indeed, Wesley is the epitome of the innovation that Whedon introduced to the sexy nerd stereotype: the ability to change and evolve. Eventually, during the course of the two series, Giles rebels against the Watchers' Council, Willow becomes a powerful witch, friend Xander finds love and becomes more mature and self-sufficient, and Fred, the nerdy librarian and Willow-like sidekick introduced in *Angel*'s second season, dies only to be reborn as the nearly omnipotent, blue demon, Illyria. Even Spike, perhaps the coolest character in Whedon's canon, is revealed to have been a nerd before he became a vampire. In the *Buffy* episode, "Fool for Love," we discover that his nickname, "William the Bloody" stems not from his murderous ways but from the fact that he was once a bloody awful poet.

As one *Firefly* fan explains, "The term 'nerd' had been used in a derogatory way, until Joss Whedon showed that nerds can contribute as much as (or more than) their less geeky peers, and are as attractive and sexy as everyone else" ("Helygen," 2005). This is certainly the case in Whedon's short-lived science fiction series *Firefly* (2002), which, thanks largely to overwhelming and outspoken fan support (Fienberg, 2005; Whittmore, 2005), received a new lease on life in its 2005 feature film incarnation, the critically acclaimed *Serenity*. The misfits and outcasts who are the series' protagonists include such sexy nerds as Kaylee (Jewel Staite), a sweet and sassy engineer with an uncanny gift for mechanics and electronics, and a pair of siblings wanted by the galactic powers that be: Simon Tam (Sean Maher), a gifted surgeon, and River Tam (Summer Glau), a genius who has escaped from a governmental program designed to develop her extraordinary abilities as a state weapon.

Whedon's ability to transcend classic nerd stereotypes resonates with fans: as one put it, "I'm very glad they didn't put Simon in wire-rimmed glasses he could whip off at an essential juncture to reveal his masculine splendor (gag) or put Kaylee's hair in a really unflattering ponytail she could pull down at an essential juncture to reveal her full blossom of womanhood (double gag) . . . . I think *Firefly* avoids this trap, and plays these people as complete and complex characters who merely happen to possess nerdly traits" (Winslow, 2005).

Conclusions: Long Live the Cerebral Hero

Although the Whedon-produced series are gone, cerebral heroes live on, most recently in *Numb3rs* (2005-present) with David Krumholtz as a sexy nerd mathematician who assists his FBI agent brother (Rob Morrow) solve cases, and in *NCIS* (2003-present) which features not one but three cerebral heroes in its ensemble. One is an eccentric Goth researcher named Abby (Paula Perrette) and one is a nerdy special agent and computer whiz named McGee (Sean Murray). The third, interestingly enough, is a nebbish chief medical examiner named Donald "Ducky" Mallard. Mallard lives with his dotty ninety-six-year old mother, lectures everyone else on the team on various arcane subjects, and, as has been observed during the series run several times, looked a lot like Illya Kuryakin when he was young. Not surprisingly, he's played by David McCallum.

The fact that the story of the televised sexy nerd begins and ends with David McCallum's portrayals suggests a full-circle journey back to where the cerebral hero first began. In fact, the sexy nerd has taken significant strides, evolving from sidekick to leading man, from accidental sex symbol to automatic heartthrob. As the self-identified nerd audience has grown larger and more vocal, and the television creators themselves have declared their own nerd allegiances, the cerebral hero has finally come into his own. While the twin inspirations of technology and education continue to drive society, sexy nerds will continue to thrive and multiply as well. No doubt, we can expect new and even more appealing innovations in the cerebral hero, whenever and wherever he (or she) may appear in the future.

Notes

1. There have been four weekly series based on Ellery Queen, one on the old Dumont Network, 1950-1951; two on ABC, 1951-1952; and two on NBC, 1958-1959 and 1975-1976. A short-lived series starring William Conrad as Nero Wolfe ran on NBC in 1981. Sherlock Holmes has appeared on television in a number of made-for-TV movies, but his most well-known and highly praised small screen incarnation was in the Granada Television series (1984-1992) starring Jeremy Brett in the title role. Produced in Great Britain, the episodes were shown on U.S. television as part of the PBS anthology series *Mystery*.

2. The number of letters to the show were variously estimated at 10,000 a week by Brossard (1965), and 2,500 a day by Brown and Shorin (1965). According to Heitland (1987), McCallum received the greatest portion, apparently more mail than any MGM star previously, including Clark Gable.

References

"Alison." (1999, November 18). Private post to the author (AHS). [published with permission]

Bacon-Smith, C. (1986, November 16). Spock among the women. *New York Times Book Review*, 1, 26, 28.

Bacon-Smith, C. (1992). Marriage and the alien male: lay-Spock. In *Enterprising women: television fandom and the creation of popular myth.* (pp. 102-110). Philadelphia: University of Pennsylvania Press.

Berner, A. (2004). The path of Wesley Wyndam-Pryce. In G. Yeffeth (Ed.). *Five seasons of Angel* (pp. 145-151). Dallas, TX: Benbella Books.

Borie, M. (1965, July). The 1,741 girls who plotted to blow up Illya. *Modern Screen*, 34-35, 67-68.

Brossard, C. (1965, July 27). U. N.C.L.E.'s Illya: new kind of TV idol. *Look*, 79-82.

Brown, L. & Shorin, R. Eds. (1964). *Topps fan magazines presents The Man From U.N.C.L.E.* New York: Topps C.G., Inc., Publishing Division, & Metro-Goldwyn-Mayer, Inc.

Cawelti, J.G. (1976). *Adventure, mystery and romance.* Chicago: University of Chicago.

Channel_D and Channel_M@onelist.com, responses from fans. (1999, March 13-16). Private posts to the author (C.W. Walker). [published with permission]

DeCandido, G.A. (2001) Rupert Giles and search tools for wisdom in *Buffy the Vampire Slayer.* Retrieved October 24, 2005, from http://www.well.com/user/ladyhawk/Giles.html

Didio, L. (1996, November 18). Where the girls aren't. *Computerworld*, 106.

DuchovnyNet. (n.d.). Retrieved on November 14, 2005, from http://www.bomis.com/rings/duchovny/13

Dunn, P. (1990). The problem of identity: was Holmes a Vulcan? In Irwin, W. & Love, G.B. (Eds.) *The best of Trek 11* (pp. 28-30). New York: Penguin Books.

Endre, T. (1980) The Adventure of the Vulcan Detective. *Stardate 7.* Orion Press. Retrieved November 14, 2005, from http://www.fastcopyinc.com/orionpress/2266-2270_The_First_Mission/theadventureofthevulcandetective.htm

Erisman, F. & Erisman, W. (1996). Data! Data! Data!: Holmesian echoes in *Star Trek: The Next Generation.* In C.R. Putney, J. A. Cutshall, & S. S. King, (Eds.), *Sherlock Holmes: Victorian sleuth to modern hero* (pp. 90-100). Lanham, MD: Scarecrow Press.

Felton, N. (1962, October). Untitled Edgar Solo synopsis. Norman Felton Collection, University of Iowa Library, Iowa City, Iowa.

Felton, N. (1968, February 14). Notes sent to Skippy Schearer of *Parade* magazine. Norman Felton Collection, University of Iowa Library, Iowa City, Iowa.

Felton, N. (1995, September 30). Speech, Spycon 11 [Video]. (Available from U.N.C.L.E. HQ, P.O. Box 8403, Rolling Meadows, IL 60008).

Fienberg, D. (2005, September 29). Whedon goes from cancellation to *Serenity. Newsday.* Retrieved November 17, 2005, from http://www.newsday.com/entertainment/movies/stv-093005ent-movies-josswhedon,0,169474.story?coll=ny-moviereview-headlines

Gates, A. (2000, April 9). David Duchovny, with facial expressions. *The New York Times*, Section II, 13-28.

Gerrold, D. (1984). *The world of Star Trek.* New York: Bluejay Books.

The Giles Appreciation Society Panters Home Page. (1997). Retrieved October 24, 2005, from Google, cached at http://64.233.161.104/search?q=cache:FxXeWjqUKZoJ:www.geocities.com/TelevisionCity/7728/gaspers.html+Giles+Appreciation+Society&hl=en

Golden, C. & Holder, N. (1998). *Buffy The Vampire Slayer: The Watcher's Guide*. New York: Pocket Books.

Golding, W. (1954). *The Lord of the Flies*. New York: Capricorn Books, GP Putnam's Sons.

Gordon, D.B. (1996, September). Future shlock. *Tikkun*, 90-92.

The greatest thing since peanut butter and jelly. (1965, April 17). *TV Guide*, 6-9.

Gwyneth. (1998). As girls go. Music by Suzanne Vega. Media Cannibals and Accomplices, Partners in Crime: Tape Collection #3.

Heitland, J. (1987). *The Man from U.N.C.L.E. book: the behind the scenes story of a television classic*. New York: St. Martin's Press.

Helmik, K. (1994, January 21). Captain Kirk looks back in time. *Christian science monitor*, 17.

"Helygen." (2005, July 25). Private post to the author (AHS). [published with permission]

Hermit.org. (n.d.) Retrieved October 13, 2005, from http://www.hermit.org/Blakes7/.

Hollandsworth, S. (1994, May). Nerdvana! *Texas Monthly*, 132-137.

Inside UNCLE. (1965, July 5). *Newsweek*, 54.

"JL." (1999, November 17). Private post to the author (AHS). [published with permission]

Jones, T. (1996, September). The dumb jock and the science nerd. *Humanist*, 44-45.

"Kathy." (1999, November 23). Private post to the author (AHS). [published with permission]

Lomrantz, T. (2005, June 9). Nerds make better lovers. *The New York Daily News*. Retrieved November 11, 2005, from http:///www.nydailynews.com/front/v-pfriendly/story/317296p-271224c

Maertens, J.W. (1996). Masculine power and the ideal reasoner: Sherlock Holmes, technician-hero. In C. R. Putney, J. A. Cutshall King and S. Sugarman (Eds.), *Sherlock Holmes: Victorian sleuth to modern hero* (pp. 296-322). Lanham, MD: Scarecrow Press.

The Man from U.N.C.L.E.: Coming on NBC TV, Fall 1964, 8:30-9:30 pm, Tuesday. (circa 1964). Promotional booklet for advertisers and affiliates. NBC Television Network.

The Man From U.N.C.L.E. 1967-1968 season information for writers. (circa 1967). Norman Felton Collection, University of Iowa Library, Iowa City, Iowa.

McNamara, M. (2002, February 13). Fan rebellion threatens *Stargate*. *Salon*. Retrieved October 24, 2005, from http://www.salon.com/ent/feature/2002/02/13/stargate_rebellion/index.html

Merriam-Webster Online Dictionary (2000). Retrieved October 24, 2005, from http://www.m-w.com/cgi- bin/dictionary?book=Dictionary&va=nerd&x=18&y=14

Muir, J.K. (2000). *A history and critical analysis of Blake's 7, the 1978-1981 British television space adventure*. London: McFarland & Company, Inc.

National Center for Education Statistics. (2005, March). Postsecondary participation rates by sex and race/ethnicity: 1974-2003. NCES 2005028. Retrieved November 17, 2005, from http://nces.ed.gov/pubsearch/pubsinfo.asp?pubid=2005028

Navarro, J. & Wells, S. (1997). *Blake's 7: the inside story*. London: Virgin Publishing.

Newcomb, H. (1974). *TV: The most popular art*. Garden City, NY: Anchor Books.

Okuda, M., Okuda, D., & Mirek, D. (1994). Number One. *The Star Trek encyclopedia: a reference guide to the future*. New York: Pocket Books, 222.

Paglia, Camille. (1995, Spring). Dear Mr. Data, you made me love you. *TV Guide Collector's Edition Star Trek: four generations of stars, stories, and strange new worlds*, 46-47.

Raddatz, L. (1966, March 19). The mystic cult of millions: the people from U.N.C.L.E. *TV Guide*, 15-18.

SaveDanielJackson.com. (n.d.) Retrieved November 11, 2005, from http://www.savedanieljackson.com/history/home/home.shtml

Schorow, S. (1997, March 12). Second fiddle, first in our hearts. *Boston Herald*, Sec. 3, 47.

Seuss, Dr. (1950). *If I ran the zoo*. New York: Random House.

Sharp, B. (1999). David Duchovny. *A cheap and evil girl*. Trauma Records.

"Siubhan." (1999, November 16). Private post to the author (AHS). [published with permission]

Skene-Melvin, D. (1996). Sherlock Holmes: The mythic hero in criminous literature. In C. R. Putney, J. A. Cutshall King and S. Sugarman (Eds.), *Sherlock Holmes: Victorian sleuth to modern hero* (pp.117-134). Lanham, MD: Scarecrow Press.

Taormino, T. (2002, April 24). Sex nerds: getting in touch with my inner nerd. *The Village Voice*. Retrieved November 11, 2005, from http://www.villagevoice.com/people/0217, taormino,34146,24.html

Thompson, H. (1965, June 20). The teen-agers cry "U.N.C.L.E." *The New York Times*, Sec. 2, 15.

Tullock, J. (1990). Brother, my soul: Spock, McCoy, and the man in the mirror. In Irwin, W. & Love, G.B. (Eds.) *The best of the best of Trek* (pp. 308-323). New York: Penguin Books.

Udovitch, M. (2000, May 11). What Makes Buffy Slay?" *Rolling Stone*, 61-64, 66. See also: http://www.smgfan.com/articles/i66.htm

Van Hise, J. (1984, July 4). Sherlock Spock. *Enterprise Incidents*, Special Collector's Edition, 62-65.

Vaughn, R. (1997, February 14). Personal interview with C. W. Walker, tape recording.

Walker, C. W. (2004). Man from/girl from U.N.C.L.E. In H. Newcomb (Ed.), Encyclopedia of Television. (2nd Ed.) Chicago: Fitzroy Dearborn, 1404-1406.

Whittmore, Sean. (2005, September 30). *Firefly*'s faithful. *The Battalion*. Retrieved on November 17, 2005, from http://www.thebatt.com/media/paper657/news/2005/09/30/Aggielife/Fireflys.Faithful-1004569.shtml?norewrite&sourcedomain=www.thebatt.com

Wikipedia (n.d.). Retrieved October 24, 2005, from http://en.wikipedia.org/wiki/nerd

Wilcox, R. & Williams, J.P. (1996). What do you think? *The X-Files*, liminality, and gender pleasure. In D. Lavery, A. Hague, & M. Cartwright (Eds.) *"Deny all knowledge": reading The X-Files* (pp. 99-120). Syracuse, NY: Syracuse University Press.

Williams, I.R. (2003). Twilight of the dorks? *Salon*. Retrieved November 11, 2005, from http://archive.salon.com/tech/feature/2003/10/29/dork/

Winslow, Dodger. (2005, August 2). Private post to the author (AHS). [published with permission]

Chapter Eight

Brains in Service of Brawn: The Scientist/Soldier Dynamic in Science Fiction Television

Christine Mains

In recent decades, some of the most popular television shows have been part of the genre of science fiction. One thing that the very different shows in this category have in common, aside from their fantastic premises, is a reliance on the character types and plot patterns of formulaic narratives. Many of these shows feature an ensemble cast of good guys—such as the reluctant and rebellious hero, the spunky heroine, the brainy sidekick, the stoic warrior companion—fighting against the forces of evil, usually agents of chaos and destruction, occasionally of order taken to extremes. While the shows are fantastic, depicting technology only dreamed of by today's audiences, writers and producers attempt to ground their narrative visions by extrapolating from real world scientific discoveries and theories. Naturally, scientists are central characters: sometimes bumbling geeks whose antics are a humorous counterpoint to the heroics of other characters, sometimes dangerously obsessed egomaniacs who regularly put the rest of the world in danger, but often larger-than-life geniuses whose creative insights and intuitive leaps of understanding are essential to the resolution of the crisis of the week.

Most science fiction television shows also include a military component; leading characters are soldiers or other agents of law and order whose presence is necessary given the ongoing battle against evil alien forces bent on the oppression or extermination of humanity. One underlying theme deals with the implications and consequences of technological advancement in times of war. Key story arcs focus on scientists torn between the desire to pursue knowledge for its own sake and on soldiers struggling with obedience to superiors who are at best foolish and at worst malicious. The interactions between soldiers and scientists reveal underlying societal attitudes about the appropriate relationship between the military, representing physical power sanctioned by law and society, and scientific research, the product of intellectual power which brings both benefits and risks. Narrative tension is generated because scientific knowledge is subordinated to urgent military needs, but often that tension remains barely examined and ultimately unresolved, the physical control of intellectual labor unquestioned.

Science Fiction Television

As a television genre, science fiction television tends to be broadly defined by fan interest in certain character types and storylines rather than strictly defined by the use of the narrative conventions of literary science fiction. Science fiction television can include, for instance, *Buffy the Vampire Slayer* and its spin-off *Angel*, which deal primarily with vampires and demons, or *The Sentinel*, featuring a cop with superhuman senses. Aside from making use of science fiction elements, these shows also share a fan base with shows more recognizably science fictional, such as *Andromeda*, *Firefly*, *Battlestar Galactica*, *Babylon 5* or the *Star Trek* franchise. Other shows are set in a time contemporary to our own, such as *The X-Files*, or *Stargate SG-1* and its spin-off, *Stargate Atlantis*. Because of its contemporary setting and its unapologetic borrowing of thematic and narrative elements from other science fiction series and films, the fictional universe of *Stargate SG-1* provides an ideal illustration of the ways in which stock characters such as the soldier and the scientist speak to our cultural concerns about military control of intellectual effort.

Stargate Command is a unit of the United States Air Force, operating from a hidden base underneath Cheyenne Mountain. While the SGC's mandate does include scientific exploration and discovery, such a pursuit of knowledge for its own sake or for the betterment of humanity is subordinated to the military's need to find technology that will aid in Earth's defensive war against the parasitic alien Goa'uld and the mechanistic Replicators. In the spin-off series *Stargate Atlantis*, an expedition composed primarily of civilian scientists and headed by diplomat and negotiator Dr. Elizabeth Weir (Torri Higginson) finds itself relying on its military support staff for more than grunt work and security when their base on the lost city of Atlantis comes under attack by the vampiric Wraith. In the gateverse, science and technology are the tools by which both heroes and villains can achieve their goals. The military is firmly in charge: all of the leading characters, whether soldiers or scientists, wear uniforms and carry weapons, and in the parent show, the science labs are located in the gray concrete bunkers of the military complex. While both soldiers and scientists are depicted heroically and have an equal share of screen time, it is, unsurprisingly, the soldier who takes the lead and the scientist who plays the sidekick. However, that does not mean that the military/science debate, a constant theme in science fiction both written and filmed, is never raised.

Stock Characters as Rhetorical Devices

The narrative conventions of science fiction include stock characters, such as the maverick soldier and the mad scientist. These stock characters function as rhetorical devices to convey the themes of the narrative, by embodying positions in the argument being made, explicitly or otherwise. A useful model of characterization which suits analysis of formulaic narratives such as those of science fiction television has been proposed by narratologist James Phelan. Phelan (1989) describes a three-component model of characterization, analyzing the mimetic, synthetic, and thematic aspects of a particular character. By mimetic, Phelan means that the character possesses recognizably realistic traits, in terms of physical and psychological characteristics which encourage readers—or viewers—to see the character as a real person. In television, the actor's appearance and acting choices have as much to do with the "personality" of the character as do the words that the show's writers put in the character's mouth; even more so than in literature, the viewer gets to know a character with a physical body and a set of psychological quirks. One method of conveying the realism of a fictional character is to draw on people and events from real life, to make a connection between what is familiar from history or current events and what is being constructed on the page or on the screen.

Of course, the lines between reality and story blur, in both directions. The more a real world person is narrativized—in documentaries, films, memoirs, historical fiction—the more like a character the person seems, and the more their real traits and actions influence the portrayal of similar characters. The figure of the mad scientist, for example, affects the perception of scientists in the real world, and the behavior of real world scientists becomes part of the literary type. This is, in a sense, what Phelan has in mind when he refers to the synthetic component of characterization. Phelan observes that a fictional character "is neither real nor the image of a real person but rather is a construct" (p. 2) created out of words and, in the case of television and film, out of images. Constructing a character out of words and images means drawing on cultural stereotypes and mythic archetypes, on the stock character types of formulaic narrative. The mad scientist is, after all, a narrative tradition and has been for as long as stories about the search for knowledge have been told. In *From Faust to Strangelove* (1994), Roslynn D. Haynes describes the typing of the scientist in western literature, outlining several categories into which typical representations of the figure of the scientist can be placed. Two of these categories are positive: the "heroic adventurer" and the idealist, "engaged in conflict with a technology-based system that fails to provide for individual human values." But such representations are outweighed by the number of negative scientist-figures in literature: the "obsessed or maniacal" scientist who never considers the human cost of research; the "absent-minded professor" alienated from the "real world of social intercourse"; the cold and unfeeling scientist who forsakes human relationships; and the helpless scientist who "has lost control either over his discovery . . . or, as frequently happens in wartime, over the direction of its implementation" (pp. 3-4). During her comprehensive discussion of these types of the scientist figure, Haynes provides examples of fictional characters, such as Victor Frankenstein and Wells' Dr. Moreau, and of real world scientists such as Albert Einstein and J. Robert Oppenheimer, director of the World War II Manhattan Project, referring not only to their real world histories but also to fictionalized versions of these real people.

The point of Haynes' discussion, and of similar works such as David J. Skal's *Screams of Reason* (1998), is to explore the ways in which the stereotypical or archetypal scientist is used to convey cultural concerns about science and technology. Skal argues that in "B movies, pulp novels, and comic books, the mad scientist has served as a lightning rod for otherwise unbearable anxieties about the meaning of scientific thinking and the uses and consequences of modern technology" (p. 18). Christopher Toumey (1992), in his article exploring the "flattening" of the mad scientist character as narratives are adapted from text to screen to sequel, states that stories about the mad scientist "include the feelings that science is downright dangerous to one's spiritual well-being and that science is too secular, in the sense that scientists have escaped the restraints of Judeo-Christian morality" (p. 434). In tying the synthetic component of the scientist character to such arguments, these critics demonstrate Phelan's third component, the thematic. According to Phelan, the character is "taken as a representative figure, as standing for a class . . . and his representativeness then supports some proposition or assertion." In other words, the character is a means of commenting on universal themes and social issues (p. 3). In science fiction shows in general and in the Stargate universe in particular, the military/science debate is conveyed largely through the construction of and interaction between the stock characters of the scientist and the soldier who instantiate positions in the argument.

The Military/Science Debate

In short, there are two poles to the military/science debate: one, the risks to humanity of allowing scientists to pursue research unchecked, and two, the dangers of allowing the military and the governments that they serve to control the direction of that research. In contemporary science fiction television, most clearly in the Stargate universe, science and scientists are dangerous, at worst evil and at best foolish or naïve, and must be guided by the military representative of the physical force of law and order. Science is a useful tool, especially in wartime, but it is also a potential threat, as Haynes' listing of negative traits makes clear. Speaking of the mad scientist in film, Skal (1998) describes "the relationship between madness and hyperintellectualism" in popular culture (p. 19), and

Toumey points out that "scientists who control the mysteries of modern secular knowledge [appear] unaccountable to standard conventions of morality" (p. 411). Thus the need for society, through its agents the military establishment, to guide and control the scientist is justified. But Haynes' description of the scientist helpless to control military appropriation of scientific research speaks to the other extreme, that both scientists and their military masters are answerable to political and corporate entities with questionable or even sinister motives.

The military/science debate is a significant theme in science fiction television, both in its presence in storylines and in its absence where it might be expected. That Phelan's model of characterization proves useful in teasing out the underlying arguments of this theme is evident in the analysis by Haynes and other critics of literary works featuring characters based, more or less, on J. Robert Oppenheimer. Clearly Oppenheimer functions mimetically, as the details of his life and his involvement with atomic weapons research resonate with viewers. As Haynes (1994) observes of narratives based on society's response to the bomb, "The real-life model . . . was almost certainly the sensitive and widely-read J. Robert Oppenheimer, whose private fears about the increasing hold by the military over the products of research in atomic physics did not prevent his public expression of delight in the H-bomb" (p. 251). Haynes also traces Oppenheimer's transformation into a character in literature, a process by which the synthetic components of such characterization is emphasized. That Oppenheimer comes to signify a character type is also evident in an article by Jules Zanger and Robert Wolfe (1986) who cite the "symbiotic relationship" between literature and science, between fictional and factual scientists, in referring to Oppenheimer and to Albert Einstein as models for literary "magicians" of post-war science fiction and fantasy (pp. 38-39). And of course the point of Haynes' use of Oppenheimer as an example of several of her character types is to demonstrate how such types convey themes about social concerns regarding scientific research. Haynes' discussion of Oppenheimer, which examines the mimetic, synthetic, and thematic components of representations of the scientist figure, points to a narrative influence on the portrayal of television scientists, including the two leading physicists in the gateverse: Colonel Samantha Carter and Dr. Rodney McKay.

The Helpless Scientist's Inner Conflict

While arguments are, naturally enough, often conveyed through dialogue and character interaction, a single character can function as a thematic locus. In *Stargate SG-1*, the argument in favor of military control over useful but potentially dangerous scientific research is embodied in the character of Major/Doctor Samantha Carter (Amanda Tapping). A theoretical astrophysicist and the world's foremost expert on wormhole theory, Carter is also an officer in the

United States Air Force, sworn to obey the commands of her superiors. In the pilot episode "Children of the Gods" (1997), then-Captain Carter confronts a roomful of fellow officers, including the general in charge of the SGC and the colonel who will be her commanding officer, demanding they show her the respect due to any officer regardless of the fact that she is a woman. But it is not her gender that Col. Jack O'Neill (Richard Dean Anderson) has a problem with, but rather the fact that she is a scientist. Carter insists on being addressed as "Captain" rather than "Doctor" by O'Neill and his men, but when she meets Dr. Daniel Jackson (Michael Shanks), the linguist and archaeologist who served with O'Neill on the first mission through the stargate, she introduces herself as "Doctor" to emphasize their bond as scientists with a shared interest in understanding the alien technology. This confusion in address, a sign within the narrative of the tension between military and science, tellingly disappears after the first few episodes, an early indication that Carter the physicist is of less importance to the narrative themes of the series than is Carter the junior officer.

That is not to say that the series' writers never make use of the character's potential to function thematically in storylines centering on the military/science debate. Particularly in the earlier episodes, Carter is occasionally torn between her desire to investigate scientific phenomena and her military duty. Raised a military brat, and much less likely than the insubordinate and irreverent O'Neill to resist the dictates of the chain of command, Carter is fascinated by the opportunities for discovery afforded by stargate travel. However, well aware of the urgent need to discover and develop a defense against the enemy, Carter seldom allows her attention to stray from military matters. On the rare occasions when her scientific enthusiasm does threaten to carry her away, her superiors are quick to slap her down. In the second season episode *A Matter of Time* (1998), the SGC's stargate connects to a stargate on a distant planet caught in the gravity well of a newly formed black hole. Through the camera of the remote probe, Carter, O'Neill, and General Hammond (Don S. Davis) are able to watch images of another SGC team, investigating the planet at the time of the event and now trapped by the effects of time dilation. Since there is nothing the SGC can do to rescue the team, Hammond orders the stargate closed, but Carter protests, clearly fascinated by the unique opportunity to study a black hole first-hand. In response to her desire to continue observing the phenomenon, O'Neill figuratively slaps her down with a reminder that what they are observing is the death of their colleagues. O'Neill's sharp voice and the emphasis on her rank rather than her name is intended to chasten Carter, to remind her of the values that should be the concern of a good officer at such a time, the respect and honor due to a fallen comrade. He sees that Carter the soldier has given way to Carter the scientist, that she is obsessed, to use Haynes' terminology, with the pursuit of astrophysical knowledge, to him an "arcane intellectual goal" (Haynes p. 3). She risks abandoning her social responsibilities, as would Haynes' absent-minded professor, and forgetting necessary human emotions, like Haynes' unfeeling scientist. This scene of interaction between the characters serves to align the scientist with the value system of the military. The scientist's contribution is important, as Carter's knowledge proves invaluable to solving the crisis later in the episode, but it is up to the soldier, who cares about the impact of such a crisis on other human beings and who is responsible to society for putting things right, to implement the intellectual solution.

Over the course of the series, the character of Carter has undergone significant changes such that she no longer functions thematically even to such a limited extent in the military/science debate. Because she is consistently written as a good soldier who willingly obeys the orders of her superiors without dissent, Carter's internal conflict between soldier and scientist is usually silenced; she is seldom free to speak her mind regarding doubts about the military's rush to press her discoveries into service against the enemy, although the actress' facial expressions often convey Carter's discomfort. In the fourth season episodes "Chain Reaction" (2000) and "Scorched Earth" (2000), for example, Carter follows orders to construct weapons despite her misgivings. In the former episode, the weapons test goes wrong, an entire alien planet is destroyed and Earth itself placed at risk; in the latter, O'Neill orders Carter to build a bomb in order to destroy an alien ship. The character has also been further distanced thematically from the military/science debate as the synthetic component of her characterization as the love interest for the hero became more prominent. When romance entered the storyline, the character's inner conflict ceased to center on her opposing vocations as scientist and soldier; Carter the scientist is reduced to spouting technobabble for expository purposes rather than questioning the uses to which her knowledge is put. To use Phelan's terms, emphasizing a different synthetic component of characterization, such as from Scientist to Love Interest, suggests a shift in the thematic component also; instead of embodying a position in the debate about the military use of scientific research, Carter's character arc centers on the tragic need for love to give way to duty.

The Civilian Scientist as Conscience

Carter's interaction with the soldier O'Neill aligned her character thematically with the military side of the debate; on occasion, her interactions with other scientists provide an opportunity to question that alignment. Because Carter the good soldier is unable to protest the military's actions, criticisms of military control of scientific research are expressed by other characters, civilian scientists freer to speak their minds. In "48 Hours" (2001) Carter meets her match in physicist Dr. Rodney McKay (David Hewlett), introduced to Carter and General Hammond as the world's foremost expert on the stargate rather than Carter who, as McKay pointedly reminds her, spends most of her time in the field, as a soldier, rather than in the lab, as a scientist. McKay's questions about Carter's competence seem at first reasonable and thoughtful; he notes that the interface she designed for the stargate is full of flaws, and observes that even when her system does report an error, Carter is likely to ignore it anyway. Faced with criticism by a fellow physicist, a defensive Carter uses her place in the military chain of command as an excuse for sloppy science, arguing that it is her superiors who decide what to do with her information. Viewers faced with the need to choose sides in the debate may find McKay's position understandable. After all, McKay is at the SGC to help Carter find a way to free her teammate trapped in the stargate's pattern buffer as a result of her inadequate understanding of the technology. Earlier in the season her decision to override the stargate's failsafe mechanism almost resulted in the destruction of a planet's entire ecosystem. Consequently, the viewer may tend to agree with McKay's argument that Carter's military mindset and loyalty to the chain of command has degraded her abilities as a scientist.

However, certain mimetic and synthetic aspects of the construction of McKay's character undercut that reading and ensure that viewer sympathies remain with the military position embodied in Carter rather than the civilian scientist position embodied in McKay. For one thing, McKay is a guest at the party; viewers of episodic television series are well aware that guest stars serve as foils for the regular leads who, according to convention, will ultimately be proved correct. More important, however, is that McKay in many ways matches the negative type of the scientist to a much greater extent than does Carter. McKay is depicted as socially inept; he tells the fair-haired Carter that he's always been attracted to dumb blondes, and he seems to be a hypochondriac, worrying about his citrus allergy and the possibility of hypoglycemic shock. He may have valuable things to say, but his tendency to say those things rudely and arrogantly demonstrates his inability or his unwillingness to practice social niceties, a combination of the absent-minded professor "out of touch with the real world of social intercourse" and the "unfeeling scientist who has reneged on human relationships" (Haynes, 1994, p. 3). In the end, his insistence that Carter's hope of saving her teammate is unfounded is proven wrong, and he is "punished" by being sent to Siberia. Because McKay is constructed as an unfavorable opponent, his thematic position is given less weight by the viewer; he is clearly on the losing side.

McKay learns his lesson, that science for its own sake must give way to military needs. In a later episode, *Redemption* (2002), McKay goes so far as to acknowledge that Carter is, in fact, a better scientist than he is. As a child, he had abandoned the piano because he was told that his technical proficiency was not the same as artistic genius; working with Carter has taught him that science is as much an art as music, and that he is no artist. His ill-mannered sniping, he admits, is based solely on his jealousy of her. In this one short scene, McKay not only dismisses his well founded criticism of Carter's out-of-the-box plans as emotional insecurity on his own part, but also, through his use of her rank rather than her name or her honorific which would indicate that they are intellectual equals, emphasizes her place in the military hierarchy. In effect, by acknowledging her military rank at the same time as her superiority to himself, he concedes the argument not only on a personal level but also at a thematic level, implicitly accepting the values of the military worldview. In the narrative of *Stargate SG-1*, McKay, as a temporary antagonist for the female lead, cannot serve as her scientific conscience, her reminder to listen to another part of herself than the good soldier; instead, it is he who must learn a lesson about the proper place of science within their world. This is a lesson reinforced in the *Stargate Atlantis* episode "Grace Under Pressure" (2005) when McKay's hallucination of Carter correctly advises him to trust in his teammates rather than in his own technological know-how. The scientist learns to agree with rather than criticize the values held by the military with which he is aligned.

The Soldier/Scholar Dynamic

Although the single character of the scientist can embody thematic concerns about military use of scientific knowledge, the soldier/scientist dynamic is more often articulated in the pairing of two characters, one a warrior relying on physical force, the other a scholar employing intellectual power. This is a common dynamic in science fiction television shows; even in ensemble shows with a fairly large cast of characters, it is not unusual for storylines to revolve around a "buddy" pairing of this type, as evident in the original *Star Trek* series of the 1960s with the dynamic between the action-oriented Captain James T. Kirk and his more cerebral Vulcan science officer Mr. Spock. Sometimes the dynamic crosses gender lines, as in the 1970s series *Space: 1999*, with the romantic pairings of Commander Koenig and Dr. Helena Russell, and in the second season with brawny Security Officer Tony Verdeschi and the brainy alien science officer Maya, or in the more recent series *Dark Angel*, pairing a wheelchair-bound male intellectual and a young woman bred to be a superhuman fighting machine. The brooding and physically strong vampire hero in *Angel* has his scholarly sidekick Wesley Wyndam-Price; in *The Sentinel*, Jim, a cop and former Army Special Forces, is partnered with anthropology grad student Blair, a long-haired hippie. The pairing of scholar and warrior concretizes the opposition between "book smarts" and "street smarts"; the warrior gains knowledge through experience, in the "school of hard knocks," while the scholar's knowledge is the product of education, learned in the "ivory tower." The pairing of the two characters suggests that both forms of knowledge are valued; it takes the combined effort of the soldier's hard-earned experience and the scientist's encyclopedic command of obscure information to resolve crisis situations. But the two characters are seldom presented as complete equals; more often than not, the soldier is the hero, the scientist his sidekick, his assistant. Thus the scientist is under the control of the soldier, implying an underlying cultural attitude that physical force is more powerful than intelligence.

This dynamic plays out in *Stargate SG-1* with the pairing of the civilian archaeologist Dr. Daniel Jackson and Col. Jack O'Neill, a career military officer. When military needs conflict with the pursuit of knowledge, Jackson is the one to voice objections, not Major Carter. On the spin-off series *Stargate Atlantis*, premiering in 2004, the soldier/scholar buddy dynamic pairs Major John Sheppard (Joe Flanigan), a helicopter pilot who becomes Atlantis' ranking military officer when his superior is killed on their first mission in the Pegasus galaxy, and Canadian physicist Dr. Rodney McKay, formerly the thorn in Samantha Carter's side and now the head of the science division of the Atlantis expedition. McKay's transformation from a recurring guest on *Stargate SG-1* into a regular lead on *Stargate Atlantis* would seem to require a certain amount of rehabilitation of the character, but in many ways McKay is the same arrogant and socially inept egomaniac. But now he is the one responsible for saving his team and the galaxy from certain doom. In effect, he becomes the Carter of Atlantis, a physicist tempered in the field, his discoveries subject to military needs. Unlike Carter, he is a civilian, and therefore free to speak his mind as does Jackson. Yet he does not take the same positions in the military/science debate.

This may, of course, be partly because in the forty episodes of *Stargate Atlantis* that have aired so far, the military/science debate has not been as explicit a theme as in earlier seasons of the parent show. The most overt exploration of this theme is in the second season episode "Trinity" (2005), an episode whose title, evoking the code name for the first test of a nuclear weapon, is only one of many references to the Manhattan Project and as such necessarily carries mimetic resonance for viewers. When Sheppard and McKay's team discover an ancient weapons array, Sheppard is pleased with the possibility of gaining a technological advantage against their enemy. McKay has another reason to be excited: a potential new power source, the greatest discovery of all time. Although it would seem that the weapon was a failure, McKay believes that the power source for the weapon was rushed into service before the proper testing of what "could have turned the tide of war." Almost more important, from McKay's point of view, is the effect such an achievement will have on his own career, moving him that much closer to his dream of a Nobel Prize. In a pivotal scene with Sheppard, he acknowledges this motivation as well the possibility, remote in his estimation, that the experiment could have drastic consequences for the entire universe. Their conversation takes place after their first attempt at test-firing the weapon has failed, resulting in the death by radiation exposure of one of McKay's subordinates. The scene is filmed as a series of alternating over-the-shoulder two-shots in close-up, a technique which emphasizes the close relationship between the two teammates. The close-up allows the viewers to witness the slightest change of expression on the actors' faces, while the two-shot, which keeps both actors in the frame together, enhances the sense of intimacy. The effect is that of eavesdropping on a private conversation between two people comfortable in each other's personal space.

The viewer cannot help but contrast this scene with earlier scenes in which the same issues were debated. Here, McKay is asking Sheppard for his support in convincing Dr. Weir to allow him to continue with his research, which has proven to be much more dangerous than anticipated. McKay presents Sheppard with several arguments: he wants his fellow scientist's death to have meaning, a reason he knows will satisfy Sheppard's military code of honor; the power source could provide humanity with a clean, safe form of energy, a reason which Weir the idealist will find difficult to resist; he wants to succeed where the Ancients, who represent the pinnacle of scientific understanding in the gateverse, could not; and he wants the scientific team to have priority in setting the agenda for the research which he knows the military will not abandon for longer than it takes the expedition to report back to the SGC. McKay has already made these arguments, unsuccessfully, at staff briefings also attended by Dr. Weir and by McKay's fellow physicists as well as Colonel Caldwell (Mitch Pileggi) representing the Air Force and the Pentagon. The difference here is that the debate is made personal, a matter of trust between two friends, two buddies.

In this pivotal scene, the buddy dynamic between the soldier and the scientist, a synthetic construct, dramatizes the thematic component of the interaction between the scientist who must ask for permission to pursue his research and the soldier who exercises control over that research. McKay needs Sheppard's approval and support to convince Weir to continue the testing. At issue is whether that control is justified, which necessitates a reminder of the dangers of scientific research. In another explicit reference to the Manhattan Project, a reference intended to add a mimetic component to McKay's characterization, McKay tells Sheppard the story of another experiment gone wrong in order to convince him that continued research is meaningful and even necessary despite the obvious risks. He recounts the real world experience of physicist Harry K. Daghlian, a young man working with the Manhattan Project whose slow death as a result of a lab accident contributed to the scientific community's understanding of radiation exposure. The parallels that McKay draws between Daghlian and his own subordinate would seem to cast McKay in the role of Oppenheimer, whom Haynes uses as an example not only of the amoral scientist but also of the helpless scientist. After the Trinity test and the deployment of the bomb to end the war, Oppenheimer, "appalled at the power and destruction that had resulted from their work, set forth on a crusade to educate the public about the dangers of unchecked research" undertaken at the behest of the military (Haynes, 1994, p. 256). McKay, in this episode the archetypal "mad scientist in all his overreaching, exultant, tragic glory" (Skal, 1998, p. 20), must be under Sheppard's control for society's protection, but the evocation of Oppenheimer at the mimetic level cannot help but recall concerns about that control.

Given that McKay's arrogance and overconfidence result in the destruction of five-sixths of a solar system and nearly kills both of them, the argument would seem to be that the soldier's control is necessary. McKay is chastened not by his colleagues but instead by his soldier buddy Sheppard, whose life and trust he has placed at risk. It is Sheppard's loss of faith in McKay's abilities, the impact of his failure on their relationship both professional and personal, that causes him the most pain; the mise-en-scène of the final scene echoes their earlier private conversation, as McKay apologizes to Sheppard and vows to work to regain his trust. It is Sheppard to whom McKay looks for guidance and approval, in a way that even Daniel Jackson seldom looks to O'Neill at the height of their recurring debate between military and scientific approaches. Jackson questions, criticizes, resists the military point of view as represented by O'Neill and the SGC; aside from token grumbling about the military's neglect of scientists until they're needed, McKay does not. In uniform, weapon in his holster, at his soldier buddy's side, McKay has become the mirror to Samantha Carter, regardless of his lack of military rank. The scientist is fully aligned with military values.

As a result of the partnership with the soldier, the scientist becomes more likely to search for a solution to a problem in physical rather than intellectual force. McKay, who once described his area of expertise as the type of science that one could do without leaving the comfort of a couch or chair ("The Brotherhood," 2004), ignores a laboratory full of potentially useful chemicals and equipment to inject himself with a drug that provides enough physical strength to overcome his captors in a fistfight ("The Hive," 2005). The transformation of the scientist into the soldier, of the intellectual into an agent of physical force, is not limited to the Stargate universe but is evident in other shows as well. In the final episode of *The Sentinel* (1996-1999), for instance, anthropology student Blair Sandburg (Garett Maggart), who has spent three years partnered with cop Jim Ellison (Richard Burgi) on the excuse of observing him as a research subject, declares his dissertation a hoax in order to protect Jim; as a result, he is rejected by the academic community and decides instead to attend the police academy and become a "real" partner. In another example, scholar Wesley Wyndam-Price (Alexis Denisof), introduced on *Buffy the Vampire Slayer* (1997-2003) as a prissy scholar, marked as a socially inept nerd by his glasses, tailored suits, fussy manners, and clumsiness with weapons, is transformed through several seasons of the spin-off *Angel* (1999-2004) into a scruffy, brooding, and physically competent rogue demon hunter, equally adept at dispatching a monster with a sword and looking up information in books. To some extent such a transformation is the result of character development consequent to plot events or of fan response to the actors' portrayal of the characters. But the increasing militarization of scientists in science fiction television also speaks to the helplessness of the scientist in the face of military control over scientific research, to the extent that the scientist becomes the soldier. The loss of type identity suggests the ultimate loss of control.

The Soldier Opposed to the Military

It is important, however, to make a distinction between the soldier in the soldier/scientist pairing and the military establishment which that soldier serves, particularly given that American science fiction television, as Jan Johnson-Smith (2005) points out, suffers from "the legacy of Vietnam" (p. 125), a legacy that must weigh even more heavily in the wake of 9/11 and the war on terror. Johnson-Smith argues that although television is an extremely conservative medium which has failed for the most part to address societal concerns arising from the Vietnam war and its effects on the American psyche, the serial and episodic nature of television "has the potential to deal with the traumas and complexity of an experience like Vietnam in more depth than a film ever could" (2005, p. 133). Despite this potential, few television shows directly address such concerns, choosing instead to highlight the heroics of individual soldiers who battle not only an external enemy but also the politically controlled military establishment. "Even within a military-based series like *Stargate SG-1*," writes Johnson-Smith, "both the military and political establishment en masse are generally cast in a negative light, whilst maverick ensembles and individuals . . . are held up as protectors and saviours" (2005, p. 125). Col. Jack O'Neill is a career soldier who has admitted to performing some pretty distasteful actions in the service of his country. He has a record of insubordination to superior officers and government officials, and has, on occasion, disobeyed orders. So has Major John Sheppard, whose "black mark," referred to three times during the pilot episode, almost keeps him from joining the Atlantis expedition despite his natural ability with Ancient technology which is desperately needed for the mission's success. Both are mavericks, irreverent and prone to cowboy heroics; both have problems with respecting the authority of military and political institutions. This distances the soldier-figures, to some extent, from the military position in the military/science debate, a distance further emphasized by what they have in common with their scientist buddies: O'Neill's hobby is astronomy and Sheppard is a math whiz who passed the Mensa test ("The Brotherhood," 2004). When Jackson and McKay take on more of the attributes of soldiers, it is as part of the bond with the maverick soldier at odds with the military establishment and its political masters, part of the emphasis on stories "about individual or small group victories, about individual honour and integrity" (Johnson-Smith, 2005, pp. 126-7).

Despite this slight nod to resistance, however, for the most part military control over the direction and application of scientific research is not questioned within the Stargate universe, reflecting a larger attitude in American science fiction television about the need for the physical forces of law and order to protect society from chaos and destruction. Despite the necessary benefits of scientific research to society, because of the dangerous consequences of allowing it to proceed unchecked, the scientist is controlled by the soldier, the representative of physical force as the agent of order. It is the scientist who must adapt to the military environment as a result of the pressures of wartime. Thus Samantha Carter is always Captain, then Major, then Lt. Colonel, almost never Doctor. Thus Dr. Rodney McKay, never afraid to speak his mind about the insufficiencies of others, seldom questions the uses to which the military desires to put his brilliant mind. Even Daniel Jackson, Peaceful Explorer, has been transformed over nine—soon to be ten—seasons into Action Jackson, with more muscles and more weapons experience than many young soldiers; gone is the long hair, in favor of a short military cut; gone is the plaid and corduroy, replaced by on-duty uniforms and off-duty tailored suits. While the characterization of the scientist and soldier in science fiction television and their role in the thematic exploration of the military/science debate does reveal a certain amount of discomfort with the political establishment and its systems for enforcing security, particularly in these difficult times of the war on terror, that discomfort is seldom sufficient to suggest that perhaps, occasionally, the soldier just might be more of a threat than the scientist under his control.

References

Bilson, D. & De Meo, P. (Executive Producers). (1996-1999). *The Sentinel* [Television series]. UPN.

Binder, C. (Writer), & Wood, M. (Director). (2005). The Hive [Television series episode]. In B. Wright & R. C. Cooper (Executive Producers), *Stargate Atlantis*. Sci-Fi Channel.

Cooper, R. C. (Writer), & Woeste, P. (Director). (2001). 48 Hours [Television series episode]. In B. Wright & J. Glassner (Executive Producers), *Stargate SG-1*. Showtime.

Cooper, R. C. (Writer), & Wood, M. (Director). (2002). Redemption Part 2 [Television series episode]. In B. Wright & J. Glassner (Executive Producers), *Stargate SG-1*. Sci-Fi Channel.

Gero, M. (Writer), & Wood, M. (Director). (2004). The Brotherhood [Television series episode]. In B. Wright & R. C. Cooper (Executive Producers), *Stargate Atlantis*. Sci-Fi Channel.

Glassner, J. & Wright, B. (Writers), & Azzopardi, M. (Director). (1997). Children of the Gods. [Television series episode]. In B. Wright & J. Glassner (Executive Producers), *Stargate SG-1*. Showtime.

Haynes, R. D. (1994). *From Faust to Strangelove: Representations of the scientist in western literature*. Baltimore & London: Johns Hopkins University Press.

Johnson-Smith, J. (2005). *American science fiction tv:* Star Trek, Stargate *and beyond*. Middletown, CT: Wesleyan University Press.

Kindler, D. (Writer), & Wood, M. (Director). (2005). Trinity [Television series episode]. In B. Wright & R. C. Cooper (Executive Producers), *Stargate Atlantis*. Sci-Fi Channel.

Mallozzi, J. & Mullie, P. (Writers), & Wood, M. (Director). (2000). Chain Reaction [Television series episode]. In B. Wright & J. Glassner (Executive Producers), *Stargate SG-1*. Showtime.

Mallozzi, J. & Mullie, P. (Writers), & Wood, M. (Director). (2000). Scorched Earth [Television series episode]. In B. Wright & J. Glassner (Executive Producers), *Stargate SG-1*. Showtime.

Mallozzi, J. & Mullie, P. (Writers), & Wood, M. (Director). (2004). The Siege Part 2 [Television series episode]. In B. Wright & R. C. Cooper (Executive Producers), *Stargate Atlantis*. Sci-Fi Channel.

Phelan, J. (1989). *Reading people, reading plots: Character, progression, and the interpretation of narrative*. Chicago and London: University of Chicago Press.

Skal, D. J. (1998). *Screams of reason: Mad science and modern culture*. New York and London: W. W. Norton & Co.

Toumey, C. P. (1992). The moral character of mad scientists: A cultural critique of science. *Science, technology, & human values*. 17.4 (Autumn, 1992): pp. 411-437.

Whedon, J. (Executive Producer). *Angel*. (1999-2004). WB.

Wright, B. (Writer), & DeLuise, P. (Director). (2000). The Other Side [Television series episode]. In B. Wright & J. Glassner (Executive Producers), *Stargate SG-1*. Showtime.

Wright, B. & Rashovich, M. (Writers), & Kaufman, J. (Director). (1998). A Matter of Time [Television series episode]. In B. Wright & J. Glassner (Executive Producers), *Stargate SG-1*. Showtime.

Zanger, J. & Wolf, R. G. (1986). The disenchantment of magic. In J. Hokenson & H. Pearce (Eds.), *Forms of the fantastic: Selected essays from the third international conference on the fantastic in literature and film* (pp. 31-40). Contributions to the study of science fiction and fantasy 20. Westport, CT: Greenwood Press.

Chapter Nine

The *CSI* Effect: Scientists and Priming on Prime Time Television

Gary R. Pettey and Cheryl Campanella Bracken

When considering Newt Minnow's *vast wasteland*, displays of intelligence may not be the first programs or character depictions that come to mind. But, periodically, intelligent, educated people doing serious and thoughtful activities have been given airtime and even found an audience. Arguably very few prime time shows have been populated with educated technicians and actual scientists doing their work as *CSI: Crime Scene Investigation*. Fewer still have been credited with having actual societal effects.

CSI: Crime Scene Investigation debuted in the fall of 2000 on CBS. In its first season it finished at the number 12 rated show. In the three seasons since it has been either the number 1 or number 2 rated show. It has two spin-offs: *CSI: Miami* and *CSI: New York*. News reports over the last four years have described the "*CSI* Effect." Lawyers report they have to explain to jurors how certain tests cannot be done and why (Willing, 2004; Rincon, 2005). Police are questioned about their forensic protocols (Lovgren, 2004). Another aspect of the "*CSI* Effect" has been the interest and growth in forensic education across the country (Lewerenz, 2003; Lemaine, 2004). This show perhaps more than most portrays the science itself as the star of the show. The science, more than the police officers, solves the crimes.

Priming has normally been associated with non-fiction political information—essentially news and opinion information (Roskos-Ewoldson, Roskos-Ewoldson, & Carpentier, 2002). Recently, priming has been expanded to fictional presentations of political information. Holbert et al. (2003) used the NBC show *The West Wing* as a priming mechanism to see if the portrayal of the fictional president Josiah Bartlet influenced individual's image of the U.S. Presidency. Further, they demonstrated that the watching of the fictional show could influence the perception of actual holders of the office. Their findings suggest that priming effects should not be limited to news and news coverage, but rather the fictional world may well have impact in non-fictional evaluations.

If portrayals in *The West Wing* have a priming effect on the evaluation of real world presidents, shouldn't the positive portrayals of science and scientists on *CSI* influence how viewers perceive actual science, scientists and their role in society?

Science, Scientists and *CSI: Crime Scene Investigation*

The presentation of science and its practitioners have periodically been the subject of research. Traditionally scientists have been portrayed as creator/destroyers (*Frankenstein*), evil geniuses (*The Fugitive*), heroes (*Apollo 13*), wizards (various Einstein portrayals) or experts (*CSI*) (Steinke & Long, 1996; Steinke, 1999; LaFollette, 1990; Basalla, 1976). Nurses have been presented as pretty and powerless in advertising (Lusk, 2000), psychoanalysts as ditsy or evil in film (Greenberg, 2000; Nairn, 1999). Televised portrayals of surgeons have been shown to be distorted and to have negatively influenced attitudes of new medical students (Kozar, et al., 2004).

Some studies suggest that these portrayals of science and scientists may lead to misunderstanding and even distrust of the science itself. Nisbet et al., (2002) concluded that general television viewing appeared, in Gerbner's terms, to cultivate reservations about science. While the presentation of science as strange and scary can lead to its distrust, Nisbet et al. (2002) do note that television is not monolithic and has many positive presentations of science and knowledge as well. Further, they speculate that as exposure to printed science and televised science leads to increase levels of trust in and knowledge about science and its procedures, then heavy users are possibly displacing real discussions of science in print and on television with heavy television use. They argue that their findings support the notion that "science-specific programming can have positive benefits for public understanding" (p. 604).

Depictions of science and scientists as problem solvers are not new. Some studies suggest that it is often the job of the televised scientist to apply reason to mystery (Hornig, 1990; Nelkin, 1995). They are often portrayed as an elite group—priests robed in white coats (Shortland, 1988; Hornig, 1990; Long & Steinke, 1996). This type of portrayal is sometimes criticized by scientists. They have argued that such distorted presentations often imply that scientists are the "truth's ultimate custodian" ("How not to respond to *The X Files*," 1998, p. 815).

CSI: Crime Scene Investigation, and its spin-offs *CSI: Miami* and *CSI: New York*, present a set of scientists and technicians as heroes in the pursuit of truth. Certainly, the impact of the CBS Show is strong. It finished as the #12 rated show in its first season (2000-2001). In the three full seasons since, it was either the #1 or #2 rated show. Almost immediately, the show went into reruns on cable networks. Currently, it runs on prime time five nights a week on *Spike T.V.*

with two episodes Monday through Thursday and four episodes on Friday. The opportunity to consume the images and messages of the show, then, are extensive. The "*CSI* Effect" is part of the cultural vernacular, even if the reality of forensics is somewhat different than its portrayal on the show ("Classes give budding CSIs a reality check," 2005). Further, the program provides viewers with information about the functioning of crime labs, and provides scientific explanations for making decisions and to interpret crime scene findings.

Priming

McCombs and Estrada (1997) revised the traditional Agenda Setting summary: "The media may not only tell us what to think about, the may also tell us how and what to think about it and even what to do about it" (p. 247). Priming is seen by many scholars as related to or a part of Agenda Setting (Iyengar & Kinder, 1987; Comstock & Sharrer, 1999; Scheufele, 2000).

Kiousis and McCombs (2004) argue that the salience of Agenda Setting operates at two levels: Object Salience and Attribute Salience. They argue that media attention toward figures, issues and organizations (Object) is the first level. These media treatments, they argue, carry with them attitudinal implications (Attitude). They found a strong relationship between the amount of media coverage of political figures and the public attitudes toward those actors.

Scheufele (2000) argues that priming relates specifically to the cognitive process of mediated information. When information is processed "individuals develop memory traces (Tulving & Watkins, 1975) or activation tags (Collins & Loftus, 1975)" (p. 299). These traces or tags are then used in the processing of future information. This is, of course, consistent with Roskos-Ewoldson, Roskos-Ewoldson and Carpentier's (2002) definition of priming in mass communication research as "the effects of the content of the media on people's later behavior or judgments related to the content" (p. 97).

Entertainment Television and Priming

While mass communication has looked at priming within a news and public affairs context, probably because of its Agenda Setting relationships, priming can be used to examine media effects in other areas. Some studies have argued that entertainment media may play a role in shaping opinion about actors and issues in the news (Holbert, Shah & Kwak, 2003; Shah, 1998). Policy and public issues are not limited to strictly political settings. Entertainment content could, indeed, shape attitudes in other important areas. Roskos-Ewoldson et al. (2002) describe how the mental model approach is a "flexible framework for an academic understanding of the media" (p. 114).

An example of priming and non-political content is a study by Segrin and Nabi (2002) who looked at consumers of romantic television programming and their attitudes toward marriage. They argue that consumption of that specific content created a mental model (tags or traces) of marriage. Other work in non-political/current events content includes an experiment by Holbert, Shah and Kwak (2003) using implicit priming/mental model argument to examine prime time media use, particularly drama and sitcoms, and attitudes toward women's rights. They found, even when controlling for situational, demographic and orientation variables, significant relationships between consumption of media portrayals and attitudes.

These studies provide evidence that use of priming as a theoretical grounding for exploring the impact of entertainment content on media audiences is not only possible but provides insightful results. The extension of priming into entertainment contexts has been encouraged by Holbert, Pillion, Tschida, Armfield, Kinder, Cherry and Daulton (2003). It has been suggested by Holbert et al. (2003) that investigation into the presentation of political communication presented in a news format versus information presented in entertainment programming may reveal priming effects.

This movement of priming research into the entertainment side of television was clear in a study of how *The West Wing* shaped attitudes toward real presidents George W. Bush and William Clinton. Holbert et al. (2003) found that subjects who were exposed to an episode of the network show generally, though not always significantly, saw Bush and Clinton as more principled, engaging and common. Providing even further evidence that the priming effect is not limited to political/news content. The authors explain that priming may work with entertainment content that allows television viewers to see fictional portrayals of positions/characters that are not usually seen by audiences or in the real world (i.e., backstage views) Meyrowitz (1985). They argue that the fictional President Josiah Bartlet, in Goffman's terms, rounds out or displays important elements in the backstage more easily than normal presentations of real presidents in the frontstage (Goffman, 1959; Meyrowitz, 1985; Holbert, Tschida, Dixon, Cherry, Steuber, & Airne, 2005). They further note that their research supports Parry-Giles and Parry-Giles (2002) where they argued that the public's exposure to shows like *The West Wing* could raise confidence in the U.S. Presidency in general.

Obviously, scientific findings and capabilities are part of public policy and public issues. From global warming to evolution to stem cell research questions of science are covered continually on all media. These issues are not limited to the news or public affairs content, but rather are featured in entertainment shows from dramas to situation comedies. Thus, the researchers feel that *CSI* provides television audiences a glimpse into the backstage (Meyrowitz, 1985) of crime scene investigation in a similar fashion to *West Wing* and the presidency.

Methodology

In a between-subject experiment, participants were randomly shown one of two programs: Subjects who received the manipulation viewed an episode of *CSI: Crime Scene Investigation*; the control group viewed an episode of *The West Wing.*

Participants

Undergraduate students were recruited from several Communication courses to participate. One hundred and twenty-six undergraduate students were given extra credit from their instructors for participation.

Stimulus

The program used for the manipulation was *CSI: Crime Scene Investigation.* The episode used was the fifth episode of the first season. "Friends and Lovers" originally aired on Nov. 3, 2000. Anticipating heavy viewership of the program among the subjects, it was decided to use an early episode of the show. The first four episodes spent more time with character introduction than using science to solve crimes. The fifth episode was the first where no new characters were introduced. An episode guide describes the program:

> Grissom investigates the death of a naked young man found in the desert. A woman who was buried the week before is no longer in her grave, but instead is found in a dumpster. A young woman claims self defense when a college dean is found dead. (http://members.aol.com/JRD203/csi-episodes.htm#S1).

For the control group, the *West Wing* pilot episode was chosen. It first aired on Sept. 22, 1999 and is described as:

> The entire White House staff bristles with activity when it's learned that the President injured himself during a bicycle accident, and his absence becomes a factor as chief of staff Leo McGarry must juggle a host of impending crises, including a mass boat lift of Cuban refugees approaching the Florida coast and the reaction of conservative Christians to a controversial televised comment by deputy chief of staff Josh Lyman. Meanwhile, Sam Seaborn, the trouble-prone deputy communications director, unknowingly spends the night with a call girl and then makes another critical error during a children's White House tour. http://www.westwingepguide.com/S1/Episodes/1_PILOT.html.

The programs were similar in length. Both are contemporary, serious dramas of similar production values. At the core of each is examination of the inner workings of complex professions that are central to our culture.

After viewing the program participants filled out a revised National Science Foundation (NSF) public survey. Participants answered questions about their interest in and attitudes about a variety of issues. Questions from the NSF study also ask about science media and fictional program use. For this study we added questions about the frequency of viewing of any of the *CSI* programs. The questionnaire took about twenty minutes to complete. Overall, the study took approximately one hour.

Measures

Nisbet et al. (2002) used measures developed by Miller et al. (1997) and Miller and Kimmel (2001) suggested by the NSF surveys. The two attitude scales each contain four-question Likert scales. This study used the 2001 NSF Public Attitudes survey to measure two conceptual domains: individuals 1) *Reservations concerning science and technology* and their 2) *Beliefs in the promise of science.* Further using measures from the NSF survey, this study will examine the attitudes toward government spending for science. As control, we measured both the level of science knowledge and the level of knowledge about scientific procedures, as well as the number high school and college science and math classes of subjects.

Reservations concerning science and technology is an index measured by combining four items: Science makes our way of life change too fast; On balance, the benefits of science research have outweighed the harmful results; We depend too much on science and not enough on faith; and It is not important for me to know about science in my daily life.

Belief in the promise of science is an index measured by combining four items: Because of science and technology, there will be more opportunities for the next generations; Science and Technology are making our lives healthier, easier and more comfortable; Most scientists want to work on things that will make life better for the average person; and With the application of science and new technology, work will become more interesting.

If participants are exposed to positive messages about scientists as regular people and science as a human endeavor, they should create positive or affirming connections or traces toward science and scientists in general.

Hypotheses and Research Questions

H1: Mean scores for Beliefs in the promise of science will be higher for viewers of the *CSI* episode than for the control group.

H2: Mean scores for the Reservations concerning science and technology will be lower for the *CSI* viewers than for the control group.

RQ1: Is there a difference between heavy viewers of *CSI* and infrequent viewers of *CSI* in their view of science or scientists?

RQ2: Do heavy *CSI* viewer use more science media the lighter viewers?

RQ3: Do heavy *CSI* viewers know more about science and science procedures than lighter viewers?

Results

Promise of Science

Table 9.1 presents the means, standard deviations and ANOVA comparing *CSI* and the control groups. The alpha for the scale (and for Reservations about Science) is low (Promise, α =.42; Reservations α =.15). Because of this, the elements of the scale are also presented. Hypothesis 1 posited that participants who watched *CSI* will report a higher belief in science than participants who watched the control. There was a significant difference between *CSI* and the control group: *CSI* M = 26.33; control M = 24.24. A single exposure to an entertainment program about scientists at work resulted in a small but significant increase in a positive evaluation of science in general. Each of the four individual items also showed that pattern, though only *Science and Technology are making our lives healthier, easier and more comfortable* showed a significant difference.

Table 9.1. ANOVA, Means and Standard Deviations for the Promise of Science and the components of the Scale: Comparing *CSI* to *West Wing*

	M	*SD*	*F*	*p*
Promise of Science				
CSI	26.33	5.25	5.22	0.024
WW	24.24	5.04		
Makes Lives Healthier, easier				
CSI	7.21	1.91	7.91	0.006
WW	6.25	1.89		

Table 9.1. continued

		M	*SD*	*F*	*p*
Opportunities					
	CSI	7.03	2.06	3.02	0.085
	WW	6.33	2.44		
Life Better					
	CSI	6.29	2.31	0.061	0.81
	WW	6.19	2.02		
Work more interesting					
	CSI	5.81	2.21	0.727	0.396
	WW	5.46	2.38		

Reservations of Science

Table 9.2 presents the means, standard deviations and ANOVA comparing *CSI* and the control group for Reservations about Science and the component items. Hypothesis 2 posited that participants who watched *CSI* will report less reservation concerning science and than for the control group. There was a significant difference between *CSI* and the control group: *CSI* $M = 17.14$; control $M = 19.54$. A single exposure to an entertainment program about scientists at work resulted in a small but significant decrease in a negative evaluation of science in general. Each of the four individual items also showed that pattern. Two items also show significant differences: *It is not important for me to know about science in my daily life* and *Science makes our way of life change too fast.*

Table 9.2. ANOVA, Means and Standard Deviations for Reservations toward Science and the components of the Scale: Comparing *CSI* to *West Wing*

		M	*SD*	*F*	*p*
Reservations toward Science					
	CSI	17.14	5.02	6.56	0.01
	WW	19.54	5.41		
Depend too much					
	CSI	4.25	2.67	1.09	0.29
	WW	4.75	2.63		
Science not important					
	CSI	3.33	2.42	4.82	0.03
	WW	4.27	2.37		

Table 9.2. continued

		M	*SD*	*F*	*p*
Makes life move too fast					
	CSI	4.21	2.16	2.93	0.00
	WW	4.98	2.89		
Provides General Benefits (Reversed)					
	CSI	5.35	2.35	0.201	0.66
	WW	5.54	2.44		

High Versus Low CSI Viewing

There was no difference between experimental and control groups in their viewing of *CSI*. Table 9.3 presents the results.

Table 9.3: Means, Standard Deviations and F tests for the Frequency of *CSI* viewing

		M	*SD*	*F*	*p*
How often do you watch CSI on network television?					
	CSI	2.33	2.35	0.001	0.97
	Control	2.32	2.42		
How often do you watch CSI in rerun on cable television?					
	CSI	2.02	2.22	0.002	0.96
	Control	2.03	1.96		
How often do you watch *CSI* spin-offs?					
	CSI	2.19	2.03	1.79	0.184
	Control	1.75	1.68		

Research Questions

To look at differences between high and low viewers of *CSI*, two questions were examined: *How often do you watch CSI on network television*? and *How often do you watch CSI in reruns on cable television?* These were measured on 1 to 10 scales with *Never* at one end and *Every Week* (*More than once a day*) on the other. When the two scales are added together, over 60 percent said they *Never* watched *CSI*. Two groups were formed. One group had those with Never watch (N=76) and the other group consisted of people who watch some amount (N=50). Only 15 subjects had a combine score over 10, only one person had a combined score of 20. Comparisons between these groups, then, are somewhat problematic. T-tests were used.

Research Questions 1 explored whether there was a difference between heavy viewers of *CSI* and infrequent viewers of *CSI* in their view of science or scientists. No mean differences were found on items that asked about opinions about scientists and their work.

In the second research question, heavy *CSI* viewers were compared to light viewers to see if there were differences in use of science media. Again, no mean differences were found between high and low CSI viewer on science media use, whether fiction (*X-Files*, *Star Trek*) or non-fiction (*NOVA*, *National Geographic*).

Research question 3 investigated any differences in knowledge about science and science procedures that might exist between heavy *CSI* viewers and lighter viewers. Only four of twenty-seven questions about science and science procedures showed significant differences between high and low *CSI* viewers. High *CSI* viewers were more likely to know what DNA is, what a molecule is, that the Earth goes around the sun. Further, they were less likely to believe that astrology had scientific merit.

Overall, there are not consistent patterns of differences between higher and lower viewers of *CSI*. Future research could seek out actual high *CSI* viewers, as this study found few of them.

Discussion

The current study provides further evidence that there is a priming effect beyond political/current events content. Specifically, the study found evidence that audiences who were exposed to *CSI* are more positive towards both science and scientist than participants who saw the control (*The West Wing)*. The results of this study extend priming into a pure entertainment context, thus strengthens the argument presented Holbert et al. (2003) that priming can be extended beyond the traditional political communication into examining the effects of entertainment programming.

The specific results of this exploratory study while limited are nonetheless promising. In only one exposure to *CSI*, and for many participants it was their only exposure, to find significant differences in attitudes towards science and scientists suggests that a cumulative effect might be even greater. As suggested by Cultivation, these results are similar to the Holbert et al. (2003) results in that the current study has similar findings regarding seeing the characters as more human and therefore as more likable.

The theoretical implications of these results are several-fold. First, these results contribute to a growing body of evidence that priming exists outside of political/current event content. Second, the mechanisms behind priming can be triggered by entertainment content or that viewers of entertainment are involved enough in the content for priming to occur. Third, and perhaps the most powerful is that priming can occur with entertainment with minimal exposure.

The implications of priming effects existing with one exposure, especially with content where the conveying of information is not the primary function of the program, has practical value for edutainment programming. This use of priming in such a context may provide a theoretical explanation for how *edutainment* may work.

Being that the current study is exploratory, the authors acknowledge that these results are limited. The use of the single episode may be problematic in that people tend to be repeat viewers of programming they enjoy. However, the design of this study was intended to investigate the possibility of *CSI* impacting audiences' perceptions of science. Since there is some support provided by this single exposure it seems likely that continuous exposure to this show and its spin-offs would continually re-enforce (and prime) the positive view of science and scientists found within this single episode manipulation.

Another issue is the lack of variance between frequent and infrequent viewers of *CSI*. This can be controlled for in future studies. The fact that 60 percent of the participants had never seen *CSI*, one of the most popular shows on network television over the last few years, reaffirms studies that college age students watch relatively little television. While the current study has expanded the use of priming to include non-political entertainment, the authors feel this is only really the first step into such priming effects (outside of political/current event content). One example includes the possible priming effect on audience of Edutainment. Other future investigations might explore the impact of the incorporation of current/recent crimes into programs such as *Law and Order* (or any of their spin-offs).

Conclusion

This exploratory study extends the use of priming into non-political content. The results suggest that television audiences can be primed to view, in this case, science and scientists more favorably depending on the treatment within an entertainment program. Further, these results suggest that future research into entertainment content and priming may prove fruitful.

References

Basalla, G. (1976). Pop science: The depiction of science in popular culture. In G. Holton & W. Blanpied (Eds.), *Science and its public* (pp. 261-278). Boston: Beacon.

"Classes give budding CSIs a reality check: Popular shows give unrealistic view of tough job" (2005) Associated Press. MSNBC. http://www.msnbc.msn.com/id/7762090/

Collins, A. M., & Loftus, E. F. (1975). A spreading-activation theory of semantic processing. *Psychological Review*, 82, 407-428.

Comstock, G. & Scharrer, E. (1999). *Television: What's on, who's watching, and what it means*. San Diego: Academic Press.

Gerbner, G., Gross, L., Morgan, M., & Signorelli, N. (1981). Scientists on the TV screen. *Culture and Society*, *42*, 51-54.

Goffman, E. (1959). *The presentation of self in everyday life*. New York: Anchor Press.

Greenberg, H. R. (2000) A field guide to cinetherapy: On celluloid psychoanalysis and its practitioners, *The American Journal of Psychoanalysis*, *60*(4), 329-339.

Holbert, R. L., Benoit, W. L., Hansen, G. J., & Wen, W. C. (2002). The role of communication in the creation of an issue-based citizenry. *Communication Monographs, 69,* 296-310.

Holbert, R. L., Kwak, N., & Shah, D. V. (2003). Environmental concern, patterns of television viewing, and pro-environmental behaviors: Integrating models of media consumption and effects. *Journal of Broadcasting & Electronic Media, 47,* 177-196.

Holbert, R. L., Shah, D. V., & Kwak, N. (2003). Political implications of prime-time drama and sitcom use: Genres of representation and opinions concerning women's rights. *Journal of Communication, 53,* 45-60.

Holbert, R. L., Pillion, O., Tschida, D. A., Armfield, G. G., Kinder, K., Cherry, K. L, & Daulton, A. R. (2003). *The West Wing* as endorsement of the U.S. Presidency: Expanding the bounds of priming in political communication, *Journal of Communication*, *53*(3), 427-443.

Holbert, R. L., Tschida, D. A., Dixon, M., Cherry, K., Steuber, K., & Airne, D. (2005). *The West Wing* and depictions of the American presidency: Expanding the domains of framing in political communication. *Communication Quarterly, 53* (4), 0146-3373.

Hornig, S. (1990). Television's *Nova* and the construction of scientific truth. *Critical Studies in Mass Communication*, *7*, 11-23.

How not to respond to *The X-Files*. (1998). *Nature, 394*, 815.

Kiousis, S. & McCombs, M. (2004). Agenda-Setting effects and attitude strength: Political figures during the 1996 presidential election. *Communication Research*, 31, 1, 36-57.

Kozar, R. A., Anderson, K. D., Escobar-Chaves, S. L., Thiel, M. A., & Brundage, S. I. (2004). Preclinical Students: Who are surgeons? *Journal of Surgical Research, 119, 113-116.*

Iyengar, S. (1987). Television news and citizens' explanations of national affairs. *American Political Science Review, 81*, 815-831.

Iyengar, S., & Kinder, D. R. (1987). *News that matters: Television and American opinion.* Chicago: University of Chicago Press.

LaFollette, M.C. (1990). *Making science our own: Public image of science 1910-1955.* Chicago: University of Chicago Press.

Lemaine, A. (2004). CSI spurs campus forensic scene. *The San Diego Union-Tribune*, September 13, 2004.

Lewerenz, D. (2003). CSI spawns interesting forensic science at college level. *Houston Chronicle*, August 28, 2003.

Lovgren, S. (2004). CSI Effect is mixed blessing for real crime labs. *National Geographic News*, September 23, 2004.

Long, M., & Steinke, J. (1996). The thrill of everyday science: Images of science and scientists on children's educational science shows in the United States. *Public Understanding of Science, 5*, 101-120.

Lusk, R. (2000). Pretty and powerless: Nurses in advertisements, 1930-1950. *Research in Nursing & Health, 23*, 229-236.

McCombs, M. & Estrada, G. (1997). The news media and the pictures in our heads. In S. Iyengar and R. Reeves (Eds.) (pp. # -#), *Do the Media Govern?* London: Sage.

Meyrowitz, J. (1985). *No sense of place: The impact of electronic media on social behavior.* New York: Oxford University Press.

Miller, J. D. & Kimmel, L. G. (2001). *Biomedical communications: Purposes, audiences, and strategies.* New York: Academic Press.

Miller, J. D., Pardo, R. & Niwa, F. (1997). *Public perceptions of science and technology: A comparative study of the European union, the United States, Japan, and Canada.* Madrid: BBV Foundation.

Mutz, D. C. (2001). The future of political communication research: Reflections on the occasion of Steve Chaffee's retirement from Stanford University. *Political Communication, 18*, 231-236.

Nairn, R. (1999). Does the use of psychiatrists as sources of information improve media depictions of mental illness? A pilot study. *Australian and New Zealand Journal of Psychiatry, 33*, 583-589.

Nelkin, D. (1995). *Selling science: How the press covers science and technology.* New York: Freeman.

Nisbet, M.C., Scheufele, D.A., Shannahan, J., Moy, P., Brossard, D., & Lewenstein, B.V. (2002). Knowledge, reservations, or promise? A media effects model for public perceptions of science and technology. *Communication Research, 29*, 584-698.

Parry-Giles, T., & Parry-Giles, S. J. (2002). *The West Wing*'s prime-time Presidentiality: Mimesis and catharsis in a postmodern romance. *Quarterly Journal of Speech, 88*, 209-27.

Price, V., Tewksbury, D. & Powers, E. (1997). Switching trains of thought: The impact of news frames on readers' cognitive responses. *Communication Research, 24*, 481-506.

Rincon, P. (2005). CSI shows give "unrealistic view." *BBC News* 2005/02/21/13:57:32 GMT. http://newsbbc.co.uk/go/pr/fr/-/1/hi/sci/tech/4284335.stm.

Roskos-Ewoldsen, D.R., B. Roskos-Ewoldsen & F.R. Dillman Carpentier (2002). Media priming: A synthesis. In J. B. Bryant & D. Zillmann (Eds.), *Media effects in theory and research* (2nd ed.) (pp. 97-129). Mahwah, NJ: Lawrence Erlbaum.

Scheufele, D. A. (2000). Agenda-setting, priming, and framing revisited: Another look at cognitive effects of political communication. *Mass Communication & Society, 3*, 297–316.

Segrin, C. & Nabi, R.L. (2002). Does television viewing cultivate unrealistic expectations about marriage? *Journal of Communication, 52*, 247-263.

Shah, D. V. (1998). Civic engagement, interpersonal trust, and television use: An individual level assessment of social capital. *Political Psychology, 19* (3), 469-496.

Shortland, M. (1988, July). *Mad scientists and regular guys: Images of the expert in Hollywood films of the 1950's.* Proceedings of the Joint Meeting of the British Society for History of Science and the History of Science Society, Manchester, England.

Steinke, J. (1999). Women scientist roles models on screen: A case study of contact. *Science Communication, 21*, 111-136.

Steinke, J. & Long, M. (1996). A lab of her own? Portrayals of female characters on children's educational science programs. *Science Communication, 18*, 91-115.

Tulving, E. & Watkins, M. J. (1975). Structure of memory traces. *Psychological Review, 82*, 261-275.

Willing, R. (2004). CSI effect has juries wanting more evidence. *USA Today* August 5, 2004

Part IV

Talk Shows and Reality Television

Chapter Ten

Media-Constructed Anti-Intellectualism: The Portrayal of Experts in Popular U.S. Television Talk Shows[1]

Lisa Holderman

In myriad ways, mass media consistently convey stories about scholarly intelligence, most of which reflect, either overtly or subtextually, an anti-intellectual philosophy (Ross, 1989, 1990). Although common sense tells us intelligence is generally regarded as an asset in our society, this may not always be the case (Hofstadter, 1964). In fact, there are important contexts in which scholarly intelligence (that is, intelligence and erudition gained through academic study) is portrayed as a social drawback. The power elite, namely those individuals and groups with social and economic power, are invested in perpetuating this view of intelligence to maintain the social order and retain dominance and control.

For example, popular lore distinguishes between "book smart" and "street smart." This distinction, which may systematically exist only in fiction, attests to the relative ineffectuality of scholarly intelligence. In both comedy and drama, for example, the intellectual is routinely shown to be a social cripple in contrast to those who are not burdened with "useless" book smarts. Various media studies and analyses (e.g., Gerbner, 1974; Thomas & Krippendorf, 1988) show fictional characters with intellectual powers are "balanced" by formidable weaknesses in other areas of their lives, particularly romance and sex.

Consider the countless, yet unquestioned, media portrayals of the "nerd" whose intelligence is invariably matched with a lack of good looks, sexual interactions, or other desirable characteristics. Studies show scientific intellectuals are often portrayed in the media as isolated and uninvolved with "real" society (Gerbner, Gross, Morgan, & Signorielli, 1981), as mad or corrupt (Tudor, 1989), powerless (Goldman, 1989), antisocial (Long & Steinke, 1996), or "geeks" (Nelkin, 1990).

In some popular portrayals, intelligent characters' deficiencies are less obvious than others. For example, FBI agents Mulder and Scully of *The X-Files* are both smart and attractive, yet are shown to have very few friends and even fewer romantic interactions. The popular film *Good Will Hunting* delivers the same anti-intellectual message in a different way as Will, the handsome natural-born genius, relinquishes an opportunity to be a noted scholar in order to pursue a romantic relationship.

Media Portrayals of Scholarly Intelligence and Social Control

It is not difficult to interpret why scholarly intelligence may be portrayed negatively in the mass media. In terms of broad social control, individuals who analyze carefully, think systematically, compile evidence, and, most importantly, meticulously survey the environment to detect both problems and solutions, may be dangerous elements to the established order. Despite apparent support for education in general, it may be argued that the social order is least challenged if citizens (*particularly* members of the lower classes most likely to be dissatisfied) are removed from and perhaps even hostile to erudition. Mass media, among other social forces and institutions, play an important role in this hegemonic process.

For decades, some media scholars have postulated that mass media shape the worldviews of their audiences. Cultivation theorists assume that television, in particular, "both (selectively) mirrors and leads society" (Morgan & Signorielli, 1990, p. 13) and that patterned representations and narratives on television operate as a form of social control and status quo maintenance (Shanahan & Morgan, 1999; Wober & Gunter, 1988). Gerbner (1990) describes these patterns as the basic "'building blocks' of the television world" which "expose large communities over long periods of time to a coherent structure of conceptions about life and the world" (p. 255). The proliferation of certain images and the relative absence of others in both fictional and fact-based media "teach" mass audiences, among other things, what and whom they should value and what they should expect from their lives.

Given this theoretical grounding, how intelligence is demonstrated, portrayed, and evaluated in the public sphere is especially significant. As Foucault (1976, 1977c) and others have argued, political power is intimately intertwined with the definition and distribution of knowledge. Knowledge, intelligence, and expertise are currency in the economics of power; issues of who knows what and how much, and who is and is not assigned to store the knowledge are crucial ingredients to understanding the political economy of intelligence. Within this power perspective intelligence and knowledge, like wealth, are often degraded or mystified in popular representations so as to make what *is* powerful less attractive or meaningful (see Thomas, 1986). It would seem, therefore, very important to explore popular instances of the definition and distribution of knowledge and intelligence.

At issue here is not if people should be wary of intellectuals nor should one assume that all intellectuals are credible sources working toward to the greater good of society. Indeed, the role of intellectuals as political and social agents has been questioned by scholars and philosophers throughout history and has resulted in a range of opinions (see, for example, Bauman, 1987; Benda, 1928; Berger & Luckman, 1966; Brym, 1980; Eyerman, Svensson, & Soderqvist, 1987; Durkheim, 1951, 1895/1982; Foucault, 1976, 1977a, 1977b; Gouldner, 1979; Gramsci, 1971; Hofstadter, 1964; Huzar, 1960; Konrad & Szelenyi, 1979; Mannheim, 1949; Martin, 1987; Marx & Engels, 1970; Shils, 1972; Wald, 1987), However, such a debate is not appropriate for this paper. What *is* at issue here is an analysis of a particular pattern of mediated portrayals that both reflects and propagates the notion that scholarly intelligence is not a condition for which one should strive.

While many contexts are available for the investigation of how intelligence is treated in the media, an examination of experts on popular television talk shows is particularly significant. While the talk show genre appears in many forms this paper focuses on the popular daytime syndicated shows described by Himmelstein (1984) as "video-talk-trial" programs. These shows combine a host, a panel of lay people and experts, and an audience to discuss social issues ranging from interpersonal, family, and sexual concerns to racial, political, and abuse problems. Munson (1993) claims these talk shows serve important functions as "advice-giver, ersatz community, entertainer, and promoter" (p. 3). Because millions of people watch talk shows daily (Carbaugh, 1990), talk-show experts have increasingly assumed a greater role as advisors and authorities in American life. The talk show forum may illustrate the most public model of how to interact with intellectual experts and to use or abuse the knowledge they produce.

Although "expertise" in a particular area and "intelligence" are not necessarily synonymous, an obvious and important linkage between the two phenomena is particularly important in this study. Popular US talk shows often define and label experts as specialists who possess superior knowledge, clearer insights, and/or better skills with respect to a given theme. More often than not, talk-show themes deal with behavioral, sociological, and/or medical matters, and the experts are, in fact, generally represented (by the host, at least) as being more intelligent about the matters on which they are being asked to comment. That is, ostensibly, experts are never invited on the shows to be made objects of derision; rather, they are brought on the shows to provide a clearer understanding and impart rationality, if not wisdom and credibility, to the show (Heaton & Wilson, 1995). Thus, while intelligence is a broader concept than expertise, TV talk-show experts are popular representations of certain important aspects of intelligence in American culture.

Talk-Show Experts

Although the popularity of talk shows over the past decade has spawned a great deal of scholarly interest (Abt & Mustazza, 1997; Brinson & Winn, 1997; Carbaugh, 1988, 1990, 1993; Cerulo, Ruane, & Chayko, 1992; Epstein & Steinberg, 1996; Greenberg, Sherry, Busselle, & Rampoldi-Hnilo, 1996; Greenberg, Sherry, Busselle, Rampoldi-Hnilo, & Smith, 1997; Hoynes & Croteau, 1991; Kurtz, 1996; Livingstone, 1994; Livingstone & Lunt, 1992, 1994; Munson, 1993; Nelson & Robinson, 1994; Nolan & Patterson, 1990; Peck, 1995; Priest, 1993, 1995; Priest & Dominick, 1994; Rapping, 1991; Scott, 1996; Shattuc, 1997; Squire, 1994), only some studies wholly or partially focus on the experts who appear on these programs (Carpignano, Andersen, Aronowitz, & DiFazio, 1990; Gamson, 1998; Heaton & Wilson, 1995; Himmelstein, 1984; Livingstone & Lunt, 1992, 1994; Munson, 1993; Nudelman, 1997; Peck, 1995; Robinson, 1980; Tulloch & Chapman, 1992).

Given that most popular syndicated talk shows have the reputation of being sleazy or even scandalous, one might wonder why experts, intellectual experts in particular, agree to participate on such programs. One explanation is that talk show appearances are an important step in the process of disseminating research to the public (Richardson, 1987). Training in techniques for television appearances is therefore considered a must for many scholars, authors, and other professionals (Mincer & Mincer, 1982; Schwager, 1986; Walker, 1990). However, experts' motives for appearing on TV talk shows may not always be scholarly or altruistic. Himmelstein (1984) argues "persons with advanced academic credentials who claim 'expert' status offer their mass-produced popular therapies in exchange for the immediate public adulation they would never achieve were their activities-of-the-mind carefully cultivated through intellectual rigor and years of observation and disciplined analysis" (p. 284).

Although experts appear on talk shows to lend credibility to the programs (Heaton & Wilson, 1995), experts' knowledge and advice is often seen as meaningless compared to the guests' emotional, real-life experiences (Munson, 1993). Robinson (1980) shows that experts frequently do not support their statements, distinguish personal opinions from scholarly research, nor do they make clear the "methodological complexities of gathering and interpreting social science data" (p. 376). According to Robinson, these issues may arise because audiences desire simple explanations, talk shows necessitate clear and concise information due to time constraints, experts serve a public relations function for their profession and behave so that the public will accept their discipline as valid. In addition, Robinson suggests that experts may feel less confident in a television studio than in a scholarly or otherwise familiar setting, and therefore not perform as well as they might.

One interpretation of the relationship between intellectuality and talk shows is that the shows' concentration on "common sense" muddies the distinction between experts and the public. Carpignano, Andersen, Aronowitz, and DiFazio

(1990) maintain that the talk show creates a new public sphere or "contested space" in which conventional representations of politics or ideology are abandoned in lieu of a new form of presentation or debate. Carpignano et al. argue that television talk shows are unique because the opinions of the public or lay people are recognized as important contributors to the discussion at hand. The content of the programs generally celebrate the ordinary and, in an illusion of a democracy, the layperson contributors are often assumed to be representative of the "general public" and common sense.

Another school of thought holds that public participation in the media may be a challenge to intelligence and expertise (Livingstone & Lunt, 1992, 1994). Although television talk shows provide a forum for experts to disseminate their knowledge, ordinary people are becoming accustomed at the same time to arguing and disagreeing with experts (Ussher, 1994). Livingstone and Lunt suggest that experts are "in conflict both with each other and with the lay public" (1992, p. 16) and that, in most cases, the layperson's personal experiences are seen as more valid or held in higher regard by the host and the audience members than are the statements made by the expert. This may be because experts are often doubted by the layperson or are viewed as too theoretical to understand the "real world" (Ross, 1989).

Certainly, the experts who appear on talk shows vary by area of expertise and credentials. Central to this paper is the distinction between intellectual and non-intellectual experts. Henceforth, intellectual experts are defined as those individuals with expertise in disciplines that require academic study, typically in traditional higher education. Non-intellectual experts are those who have acquired their expertise in other ways, most likely through life experience.

Through a content analysis of popular television talk shows, this study investigates how experts on these programs are scheduled, positioned, questioned, and otherwise treated, in relation to the host, other experts, and lay audience members also appearing on the shows. In addition, the ways experts give and support their own opinions will also be measured. In particular, this analysis will focus on the social value placed on intellectual experts in this context and the differences in treatment between intellectual experts and non-intellectual experts[2].

Method

This study analyzes the ten highest-rated syndicated talk shows as of June, 1995, according to their Nielsen ratings as reported in *Variety* (Benson, 1995). In descending order, these were: *Oprah Winfrey*, *Sally Jessy Raphael*, *Maury Povich*, *Donahue*, *Montel Williams*, *Ricki Lake*, *Jenny Jones*, *Jerry Springer*, *Rolanda*, and *Geraldo*. Twenty episodes of each program were videotaped from television over a period of four weeks for a total sample of 200 programs.

Two coding instruments were used to analyze two distinct units of analysis. The first instrument of analysis was applied to the hour-long "episode" of every program sampled. That is, one part of the analysis considered the hour-long show as social context for the representation of experts. The second coding instrument was applied to each expert appearing in the sampled programs. Together, these two instruments examined not only the characteristics and style of each expert featured, but also the fuller context in which experts were portrayed.

An individual appearing on any of the sampled talk shows who is introduced by the host or announcer as an expert was coded as an expert for this analysis. Typically, when the host introduces or regards an individual as an expert it is almost invariably true that, in the given field or context, this individual will conform to most standard definitions of expertise. In rare events, atypical "experts" are introduced (e.g., on *Ricki Lake*, laypersons are sometimes brought on to be mediators or "judges" of the show's guests). The coding instrument differentiated among all types of individuals introduced as experts so the kinds of people represented *as* experts could be precisely determined.

One final distinction is important to the identification of experts. As noted, experts are almost invariably introduced as being separate from "guests" and the only people to be examined in this study *are introduced* as being distinct from guests. Guests are the central players on the talk shows—the subjects or "experiencers" of the problem covered in that episode. Sometimes, talk shows have only specialists as guests, i.e., a panel of lawyers discussing a popular case or a panel of gossip columnists discussing celebrities. However, the most common format provides lay guests with, and sometimes without, expert commentators. For purposes of this study, an individual was coded as an expert only when he or she was featured as a separate onlooker, critic, or advice-giver.

The author served as the primary coder and two trained secondary coders re-coded 10% of the sample for reliability. Inter-coder agreement was determined by utilizing Krippendorff's (1970, 1980) alpha; there was a .92 overall agreement among the three coders. All individual variables with a reliability level above .70 were accepted for analysis. Because the sample in this study was purposeful and non-random, inferential statistics were inappropriate (see also Grabe, 1999). Therefore, all variables were subject to simple frequency counts and crosstabulations.

Results

The results indicate that experts are a consistent and predictable presence on popular television talk shows. Experts appeared in 59% (118) of the 200 programs analyzed. In total, 158 experts appeared. Several experts appeared on more than one of the sampled shows. Fifteen experts appeared on two of the shows in the sample and three appeared on three of the sampled programs. The number of experts among all other people who speak on the shows is somewhat predictable. On average, guests outnumbered experts by a ratio of seven to one and audience speakers outnumbered experts by a ratio of nearly eight to one.

Expert Demographics

As portrayed on popular television talk shows, experts are middle-aged (70%), Caucasian (87%), female (71%), therapists/psychologists (58%). The majority of the experts appearing on the talk shows were not described as having authored a book or having any teaching experience; of the 158 experts in the sample, 32% were described as having authored or co-authored a published book, and 8% were described as having had some teaching experience.

The term "expert" is used to identify a variety of people and qualifications (see Table 10.1). For example, while the majority of the experts (58%) fell into the general category of therapist/psychologist, 13% of the experts fell into an "instructional entertainment" category that included psychic readings, dance, and matchmaking, nine percent were described as experts in beauty/fashion, four percent in medicine, four percent in law, and 3% in fitness. The remaining 9% of experts represented a variety of disciplines including advertising, social work, fitness, and men's rights.

Table 10.1. Areas of Expertise

Area of Expertise	% of Experts (N = 158)
Psychology/Therapy	58%
Instructional Entertainment	13%
Beauty/Fashion	9%
Medicine	4%
Law	4%
Food/Nutrition	3%
Fitness	2%
Social Work	2%
Communication	1%
Advertising/Media Studies	1%
Specific Products	1%
Men's Rights/Masculinity	1%
Photography	1%

Experts' Use of Evidentiary Support

Experts' citation of evidence, such as statistics, studies, and theories, to support their opinions was coded. Results indicate that the experts in the sampled shows rarely used evidence to support their claims. Of the 158 experts in the sample, only seven percent cited statistics in their statements and/or explanations (e.g. "seventy percent of women are unhappy with their home life"), 10 percent named studies as evidence (e.g., "a recent study shows that most men are unfaithful to their spouses"), and 15 percent cited theories (e.g., "a theory of intimacy avoidance suggests . . . ").

The Treatment of Experts

Seven variables were coded to measure the treatment of experts: the timing of experts, who had the final word on the show, physical placement of experts, how often experts were questioned, how often experts were interrupted, the number of disagreements with experts, and the number of positive reactions (e.g., applause) to experts. In addition, any differences between the treatment of intellectual experts and non-intellectual experts were measured.

The Timing of Experts

Two timing variables were coded: at what point in the programs the expert is introduced and the amount of speaking time allotted for each expert. Results showed that, as a rule, experts were introduced late in the programs. Of the total 158 experts that appeared in this sample, 73 percent were introduced after 30 minutes had elapsed. More specifically, 51 percent of the experts were introduced between 41 and 53 minutes into the programs.

Generally, experts on the sampled talk shows were allotted relatively little time to speak. On average, experts were allotted 2.5 minutes to speak. As Table 10.2 demonstrates, very few of the 158 experts spoke for more than three minutes of combined speaking time. Only two of the 158 experts spoke for more than eight minutes: one for nine minutes and one for ten. In addition, only very small variations were found when speaking time was crosstabulated with area of expertise.

Table 10.2. Experts' Speaking Time

Amount of Speaking Time	Percent of Experts (N = 158)
Less than 30 Seconds	5%
One Minute	28%
Two Minutes	25%

Table 10.2. continued

Amount of Speaking Time	Percent of Experts (N = 158)
Three Minutes	23%
Four Minutes	7%
Five Minutes	5%
Six Minutes	3%
Seven Minutes	2%
Eight Minutes	1%
Nine Minutes	.5%
Ten Minutes	.5%

The Final Word

In nearly every case (90%), the host of the talk-show issued a formal good-bye to the studio and home audiences. Apart from this formal good-bye the host was also the last person to provide concluding remarks (this in 79% of the 118 cases). In the small percentage of shows where someone other than the host provided closure, experts concluded in 9% (11) of the cases, guests in 7% (8), and 5% (6) ended in the middle of a conversation or argument. On average, experts are allowed the last word once every ten shows.

Given that the closure provided by hosts is sometimes separated in time (often directly after a commercial) and/or in space, against a quiet, non-audience backdrop, the program's last words *prior* to the hosts' closures were also examined. For 79 percent (93) of the cases, the person who spoke *before* the host's last word was as follows: guests in 52 percent (48) of the cases, experts in 38 percent (36) cases, and audience members in 10 percent (9). When all of the instances of the host making final remarks were excluded from the data set and the person who spoke last other than the host was assessed, guests conclude 47 percent of the time, experts speak in this position 40 percent of the time, and audience members conclude 13 percent of the time.

Physical Separation of Expertise

When initially introduced, the majority of the 158 experts in the sample were seated on the stage. Sixty-five percent (102) of the experts were situated on the stage when first introduced and 33 percent (53) were seated in the audience. The remaining two percent of the experts (3) first appeared via satellite or videotape.

When experts appeared on stage, where they were positioned was recorded. Of the 102 experts situated on the stage when first introduced, 35 percent (36) were seated on stage in the middle of non-expert guests, 32 percent (33) at the end of non-expert guests, 20 percent (21) were on stage yet physically separated from non-expert guests and 13 percent (13) were seated on stage alone or with other experts only.

Questioning the Experts

Nearly every expert was asked a question by the host: 150 of the experts (95%) were asked at least one question throughout the program. The majority of experts were asked one to four questions: 18 percent (28) were asked one question, 25 percent (40) were asked two questions, 15 percent (23) were asked three questions, 22 percent (35) were asked four questions. Fifteen percent of experts were asked five or more questions. The majority of the questions were asked in a neutral, fact-finding tone. Of the 150 experts who were asked at least one question, 98 percent (147) were asked in a relatively neutral tone of voice while only 2 percent (3) were questioned in an obviously deferential manner. No hostility was expressed in the questioning of experts.

Interrupting the Experts

Of the 158 experts in this sample, 89 percent (141) were interrupted (that is, cut off in mid-sentence) by a participant of the show. Of all the personalities on the programs, the host was the most likely to interrupt the expert. Of the 158 experts in the sample, 63 percent (99) were interrupted by the host; 33 percent of all experts were interrupted once by the host, 20 percent twice, and 13 percent three times or more.

Only six (4%) of the 158 experts in the sample were interrupted by other experts appearing on the show. In contrast, 53 (34%) of the 158 experts in the sample were interrupted by non-expert guests. Only four (3%) of the 158 experts were interrupted by an audience member. With all the talk-show personalities combined, experts were interrupted 331 times. On average, each expert was interrupted twice.

Of the total 331 interruptions invoked on experts' speech, in only five percent (15) of the cases was the floor returned to the expert. While hosts interrupted experts a total of 210 times, they only returned the floor to the expert in two percent (4) of the cases. Of 108 interruptions by guests, the host returned the floor to the expert in nine percent (10) of the cases. In short, once the expert was interrupted, it was unlikely that he or she would regain the floor.

Disagreeing with the Experts

Of the 158 experts in the sample, 22 percent (35) met with disagreements and/or challenges: 19 percent (30) encountered disagreements by guests, 4 percent (7) by other experts, 4 percent (6) by an audience member, and 2 percent (3) encountered host disagreement.

Positive Reactions to Experts

Of the 158 experts in the sample, 46 percent (73) received at least one burst of applause after a statement. A total of 139 bursts of applause for experts were found: on average, less than one incident per expert. Far fewer of the 158 experts in the sample received positive hoots or praise; 4 percent (6) of the experts received positive hoots from the audience, 9 percent (15) were praised by the host. Although not coded, applause typically followed an expert's statement when it was a repetition of opinions already voiced by audience members.

Differences in the Treatment of Intellectual and Non-Intellectual Experts
Seventy percent of the experts fell into the intellectual category as defined above. In terms of host questions to experts, the presence of a scholarly, intellectual background was inconsequential. The data showed that 51% of those experts questioned at least once by the host were intellectual experts, while 48% were non-intellectual. Thus, no apparent hierarchy emerged based on relevant educational factors.

The degree to which an expert was intellectual sometimes played an important role in whether or not he or she was interrupted. While interruptions by hosts and interruptions by audience members did not differ greatly, this was not the case for interruptions by guests and other experts. Of the experts interrupted by guests, 77% were intellectual experts. Similarly, 68% of interruptions by other experts were intellectual experts. Thus, guests clearly were more inclined to interrupt intellectual experts than non-intellectual experts.

Whether or not an expert was intellectual played an important role in whether or not he or she was challenged. *All* of the experts with whom the host and audience members disagreed or challenged were intellectual experts. In addition, 79% of the experts who encountered disagreements by guests were intellectual. Similarly, 72% of the experts challenged by other experts were intellectual experts.

The degree to which an expert was intellectual played an important role in whether or not he or she received certain positive reactions. Of those experts who received applause, 70% were intellectual experts. In stark contrast, of those who received praise from the host, 73% were non-intellectual. Of those who received positive hoots from the audience, 54% were non-intellectual.

Discussion

Results of this content analysis revealed several important patterns. The data seem to indicate a multifaceted "leveling" phenomenon in which talk-show experts, intellectual experts in particular, are constructed in such a way that their power is lessened and/or intellectuality is seen as undesirable.

First, even though experts are a consistent and predictable presence on popular television talk shows, lay guests outnumber them by a large margin. Clearly, producers presume that the viewing public is more interested in the problems and/or personally-narrated experiences of regular folks with relatively little intervention and analysis from experts.

Second, as noted earlier, every individual introduced as an expert was coded as such, this ranged from neurosurgeons to psychics. While technically the semantics are correct—someone can be just as expert in astrology as astrophysics—the presentation of that kind of variety under the same epithet in the same forum may give rise, at the very least, to a definitional confusion about expertise. More importantly, the "jumbling" of experts may lend support to the notion that all experts are of the same stature regardless of subject matter.

Third, since the sampled experts rarely used evidence such as statistics, studies, or theories to support their claims or advice, it may appear quite reasonable to a normal audience that the expert is giving his or her opinions on the matter. The general failure to include research and scholarly-based ideas may contribute to the impression that the expert is not speaking from a position of scientific authority, but as just another voice on the stage. Moreover, this failure to integrate external substantiation with opinions certainly does nothing to teach viewers about the discourse of empirical justification in general. Any disregard that the average person has for strong empirical or scholarly discourse may be related to his or her lack of exposure to this discourse. In fact, talk shows might be a perfectly easy and benign place to acquaint people with the use of evidence and critical argumentation.

Fourth, the timing measurements lead to several conclusions. In the first place, the amount of speaking time devoted to experts did not vary greatly across area of expertise; the time allotted to psychologists, psychics, medical doctors, and beauty/fashion experts typically does not differ by more than one or two minutes. In other words, no representative of an area of expertise (intellectual or non-intellectual) seems to be given more special treatment than any other. In the second place, experts tend to appear at the end of the programs. In terms of the narrative flow of the program it may make sense that experts appear at the end since exposition precedes analysis; i.e., the audience must first hear the stories upon which the experts will comment. However, what these timing data do not reveal is that both hosts and audience members typically make comments and ask questions throughout the entire show; thus, in talk shows, analysis tends to accompany the expositions from early on in the hour. Yet, when the analyst is the expert, he or she is generally required to wait until others have had their say. Finally, in most cases experts are given very little time to speak.

Therefore, when the brevity of appearance is combined with the late position in the show one may be left with the impression of the expert as an afterthought, the expert as filler, and/or the expert as a socially-redeeming (but unnecessary) symbol. By essentially sticking experts in for a quick appearance at the end, it might seem that the expert is not essential to the integrity of the show. Especially since talk shows get a lot of bad press because of alleged tawdriness, producers may schedule experts to provide a patina or façade of serious, redeeming context to the real entertainment.

Fifth, where experts are placed within the studio is telling in terms of how they are considered. Since nearly one-third of experts are positioned in the audience sector and given that audience members are typically called upon to offer advice, analysis, and comments, the placement of the expert alongside of these anonymous audience members may tend to blur the distinction between expert and amateur. By reasonable extension, it may even suggest that the opinions from both sources are of the same value.

Where experts are seated on stage may be similarly meaningful. Since the majority of stage-seated experts are placed immediately among the lay guests they become, in a sense, part of the tableau, part of the spectacle. This positioning may be contrasted with where the host typically appears. Although hosts occasionally go to the stage to comfort or whisper to a guest, they typically remain standing in the space between stage and audience or in the audience aisles, thereby remaining apart from the central spectacle and maintaining a somewhat detached persona. Although a certain degree of a detachment might also be expected from an expert, *talk-show* experts are typically not positioned for that role.

Sixth, in looking at how experts are questioned, interrupted, and challenged on talk shows, patterns emerge which reinforce the notion of the leveling of the status of experts. To begin, were it not for the hosts, experts would not speak much; generally, in this context at least, guests and audience members do not seem interested in providing the experts with more speaking time or in hearing more expert insights. Although the hosts almost always ask the expert at least one question, they typically ask only one or two questions over the course of the experts' presence on stage. Moreover, the hosts' tone and phrasing of questions does not differ much in terms of whether an expert or a regular guest is being questioned.

Given that the discursive style of talk shows engenders considerable interruption, interruptions of experts may not seem to be exceptional. However, there are two caveats to this line of thinking. First, Greenberg et al. (1996) find that conflict on talk shows is not as great as popularly presumed and so perhaps it is wrong to characterize the genre as argumentative. Second, even if the present sample provided more confrontation than that of Greenberg et al., the data strongly suggest that experts, intellectual experts in particular, are treated differently in this context. Therefore, the fact that experts are interrupted as frequently as they are questioned is meaningful. Moreover, the data also show that experts rarely recover the floor after being interrupted; neither guests nor hosts generally channel the discussion back to the expert after such an interruption has occurred.

In addition to infrequent questions and frequent interruptions, experts often encountered disagreements or challenges from non-experts. In many cases, after an expert gave a fairly long explanation, the regular guest dismissed or challenged an expert's analysis with a statement such as "Oh, please—you don't really know me," "Well, that's your opinion," or "I don't agree with that."

Finally, it is noteworthy that intellectual experts received the majority of negative reactions from other talk show participants and more applause than non-scholars. In the context of all treatment variables, it might then be argued that intellectual experts generally engender more reactions than any other type of guest. However, before too much is inferred from the applause measurement, it must also be repeated that scholars did *not* receive the majority of praise offered by hosts; indeed host praise was largely directed to non-scholars.

What these data on performance interaction seem to indicate is a lack of special attention to the opinions of talk show experts. While some may argue that experts are not owed any deference, they are structurally a separate unit of talk-show performers, and, in this context alone, one might imagine that their treatment might be different. When one adds to this structural distinction the higher hierarchical status supposedly associated with experts in our culture, one might imagine an elevated degree of respect to be shown them.

Certainly the data on positive reactions must also be considered here; positive reactions to experts may seem to balance the negative responses. Nonetheless, it is important to understand that applause—by far, the most common positive reaction—generally occurred when the expert's remark was a reiteration of audience members' opinions. In this context, it should also be pointed out, then, that audience members often applaud and cheer each other in the same way. Also, applause may not be the best measurement of response in that it is often orchestrated by (invisible-to-the-home-viewer) signs and/or production assistants. More importantly, when the raw number of positive reactions (i.e., applause, hoots, and praise) is compared to the raw number of negative reactions to experts (i.e., interruptions and disagreements), incidents of negative reactions outweighed incidents of positive reactions by more than two to one.

Conclusion

As previously discussed, intelligence, intellectuality and expertise would seem, in theory, to be regarded as positively elite and socially beneficial. However, in everyday practice, these qualities are not all that well regarded. That experts, intellectuals in particular, appear mundane and inconsequential may both reflect and propagate ideas important to the maintenance of the social order. More specifically, the social order may be more easily sustained if individuals are apathetic or antagonistic toward most intellectuals, experts, and scholars insofar as these individuals breed critical examination and sometimes even rebellion and revolution. To admire and emulate scholars and intellectuals would more likely entail pressure to examine facts carefully and to search for meaningful, underlying patterns in life. Although some may argue against this elitist interpretation, it nonetheless seems fair to suggest that an electorate and a workforce comprised of such thorough and systematic analysts might be a threat to traditionally established power.

The results of this study support the work of Carpignano et al. (1990), Robinson (1980), and Livingstone and Lunt (1992, 1994) that show popular talk shows blur the separation between expert and lay knowledge and that the celebration of lay knowledge in the media is a challenge to expertise and intelligence. What seems to be happening on talk shows is somewhat contradictory; when experts are summoned onto the stage it is seemingly, at least in part, to recognize their potential insights, however, as their role actually evolves, it becomes less clear that tribute is to be paid at all. It is not that experts are mistreated or suffer gross disrespect on talk shows, the effect of their involvement is more subtle. Instead talk shows generate an equalization so that "expert opinion" is often represented as no different from anyone else's opinion. Moreover, the more intellectual the experts, the more likely they are to receive what might be regarded as negative treatment. This connects to the more fundamental premise that television talk shows contribute to social-order maintenance by leveling, if not weakening, the image of intellectual expertise. These findings have implications, in particular, for individuals who watch talk shows regularly; cultivation theory argues that television shapes heavy viewers' perceptions such that they are concordant with the patterned portrayals they see.

Knowledge and intelligence are limited commodities that are tied importantly to power and prestige. The maintenance of the social order is dependent upon a certain amount of resentment or antagonism by the masses toward both intelligence and the individuals that possess intelligence. This resentment is functional because it defers most individuals from acquiring the erudition and intellectual skills that may contribute significantly to social mobility. Moreover, the relative absence of erudition and honed critical skills among the mass workforce and electorate may be seen as providing for a more malleable public. In sum, the leveling of experts on talk shows contributes to this apathy and/or resentment toward experts, especially intellectual experts and may thereby be viewed as threatening to social growth and development.

Notes

1. This chapter was originally published in *The New Jersey Journal of Communication*, Volume 11 (1), 45-62, and is reprinted with permission.

2. To assess any differences between "intellectual" and "non-intellectual" experts areas of expertise were collapsed into the two categories: "intellectual" areas of expertise are those that require academic study in traditional higher education while "non-intellectual" are those areas in which expertise is otherwise acquired—most likely through life experience. Areas placed in the "intellectual" category were psychology/therapy, communication, law, medicine, nutrition, media studies, and social

work. Areas of expertise placed in the "non-intellectual" category included areas such as dance, cheerleading, psychic readings, photography, fitness, and fashion.

References

Abt, V., & Mustazza, L. (1997). *Coming after Oprah: Cultural fallout in the age of the TV talk show.* Bowling Green: Bowling Green State University Popular Press.

Aronowitz, S. (1990). On intellectuals. In B. Robbins (Ed.), *Intellectuals: Aesthetics, politics, academics.* Minneapolis: University of Minnesota Press.

Bauman, Z. (1987). *Legislators and interpreters: On modernity, post-modernity, and intellectuals.* Ithaca, NY: Cornell University Press.

Benda, J. (1928). *The betrayal of the intellectuals.* (R. Aldington, Trans.). Boston: Beacon Press.

Benson, J. (June 26, 1995). Sweeps ratings show mellow Oprah sliding. *Variety, 359* (9), 21, 26.

Berger, P. L., & Luckmann, T. (1966). *The social construction of reality: A treatise in the sociology of knowledge.* New York: Doubleday.

Brinson, S. L., & Winn, E. J. (1997). Talk shows' representations of interpersonal conflicts. *Journal of Broadcasting and Electronic Media, 41*, 25-39.

Brym, R. J. (1980). *Intellectuals and politics.* London: George Allen and Unwin.

Carbaugh, D. (1988-89). Deep agony: "Self" vs. "society" in Donahue discourse. *Research on Language and Social Interaction, 22*, 179-212.

Carbaugh, D. (1990). Communication rules in *Donahue* discourse. In D. Carbaugh (Ed.), *Cultural communication and intercultural contact* (pp. 119-149). Hillsdale, NJ: Erlbaum.

Carbaugh, D. (1993). "Soul" and "self": Soviet and American cultures in conversation. *Quarterly Journal of Speech, 79*, 182-200.

Carpignano, P., Andersen, R., Aronowitz, S., & DiFazio, W. (1990). Chatter in the age of electronic reproduction: Talk television and the "public mind." *Social Text, 9*, 33-55.

Cerulo, K. A., Ruane, J. M., & Chayko, M. (1992). Technological ties that bind: Media-generated primary groups. *Communication Research, 19*, 109-129.

Cole, B. J. (1975). Trends in science and conflict coverage in four metropolitan newspapers. *Journalism Quarterly, 52*, 465-471.

Durkheim, E. [1895] (1982). *The rules of sociological method.* New York: Free Press.

Durkheim, E. (1951). *Suicide.* (J. A. Spaulding & G. Simpson, Trans.) Glencoe, IL: The Free Press.

Epstein, D., & Steinberg, D. L. (1996). All het up! Rescuing heterosexuality on the Oprah Winfrey show. *Feminist Review, 54*, 88-115.

Eyerman, R., Svensson, L. G., & Soderqvist, T. (1987). *Intellectuals, universities, and the state in western modern societies.* Los Angeles: University of California Press.

Foucault, M. (1976). The political function of the intellectual. *Radical Philosophy*, 17, 12-14.

Foucault, M. (1977a). *The archeology of knowledge.* London: Tavistock Publications.

Foucault, M. (1977b). *Language, counter-memory, practice: Selected essays and interviews.* Ithaca, NY: Cornell University Press.

Foucault, M. (1977c). *Power/knowledge: Selected interviews and other writings 1972-1977.* New York: Pantheon Books.

Gamson, J. (1998). *Freaks talk back: Tabloid talk shows and sexual nonconformity.* Chicago: The University of Chicago Press.
Gerbner, G. (1974). Teacher image in mass culture: Symbolic functions of the "hidden curriculum." In P. Olson (Ed.), *Media and symbols: The forms of expression, communication, and education* (pp. 470-497). Chicago: The National Society for the Study of Education.
Gerbner, G. (1990). Epilogue: Advancing on the path of righteousness (maybe). In N. Signoriclli & M. Morgan (Eds.), *Cultivation analysis: New directions in media effects research* (pp. 249-262). Newbury Park, CA: Sage.
Gerbner, G., Gross, L., Morgan, M., & Signorielli, N. (1979). The mainstreaming of America: Violence profile no. 11. *Mass Communication Review Yearbook*, 509-529.
Gerbner, G., Gross, L., Morgan, M., & Signorielli, N. (1981). Scientists on the TV screen. *Society, 18* (4), 41-44.
Goldman, S. L. (1989). Images of Technology in Popular Films: Discussion and Filmography. *Science, Technology and Human Values, 14*(3), 275-301.
Gouldner, A. W. (1979). *The future of intellectuals and the rise of the new class.* New York: The Seabury Press.
Grabe, M. E. (1999). Television news magazine crime stories: A functionalist perspective. *Critical Studies in Mass Communication, 16,* 155-171.
Gramsci, A. (1971). *Selections from the prison notebooks.* (Quintin Hoare, Trans.) New York: International Publishing.
Greenberg, B. S., Sherry, J. L., Busselle, R. W., & Rampoldi-Hnilo, L. (August, 1996). *Daytime television talk shows: topics, guests, and reactions.* Paper presented at the convention of the Association for Education in Journalism and Mass Communication, Anaheim, CA.
Greenberg, B. S., Sherry, J. L., Busselle, R. W., Rampoldi-Hnilo, L., & Smith, S. W. (1997). Daytime television talk shows: Guests, content, and interactions. *Journal of Broadcasting and Electronic Media, 41,* 412-426.
Heaton, J. A., & Wilson, N. L. (1995). *Tuning in trouble: Talk TV's destructive impact on mental health.* San Francisco: Jossey-Bass.
Himmelstein, H. (1984). *Television, myth, and the American mind.* New York: Praeger.
Hofstadter, R. (1964). *Anti-intellectualism in American life.* New York: Alfred A. Knopf.
Hoynes, W., & Croteau, D. (1991). The chosen few: Nightline and the politics of public affairs television. *Critical Sociology, 18,* 19-34.
Huzar, G. B. (1960). *The intellectuals: A controversial portrait.* Glencoe, IL: The Free Press.
Konrad, G., & Szelenyi, I. (1979). *The intellectuals on the road to class power.* New York: Harcourt Brace Javanovich, Inc.
Krippendorff, K. (1970). Bivariate agreement coefficients for the reliability of data. In E. F. Borgatta and G. W. Bohrnstedt (Eds.), *Sociological methodology* (pp. 139-150). San Francisco: Jossey-Bass.
Krippendorff, K. (1980). *Content analysis: An introduction to its methodology*, Beverly Hills: Sage.
Kurtz, H. (1996). *Hot air: All talk, all the time.* New York: Random House.
Livingstone, S. M. (1994). Watching talk: Gender and engagement in the viewing of audience discussion programmes. *Media, Culture, and Society, 16,* 429- 447.
Livingstone, S. M., & Lunt, P. K. (1992). Expert and lay participation in television debates: An analysis of audience discussion programmes. *European Journal of Communication, 7,* 9-35.

Livingstone, S. M., & Lunt, P. K. (1994). *Talk on television: Audience participation and public debate*. New York: Routledge.

Long, M., & Steinke, J. (1996). The thrill of everyday science: images of science and scientists on children's educational science shows in the United States. *Public Understanding of Science, 5* , 101-120.

Mannheim, K. (1949). *Ideology and utopia*. New York: Brace.

Mannheim, K. (1971). *From Karl Mannheim*. (K. H. Wolff, Ed.). New York: Oxford University Press.

Martin, A. (1976). On some neglected points of view. *Communication et Information, 1*, 13-50.

Martin, W. C. (1987). The role of the intellectual in revolutionary instituions. In R. P. Mohan (Ed.), *The mythmakers: Intellectuals and the intelligentia in perspective*. Westport, CT: Greenwood Press.

Marx, K., & Engels, F. (1970). *The German ideology* (C. J. Arthur, ed. & introduction). London: Lawrence & Wishart.

Merton, R. K. (1973). *The sociology of science*. Chicago: University of Chicago Press.

Mincer, R., & Mincer, D. (1982). *The talk show book: An engaging primer on how to talk your way to success*. New York: Facts on File Publications.

Morgan, M., & Signorielli, N. (1990). *Cultivation Analysis: New Directions in Media Effects Research*. Thousand Oaks, CA: Sage.

Munson, W. (1993). *All talk: The talk show in media culture*. Philadelphia: Temple University Press.

Nelkin, D. (1990). Selling science. *Physics Today*, 43, 41-46.

Nelson, E. D., & Robinson, B. W. (1994). "Reality talk" or "telling tales"? The social construction of sexual and gender deviance on a television talk show. *Journal of Contemporary Ethnography, 23*, 51-78.

Nolan, L. L., & Patterson, S. J. (1990). The active audience: Personality type as an indicator of TV program preference. *Journal of Social Behavior and Personality, 5*, 697-710.

Nudelman, F. (1997). Beyond the talking cure: Listening to female testimony on The Oprah Winfrey Show. In J. Pfister & N. Schnog (Eds.), *Inventing the psychological: toward a history of emotional life in America* (pp. 297-315). New Haven: Yale University Press.

Peck, J. (1995). TV talk shows as therapeutic discourse: The ideological labor of the televised talking cure. *Communication Theory, 5* (1), 58-81.

Priest, P. J. (1993). Self-disclosure on television: The counter-hegemonic struggle of marginalized groups on "Donahue." *Dissertation Abstracts International, 53*, 2147.

Priest, P. J. (1995). *Public intimacies: Talk show participants and tell-all TV*. Cresskill, NJ: Hampton Press, Inc.

Priest, P. J., & Dominick, J. R. (1994). Pulp pulpits: Self disclosure on "Donahue." *Journal of Communication, 44* (4), 74-97.

Rapping, E. (1991, October). Daytime inquiries. *Progressive*, pp. 36-38.

Richardson, L. (1987). Disseminating research to popular audiences: The book tour. *Qualitative Sociology, 10*, 164-176.

Robinson, B. E. (1980). Family experts on television talk shows: Facts, values, and half-truths. *Family Relations, 31*, 369-378.

Ross, A. (1989). *No respect: Intellectuals and popular culture*. Routledge: Chapman and Hall.

Ross, A. (1990). Defenders of the faith and the new class. In B. Robbins (Ed.), *Intellectuals: Aesthetics, politics, academics* (pp. 101-132). Minneapolis: University of Minnesota Press.

Schwager, M. (1986). Training for television. Special issue: Communications. *Training and Development Journal, 40*, 62-65.

Scott, G. G. (1996). *Can we talk? The power and influence of talk shows.* New York: Plenum Press.

Shanahan, J., & Morgan, M. (1999). *Television and its viewers: Cultivation theory and research*. Cambridge: Cambridge University Press.

Shattuc, J. M. (1997). *The talking cure: TV talk shows and women*. New York: Routledge.

Shils, E. (1972). *The intellectuals and the powers and other essays*. Chicago: University of Chicago Press.

Squire, C. (1994). Empowering women? The Oprah Winfrey show. *Feminism and Psychology, 4* (1), 63-79.

Stocking, S. H., & Dunwoody, S. L. (1982). Social science in the mass media: Images and evidence. In J. Sieber (Ed.), *The ethics of social research: Fieldwork, regulation, and publication*. New York: Springer-Verlag.

Thomas, S. (1986). Mass media and the social order. In G. Gumpert & R. Cathcart (Eds.), *Inter/Media: Interpersonal communication in a media world* (3rd ed.) (pp. 611-627). New York: Oxford University Press.

Thomas, S., & Krippendorf, M. (1988). *Beauty and Brains: The sociopolitical concomitants of sexuality on television*. A paper presented at the Convention of the International Communication Association, New Orleans.

Tudor, A. (1989). *Monsters and mad scientists: A cultural history of the horror movie.* Cambridge, MA: Basil Blackwell, Inc.

Tulloch, J., & Chapman, S. (1992). Experts in crisis: The framing of radio debate about the risk of AIDS to heterosexuals. *Discourse and Society, 3* (4), 437-467.

Ussher, J. (1994). Media representation of psychology: Denigration and popularization, or worthy dissemination of knowledge? In C. Haslam and A. Bryman (Eds.), *Social scientists meet the media* (pp. 123-137). New York: Routledge.

Wald, A. M. (1987). *The New York intellectuals: The rise and decline of the anti-Stalinist left from the 1930s to the 1980s*. Chapel Hill: The University of North Carolina Press.

Walker, K. B. (1990). Confrontational media: Training for administrators: Performance and practice. *Public Personnel Management, 19*, 419-427.

Wober, M., & Gunter, B. (1988). *Television and social control*. New York: St. Martin's Press.

Chapter Eleven

Portrayals of Intelligence in Reality Television

Marilyn Ellzey and Alison Miller

Though not a new phenomenon in American television programming or popular culture, reality television is a genre that provides a form of cultural transmission. The genre is a means by which television programmers attempt to portray a constructed "reality" that will gratify the needs of the audience, a similar goal of much—if not all—television programming. "The most popular shows are those that succeed in speaking simultaneously to audiences that diverge in social class, race, gender, region, and ideology . . . appealing to a multiplicity of social types at once" (Gitlin, 1982, p. 248) in order to satisfy market demands.

An important question regarding reality television is how it reflects intelligence as an ideology in contemporary culture. Recent reality programs such as NBC's *The Apprentice*, *Beauty and the Geek* on WB, and ABC's *The Scholar* have focused specifically on various aspects of intelligence. However, contemporary reality television portrayals of intelligence, when measured by contestant and ratings success—or lack thereof—may be more incumbent on audience appeal, physical/personality characteristics, and medium production requirements than more traditional indicators such as IQ or education level. This chapter examines portrayals of intelligence in reality television programming. Content analysis was used to compare and contrast portrayals within and among selected programs.

As many as 200 reality programs can currently be viewed on network and cable television; and the genre has been widely criticized as an indicator of the continued decline of "quality television." However, "whenever an entire nation is riveted by such a spectacle" as *Survivor*, we can learn something about our society" (Rieder, 2000, p. 6).

Literature Review

The portrayal of intelligence in reality programming is related to three important components: 1) the audience, 2) actual content, and 3) production requirements. The diverse audience garnered by reality programming is separated not only by demographic, but cultural, gaps. It can be argued reality programming represents a significant rhetorical shift for television programming away from traditional values based in "enduring family unity, mutual regard among human beings, satisfaction from simple pleasures, integrity of motive and purpose together with familial and social cooperation" (Brown, 1976, p. 390) to those of an urbanized, disenfranchised culture based in technology.

The Audience

According to Nachbar and Lause (1992), "the popularity of a given cultural element (object, person, event) is directly proportional to the degree to which that element is reflective of audience beliefs and values" (p. 5). For example, television programming of the 1950s, known as the decade of conformity, often reflected the myth of the nuclear family (e.g., *The Donna Reed Show*, *Ozzie and Harriet*) as most desirable. Accordingly, reality programming is a reflection of current audience beliefs and values—the beliefs and values, that is, of the demographic most desirable to advertisers.

Specifically targeted at the "MTV generation," *Survivor* was the first programming move CBS hoped would pull the network out of the ratings basement and provide it with an infusion of younger viewers. When the prototypical reality program debuted in the summer of 2000, CBS was ranked fourth among the four networks and known unofficially as "the Geritol network," relying on prime time offerings such as *Walker, Texas Ranger* and *Touched by an Angel.* While ultimately—and more importantly to the network—expressed in terms of advertising revenue, the differences between the older and younger demographics are more complex than merely available disposable income. One is a generation of newspaper readers, the other a generation of sometimes barely literate computer experts. The distinction is particularly important in the context of knowledge and skills deemed important and valuable. Witness the plight of displaced older workers suddenly unemployed in a job market where their knowledge base and skills are hopelessly outdated, their competition in an even more competitive job market half their ages and light years ahead of them in technology expertise. Reality television reflects this dichotomy in generational values and sends the message that in contemporary culture, the wit and cunning of a *Survivor* is a better indicator of intelligence than traditional education.

Content

Traditional education has fared little or no better as setting or subject matter for television programs. Mayerle and Rarick (1989) analyzed series centering on education over a forty-year period from the beginnings of television until 1988. Out of the forty series fitting the category, only six lasted three or more seasons. The researchers attributed this poor track record to a lack of inherent drama in education or the educational setting.

The study also led to the development of five thematic categories: "teacher as bumbler, teacher/student personal or family comedy, non-traditional educator comedy, process of education/comedy, and process of education/drama" (Mayerle & Rarick, 1989, p. 142). Three of these categories are relevant to the current topic. The "teacher as bumbler" was either academically or socially inept or both. The "teacher/student personal or family life" category included the 1980s series *Square Pegs* in which two freshman girls, labeled "losers" because they excel academically, attempt to gain acceptance into the high school's exclusive cliques.

The "non-traditional educator comedy" included the 1980s situation comedies *Welcome Back, Kotter* and *Head of the Class.* These series had in common educators/students who were "out of the mainstream or different because of age, life-style [sic], or teaching/learning method . . . . and both feature a rather bumbling male principal" (Mayerle & Rarick, 1989, p. 150). *Kotter's* "sweat hogs" were academic underachievers, but "street smart." *Head*'s teacher Charlie Moore was "cool" but his students were academically gifted and labeled "nerds" by the rest of the student body. *Kotter* ran for four seasons on (1975-1979), while *Head* lasted five seasons (1986-1991), both on ABC.

In summary, television portrays *traditional* educators and/or administrators as inept and students who achieve in the *traditional* academic setting and according to those standards as "nerds" and "losers." Television tells us that only those educators "out of the mainstream" and "street smart" students are worthy of attention.

Recent reality programming has focused specifically on intelligence, with mixed results. The ratings success of NBC's *The Apprentice* is obvious in its longevity. *Beauty and the Geek* (WB) was renewed for a second season based on its first season numbers ("ABC Contemplates . . . " 2005). However, ABC's *The Scholar* was, according to Broadcasting and Cable magazine, "a miss" ("ABC Pulls . . . " 2005). The following is a synopsis of the first season of *The Apprentice*:

> Candidates on *The Apprentice* were chosen from all walks of life. Some have Ivy League educations and some never went to college. They will compete for a chance to become the head of a division of the Trump Organization for one year at a salary of $250,000. Men will compete against women in challenges that, according to NBC.com will test their "intelligence, chutzpah, and street smarts." Each week, head of the Trump Organization Donald Trump will fire one candidate (McKay, 2005).

While *Apprentice* candidates were chosen based on their strengths, *Beauty and the Geek* participants were chosen based on weakness in an area specified by the program's premise. Realitytvcalendar.com called *Beauty and the Geek* "reality television's latest social experiment."

> It all starts with seven women who are academically impaired. Next, add seven men who are brilliant but socially challenged. The concept is to pair up couples for a chance to win a $250,000 grand prize. Each mismatched pair competes in various activities designed to test intellect, fashion savvy, and even dance moves. There's a spelling bee for the girls, massage lessons for the guys, and an introduction to actual rocket science when the girls compete to see who can build a working rocket. During these competitions, the geek must try to pass his brains on to the beauty, while the beauty tries to pull the game out of the geek. They're so far apart on the social spectrum that they're practically different species, but if they make it to the end, they could both walk away gifted and gorgeous (Killough, 2005).

All ten contestants on *The Scholar* boasted GPA's of 3.3 or higher. The prize was a full scholarship to the college of the student's choice, donated by philanthropist Eli Broad. These contestants, all obviously academically capable and deserving, "might not have otherwise had the opportunity to attend one of America's top universities." The students must show that they are deserving of the scholarship by demonstrating their book smarts, leadership ability, creativity, and community service" (West, 2005).

Production Requirements

The traditional education accomplishments of some participants notwithstanding, reality television fashions non-mainstream themes of intelligence not into a script, but a scenario. Pecora (2002) defines reality programming as "largely unscripted, though heavily edited, programs" (p. 345) filled with non-professional actors that focus on some element of group dynamics. Along with the audience and content, production is the third factor in the intelligence equation. Traditional, scripted television situation comedies and dramas are extraordinarily expensive to produce. Actors and technicians must be hired, sets must be built, costumes designed, makeup applied. But before any of that can happen, writers must write and be paid. The argument can certainly be made that it is easier (i.e. requires less mental capability) to come up with a "what can we make them do next" scenario than with interesting story lines and clever, witty, intriguing, or otherwise absorbing dialogue to tell those stories week after week. The Writers Guild of America (WGA) West recently raised concerns over pay discrepancies between traditional scriptwriters and writers and editors of reality programs (Lasswell, 2005). Just as the cost of production of traditional television fare multiplies exponentially with the level of writing/technical/acting expertise hired, so does the cost go down when the reverse is true. Again the question is one of what is valued. The answer in terms of the current competitive television environment is—at all levels of production, both news and entertainment—get more return for less investment. The distinction between traditional scripted fare and reality scenarios is an important one: reality scenarios seek and cast "real people" to fill predetermined roles in the program, further reinforcing existing stereotypes without even the limited mitigation of a fictional premise. Programs are then edited to portray the dynamics of the predetermined scenario in the most dramatic fashion.

Insufficient research exists on this topic to allow for the development of formal hypotheses. The authors were interested in answering the following research questions regarding the portrayal of intelligence in contemporary reality programming:

> RQ^1: Are physical attributes more important to contestant success than traditional education qualifications?
>
> RQ^2: Are personality traits more important to contestant success than traditional education qualifications?
>
> RQ^3: How are contestants portrayed as a result of differences in physical attributes, personalities, education/knowledge/skill levels?

Method

Content analysis was used to examine content and production in NBC's *The Apprentice*, WB's *Beauty and the Geek*, and ABC's *The Scholar*. The first seasons of *The Apprentice* (15 episodes), *Beauty and the Geek* (6 episodes), and the single-season run of *The Scholar* (6 episodes) yielded a finite amount of data that allowed for a census of content. The programs were chosen based on focus on desired comparisons of intelligence as a specified program element, attributes of contestants, success of contestants, and program longevity/success. At the time of this writing, *The Apprentice* was in its sixth season and a consistent ratings leader for NBC. *Beauty and the Geek* had been renewed for a second season based on ratings from the first season, but *The Scholar* was cancelled after one season due to poor ratings.

The unit of analysis was the program episode. The first season of *The Apprentice* was available on DVD. *Beauty and the Geek* and *The Scholar* were videotaped by the authors.

Categories of content were devised to address traditional intelligence measures, non-traditional skills/knowledge, physical attributes, and personality traits.

Traditional intelligence measures were defined as highest level of education obtained for *The Apprentice* and *Beauty and the Geek*. For *The Scholar*, grade point averages were indicated. Non-traditional skills and/or knowledge were defined as unskilled labor, professional experience, musical talent, arts and crafts, specialized knowledge (i.e. fashion sense, sports trivia knowledge, etc.). Coders were also asked to indicate gender/ethnicity of contestants and contestant performance within and status at the end of the episode. Contestant performance within an episode was defined as whether a contestant was successful within the context of an episode, e.g. succeeding at a particular task or winning a specified contest. Contestant status at the end of an episode was defined as whether the contestant was allowed to progress to the next episode. Coders also determined a contestant's dominant appeal and dominant slant of a contestant's portrayal. Dominant appeal was defined as the overall portrayal of the participant as logical, emotional, or ethical. Dominant slant was defined as the overall portrayal of the participant as positive, neutral, or negative.

Program episodes were divided evenly between two independent coders. The coders were trained in a session conducted by the authors. The coders were informed of the background, nature, and scope of the analysis and then provided with copies of the codebook and coding sheet. Following an explanation of the codebook and coding sheet, the coders were allowed to ask questions about those items or the analysis itself.

In order to determine the intercoder reliability coefficient, coders analyzed two episodes from the census. Intercoder reliability was calculated using Holsti's (1969) formula (total agreement)-(total disagreement) ÷ (total decisions). The resulting equation (74)-(20) ÷ (68), yielded a reliability coefficient of 79 percent.

Results

Technical difficulties resulted in the loss of two episodes of *Beauty and the Geek* and one episode of *The Scholar*. Therefore the total number of episodes (n=24) in the data census was comprised of fifteen episodes of *The Apprentice* (complete first season), four episodes of *Beauty and the Geek*, and five episodes of *The Scholar*.

Physical Attributes

Sex of contestants was divided evenly between male and female. Ethnicity of contestants was overwhelmingly Caucasian, both within and among programs. Over 80 percent of contestants in *The Apprentice* were Caucasian (13 out of 16), and all contestants in *Beauty and the Geek* (14) were Caucasian. *The Scholar* was most evenly divided with four out of ten Caucasian contestants (40%). Only two African Americans (12.5%) competed on *The Apprentice*, none on *Beauty and the Geek*, and four (40%) on *The Scholar*. One contestant of Asian descent (6.25%) appeared in *The Apprentice* and one (10%) in *The Scholar*. No Hispanics appeared in any of the programs. One contestant in *The Scholar* was coded as something other than the four ethnic choices provided on the code sheet.

For all programs, 31 out of 40 contestants (77.5%) were Caucasian, while only six out of 40 (15%) were African American. No Hispanics competed, while Asians accounted for only five percent of total contestants.

Across all programs, dark hair was most common for males, while medium hair color was most common in females. Close-cropped hair was most common for males while long hair was most common for females. No contestants on *The Apprentice* wore eyeglasses. Four *Beauty and the Geek* contestants wore eyeglasses (all men), while two female contestants on *The Scholar* wore eyeglasses. No *Apprentice* contestants had facial hair, while two *Beauty and the Geek* contestants had a mustache with a beard or goatee. One *Scholar* contestant also had a mustache with a beard or goatee, while one had a mustache only. The most common body type for both men and women across programs was muscular/athletic.

Education/Other Skills and Qualifications

Education credentials varied within and among programs. Nine contestants (56.25%) on *The Apprentice* held a bachelor's degree, three (18.75%) had earned a master's degree, and two (12.5%) had only completed high school. One contestant (6.25%) held a Ph.D., while one contestant was coded as having education credentials other than the choices offered. None of the contestants held vocational or technical degrees.

The contestants on *Beauty and the Geek* were overwhelmingly coded as having education credentials other than the choices provided—11 out of 14 (78.57%). Two contestants (14.29%) held a bachelor's degree, and one (7.14%) had completed high school. No other education (vocational/technical, master's degree, Ph.D.) was indicated by coders.

All contestants on *The Scholar* were still in high school. These contestants were distinguished from one another by grade point average. On a four-point scale, these scores ranged from a low of 3.36 (one contestant or 10%) to a high of 4.0. Five students (50%) had perfect 4.0 GPA's.

Formal education was referred to or mentioned sixteen times in *The Apprentice*: thirteen times by contestants when talking about themselves, six times by other contestants, and three times by Donald Trump. *Beauty and the Geek* and *The Scholar* each included only two references to formal education, twice by contestants talking about themselves and once by another contestant.

Coders determined professional experience and/or specialized knowledge to be the most common types of skills or qualifications other than formal education. All 16 contestants on *The Apprentice* had professional experience, while four had some type of specialized knowledge. Coders indicated two *Apprentice* contestants had skills or qualifications (i.e. hobbies, club or group membership) other than the choices provided.

Eleven *Beauty and the Geek* contestants had some type of job experience, four had specialized knowledge, while eight were judged as having skills or qualifications other than the choices provided. No contestants on *The Scholar* had any professional experience, but six had some type of musical talent, four had specialized knowledge, while one was judged to have other skills or qualifications. These other skills and qualifications were referenced even less frequently than formal education.

Female contestants on *The Apprentice* most frequently wore medium makeup, while those on *Beauty and the Geek* wore heavy makeup. *Scholar* contestants most frequently wore light makeup.

Statistical Analysis

Cross-tabulations and Pearson chi-squares were used to test all relationships proposed in the research questions. These relationships were tested separately for statistical significance but will be combined in the discussion that follows.

Physical Attributes

A significant relationship (x^2=.000) was indicated for performance within an episode and sex of contestant in *The Apprentice.* Females tended to be more successful (70.8%) than males (29.2%). Males were successful in less than one-third (29.2%) of instances coded, and unsuccessful within episodes in more than two-thirds (69.5%), nearly a direct inversion of female success rates.

There was no significant relationship between performance within an episode and sex of contestant in *Beauty and the Geek* or *The Scholar.* No significance between status at the end of an episode (retained or released) and sex of contestants, ethnicity, or education was indicated for any of the programs.

Other physical attributes were for the most part not significant. However, there were some exceptions. A significant relationship (x^2=.000) was indicated between hair length and performance within an episode for *The Apprentice.* Contestants with long hair were more successful within episodes (55.6%) than those with short or close-cropped hair.

Significance was also indicated for the relationship between overall body type (thin, muscular/athletic, full-figured/stocky) and performance within an episode. In *The Apprentice* (x^2=.007), full-figured or stocky contestants had no success. Thin contestants were successful slightly more than 40 percent of the time, while those judged muscular or athletic had the most (59.7%) success.

No significance for overall body type was indicated for *Beauty and the Geek* or *The Scholar*. Overall body type was not significant for status at the end of an episode for any of the programs.

Significance was indicated (x^2=.001) in the relationship between composite body type and success within an episode. Composites were derived using common body shapes (rectangle, square, apple, pear-shaped, hour-glass, triangle) that emerged from coders' indications on individual body parts (shoulders, chest, waist, hips). Contestants with rectangular figures (uniform size of individual body parts) were more successful than triangular body types, e.g. broad shoulders, medium waist, and narrow hips.

Makeup application was significant (x^2=.000) for female contestants on *The Apprentice*, as was style of dress (x^2=.037). Medium makeup application and a sexy style of dress were indicators of a successful performance within an episode.

Personality Traits

For *The Apprentice* only, several significant relationships emerged between success within an episode and agreement or disagreement between self-described personality traits and the assessments of fellow contestants. A significant relationship (x^2=.000) exists between a contestant's self-described passivity and disagreement by fellow contestants. Similarly, contestants who described themselves as aggressive but were not described similarly by their fellow contestants were more successful (x^2=.005). Furthermore, contestants who were not described by others as emotional (x^2=.000), arrogant (x^2=.001), self-centered (x^2=.000), timid (x^2=.000), blaming others (x^2=.000), irresponsible (x^2=.001), or disorganized (x^2=.000) were successful despite describing themselves as having these personality traits. Conversely, contestants whose self-descriptions as sympathetic (x^2=.000), a leader (x^2=.001), or a risk-taker (x^2=.000) did not match their fellow contestants' evaluations were not successful. No similar significant relationships were indicated for *Beauty and the Geek* or *The Scholar*. No significant relationships were indicated for status at the end of the episode for any of the programs.

Dominant Appeal and Slant

Dominant appeal of a contestant was significant (x^2=.008) for performance within an episode in *Beauty and the Geek*, and approached (x^2=.051) in *The Apprentice*. Emotional appeals were indicators of success within both programs, 84.2% and 68.1% respectively. The relationship between dominant slant and performance within an episode was also significant for *The Apprentice* (x^2=.000) and *Beauty and the Geek* (x^2=.011). Contestants whose portrayal during an episode was judged negatively slanted were unsuccessful 50 percent of the time. Neutral slants were most successful (54.2%). Conversely, positively slanted contestants were most unsuccessful (52.6%) on *Beauty and the Geek*.

Dominant slant was also significant for status at the end of an episode: *The Apprentice* (x^2=.000), *Beauty and the Geek* (x^2=.047). Positive or neutral slants accounted for over 70 percent of *Apprentice* contestants retained at the end of an episode, while negative slants were released 80 percent of the time. Conversely, *Beauty and the Geek* contestants judged to be positive or neutral were released nearly 90 percent of the time.

Discussion

Research question one proposed relationship between contestant success and physical attributes in which physical attributes were more important than traditional education qualifications such as advanced degrees or grade point averages. The findings suggest that this is indeed the case.

The statistically significant relationship between hair length and success within an *Apprentice* episode is simply due to the fact that most female contestants had long hair. However, the significant relationships between performance within an episode and sex, overall body type, composite body type, and style of dress are more important. Women with thin, athletic bodies who wore sexy clothes were more successful in the context of the tasks assigned each week. However, Donald Trump reprimanded female team members on more than one occasion for relying on sex appeal for success. Ultimately however, the lack of a similar significant relationship between status at the end of an episode (being retained or released) and sex, overall body type, composite body type, and style of dress indicates the women won the battles but lost the war. Furthermore, both finalists—and the eventual winner—were male.

As expected, formal education had very little to do with the original contestant pool—or the eventual outcome—of any of the programs. On *The Apprentice*, a bachelor's degree was the most commonly held credential. However, the two contestants with only high school educations, Troy and Jessie, actually fared better than their better-educated counterparts, including a medical doctor. Jessie was released after the sixth episode, while Troy survived until episode fourteen. The doctor, David, was the first to be released.

As stated earlier, female contestant selection on *Beauty and the Geek* was actually based on unusual hobbies or occupations and lack of education. The beauties included a cocktail waitress, an NBA dancer, a lingerie model, and a beer spokesmodel. The geeks included a Boy Scout master, a Dukes of Hazzard Fan Club vice-president, and a member of Mensa.

On *The Scholar*, the overall winner actually had the lowest grade point average (3.36) of the ten contestants. One 4.0 contestant also won a scholarship, and two others finished in the top three. However, the low incidence of references to education and the lack of any statistically significant relationship between education and within-episode performance or overall success is evidence of the minor role traditional measures of intelligence ultimately played on all three programs.

Research question two proposed a relationship between personality traits and contestant success. The significance emerging from this analysis is more complex than that of physical attributes and success. Contestant success within an *Apprentice* episode was related to the level of agreement or disagreement by fellow contestants on positive or negative personality traits. Successful contestants were those who described themselves as possessing negative personality traits (passive, emotional, arrogant, self-centered, timid, blaming others, irresponsible, disorganized) but were not described by their fellow contestants as having those traits. Accordingly, unsuccessful contestants described themselves as having positive personality traits (sympathetic, leadership, risk-taking) but were not viewed similarly by fellow contestants. Stated another way, contestants who were more self-deprecating tended to be more successful at getting within-episode tasks accomplished than those whose style was more aggressive. For example, Omarosa—arguably the first season's most obnoxious and abrasive contestant—described herself as being a sympathetic leader, but her fellow contestants thought otherwise. Omarosa was successful in her first four in-episode tasks, but failed three out of four times in the final four weeks leading up to her release from the program.

Research question three asked how contestants were portrayed in terms of physical attributes, personality traits, and education. The significant relationship between dominant appeal (emotional, logical, ethical) and success within an episode on both *The Apprentice* and *Beauty and the Geek*—particularly when taken together with the lack of significance for formal education—indicates the importance of an emotional connection on the part of the audience. This is what motivates the audience to like or dislike contestants, fueling the scenarios that drive the programs.

Dominant slant (positive, neutral, negative) was also significant for success within an episode. Neutral contestants were able to play the game and get along with fellow contestants. This is also true for ultimate retention or release at the end of an episode. Contestants judged neutral—neither too positive (about themselves) nor too negative (toward or about others) —in their overall portrayals were retained more often than their more negative counterparts.

Conclusions

Outwit. Outlast. Outplay. Those words expressed the philosophy for success in the prototypical reality program, *Survivor*. The program emphasized physical prowess, endurance, and gamesmanship over intelligence as defined by formal education. The same is true for the programs examined here, and to draw some meaningful conclusions it is useful to return to the three components of reality programming stated earlier: the audience, actual content, and production requirements.

The target audience for reality television is eighteen to thirty-four years old. It can be argued this demographic is indeed the product of an urbanized, disenfranchised culture based in technology (Brown, 1976). It is skeptical, even disdainful, of traditional measures of success such as high GPAs or college degrees, valuing more highly the measures touted by a technological consumer culture: computer prowess, instant celebrity, possessions. Reality television provides its audience access to culture as a commodity (Rifkin, 2000). While eschewing traditional measures of success, this audience is less willing, perhaps even incapable, of abandoning the stereotypical television portrayals on which it has been raised (Gerbner, 1997).

While education ostensibly was used as a criterion on *The Apprentice*, it played no part in a contestant's ultimate success or failure. *Beauty and the Geek* never even pretended to be about education except for the discrepancy that existed between the members of each couple. Contestants on *The Scholar* all had GPAs that were above average, but based on the program's other stated criteria, grade point average was not the only—or most important—factor in contestant selection.

The even sex distribution of the original contestant pools is not surprising. All three programs were by design half male and half female. However, the predominance of Caucasian participants is consistent with the under representation of minorities in television programming (Gerbner, 1997). Ironically *The Scholar*, the most ethnically-balanced of the three programs, was also the least successful in terms of ratings success.

The prevalence of dark hair, above average height, and muscular/athletic builds in male contestants—particularly in *The Apprentice*—is consistent with the virile male stereotype. It can also be argued that female stereotypes were perpetuated as well. A dearth of female contestants with light-colored hair—even in *Beauty and the Geek*—and a complete absence of any females conforming to the hourglass body type suggest at least a nod to the "dumb blonde" stereotype.

Four out of seven male contestants on *Beauty and the Geek* wore eyeglasses in keeping with the stereotypical "nerd" image. Conversely, their female partners had long, dark hair and wore heavy makeup, arguably an exaggeration of the polar opposites premise of the program

For all three programs, contestants who were least successful were those unable to distinguish themselves on any level, essentially fading into the woodwork, and therefore not appropriate to further the dramatic dynamics required in reality television. Moderately successful contestants (i.e., Omarosa) made for good drama, but eventually outlived their usefulness and wore out their welcomes. Contestants who were most successful were those who managed to distinguish themselves enough to contribute to the storyline, but not create too much tension between themselves and fellow contestants.

At this writing, *The Apprentice* was in its sixth season. *Beauty and the Geek* had been renewed for a fourth season, while *The Scholar* was cancelled following its single-season debut. Based on the findings of this study, it can be argued reality television is far from the evolutionary programming touted by its creators and promoters. More accurately perhaps is that reality programming merely repackages television's stereotypical portrayals of the physical attributes and personality traits of males, females, and minorities in a form palatable to its target audience.

References

ABC contemplates next step for "Dancing" (2005). *Broadcasting & Cable*, July 4, 6.

ABC pulls "Welcome to the neighborhood" (2005). *Broadcasting & Cable*, July 4, 4.

Brown, W. R. (1976). Prime-time television environment and emerging rhetorical visions. *Quarterly Journal of Speech*, 62, 389-399.

Ellis, D.G. & Armstrong, G.B. (1989). Class, gender, and code on prime-time television. *Communication Quarterly*, 37, 157-169.

Gerbner, G. (1997). *The electronic storyteller: Television and the cultivation of values.* Northampton, MA: Media Education Foundation.

Gitlin, T. (1982). *Television's screens: Hegemony in transition.* In American Media and Mass Culture, D. Lazere, Ed. Berkeley: University of California Press.

Holsti, O.R. (1969). *Content analysis for the social sciences and humanities.* Reading, MA: Addison-Wesley.

Killough, G. (2005). Beauty and the geek. *Reality TV calendar* (on-line). Available: http://www.realitytvcalendar.com/shows/beauty-geek.html

Lasswell, M. (2005). Thursday, July 7: B & C Week. *Broadcasting & Cable*, July 4, 3.

Mayerle, J. & Rarick, D. (1989). The image of education in primetime network television Series. *Journal of Broadcasting & Electronic Media*, 33, 139-157.

McKay, D. R. (2005). The Apprentice: Reality TV at work. *Career planning* (on-line). Available: http://careerplanning.about.com/cs/jobsearch/a/apprentice.htm

Nachbar, J. & Lause, K. (1992). *Popular culture: An introductory text.* Bowling Green: Bowling Green State University Popular Press.

Pecora, V. (2002). The culture of surveillance. *Qualitative Sociology*, 25. 345-358.

Rieder, R. (2000). Surviving reality television. *American Journalism Review*, 22, 6-7.

Rifkin, J. (2000). *The age of access.* New York: Putnam.

Selnow, G. (1986). Solving problems on prime-time television. *Journal of Communication*, 36, 63-72.

West, L. (2005). The scholar. *Reality TV* (on-line). Available: http://realitytv.about.com/od/summer2005realityshows/p/TheScholar.htm

Chapter Twelve

"Faking" Intelligence? Representing Intelligence in TLC's *Faking It*

Joan L. Conners

> "After just a few weeks of being coached in the ways of a new life, our fakers must try and pass the muster in their new identities before a panel of experts. Did they fake it and make it?" (The Learning Channel's description of the reality program *Faking It*)

The Learning Channel's *Faking It* (aired in 2003 and 2004) is a reality program that removes a person from his or her normal environment or occupation and trains him or her for a different occupation or social setting. After four weeks of training, which may involve intellectual, emotional, or physical development, the "faker" is put to a test to prove his or her ability. In these experiences, participants set aside their past experiences, as well as educational and intellectual history, to learn new skills.

Faking It is among other reality programs since 2000 that indirectly address issues of intelligence. For example, CBS's *Survivor*'s slogan is "Outwit, Outplay, Outlast," in which contestants face challenges in competitions, some of which are physical contests while others are intellectual games. Many end-of-the-season episodes of *Survivor* focus on how participants were able to outsmart others to survive longer in the game. Teven's (2004) analysis of *Survivor*'s Amazon season focuses on players' use of deception to outwit their competition, and in the case of one of the final four contestants, to go so far as to feign incompetence to reduce others' perceptions of that contestant being an intellectual threat in the game.

Another TLC program, *What Not to Wear*, addresses intelligence from a different angle: these participants might have "professional smarts" in their jobs as lawyers, college professors, music industry executives, but they lack "fashion savvy," which is provided through instruction by the hosts of the program.

Faking It may most closely parallel reality programs that address lifestyle alterations, although most may be done for some improvement from the transformation. Hill's (2005) analysis of British reality television categorizes the BBC's *Faking It* (the predecessor to the program studied here) as a life experiment program. For example, MTV's *Made* features young people coached to pursue a variety of transformations; NBC's T*he Biggest Loser* does pose a contest, but the focus is on weight loss and learning healthy living behaviors. More serious attempts at transformation are the focus of ABC's *Brat Camp*, which sends uncontrollable teens to a remote behavior program, and A&E's *Intervention*, which involves family and friends confronting a loved one who is dealing with some type of addictive behavior.

Faking It differs from these reality programs, however, in that its goal is not to instill new lifestyles, but to train someone for a short-term change. However, there have been accounts of show participants continuing to pursue their newly acquired interests and talents after the program is completed.[1] But what types of knowledge is acquired through this "crash-course" training? Hill (2005) suggests that with BBC's *Faking It*, participants are matched with employment that they should be successful in learning and demonstrating at the conclusion of the training. Producers are quite deliberate in matching participants to learning particular skills and background that they should be capable of achieving. Hill also says "If this means the program makers have to work hard to ensure a likely positive outcome by pre-selecting someone who has a high chance of succeeding, then many viewers would accept this constructed element of the program in return for a successful outcome" (p. 177).

This chapter will review approaches to conceptualizing intelligence, how past television research has examined representations of intelligence, and analyze how intelligence was portrayed in different ways on the reality program *Faking It*.

Conceptions of Intelligence

Sternberg (1996) summarized various expert definitions of intelligence as containing the following common themes: "(1) the capacity to learn from experience, and (2) the ability to adapt to the surrounding environment" (p. 91). Beyond these commonalities, there are numerous approaches to and interpretations of intelligence. This section will discuss a number of approaches taken in understanding intelligence, which will provide a framework for the analysis reported in this chapter.

One conceptualization of intelligence makes the distinction between social and intellectual competence (Eagly, Ashmore, Makhijani, & Longo, 1991). These authors describe *social competence* as "interpersonal skills and traits concerned with sociability . . . as well as the outcomes of such skills and inclinations" (p. 111), and *intellectual competence* as "intellectual and task relevant ability . . . and the results of such ability and motivation" (p. 111). Various skills and possible occupations may come to mind when considering individuals with social competence as opposed to those with greater intellectual competence.

Sternberg's (1990) analysis also involves aspects of social competence, but a different distinction between types of intelligence. He reviews past research and identifies overlapping factors in the distinction of *academic* intelligence vs. *everyday* intelligence. His research found academic intelligence was reflected by one's "verbal ability, problem solving ability, and social competence" (p. 60), while everyday intelligence was reflected by one's "practical problem solving ability, social competence, character, and interest in learning and culture" (p. 60). Given these descriptions, some aspect of social competence, having skills in relating to and understanding others, has a role in academic as well as everyday intelligence.

Another conceptualization for everyday intelligence is described by Sternberg, Wagner, Williams, and Horvath (1995) as *practical* intelligence, which is derived from Neisser's (1976) earlier distinction between academic and practical intelligence. Sternberg et al. (1995) refer to academic intelligence as "book smarts" and practical intelligence and "street smarts," "learning the ropes," and "getting your feet wet" (p. 913). In relationship to age, they speculate that academic intelligence may decline from early to late adulthood, but practical intelligence remains the same or potentially increases through late adulthood. In presenting numerous definitions of intelligence, Sternberg (1990) acknowledges that these definitions often differ by approach to cognition, motivation, and behavior.

These conceptualizations are relevant to consider in the context of *Faking It* participants, who have some potential motivation to try something different or to learn new things. One might argue the participants have a curiosity to learn about themselves or they would not choose to participate on the program. Sternberg also discusses the behavioral analysis of intelligence (in contrast to cognitive or motivational), which he says "looks not 'inside' the head, but outside it—at what the person does rather than at what he or she thinks" (Sternberg, 1990, p. 40). In facing a final challenge of their skills in *Faking It* episodes, participants clearly demonstrate their intelligence in their behavior rather than in more conventional measures of cognitive intelligence.

Sternberg (1990) further identifies three domains of behavior and intelligence that are relevant to consider in the context of representations of intelligence in television images: academic ("behavior existing in school work"), social ("behavior exhibited in between as well as within person interactions"), and practical ("behavior exhibited in one's occupation and in one's daily living") (p. 41). These domains of intelligence are relevant in considering the reality program *Faking It*. While many participants may come from primarily academic behaviors of intelligence, the training they face often involves a mix of academic skills (studying information relevant for an occupation) and social skills (learning to deal with people in a particular social context). Furthermore, their challenge at the conclusion of the training is primarily a test of practical intelligence to see if they have learned the skills necessary to succeed in this challenge.

One approach that is particularly relevant for consideration in this study is Sternberg's (1990) summary of Gardner's (1983) theory of multiple intelligences. Gardner defines intelligence as "an ability or set of abilities that permits an individual to solve problems or fashion products that are of consequence in a particular cultural setting" (Walters & Gardner, 1986, p. 165, as cited in Sternberg, 1990, p. 262). This approach suggests a broader perspective of conceptualizing intelligence than those offered above, and may reflect the variety of experiences that compose some aspect of intelligence. The following is a summary of these various intelligences from Sternberg (1990) that will be used in this analysis:

- Linguistic intelligence: skills in reading, writing, listening and talking
- Logical-mathematical intelligence: numerical computation, solving logical puzzles, scientific thinking
- Spatial intelligence: geographic navigation, and used in arts to image what something will look like
- Musical intelligence: skills in singing, playing an instrument, composing, and appreciating music
- Bodily-kinesthetic intelligence: ability to use one's body in solution of problems or in construction of displays
- Interpersonal intelligence: understanding and acting upon one's understanding of others
- Intrapersonal intelligence: ability to understand oneself (pp. 264-5)

Representing Intelligence in Television

Given the number of conceptualizations of intelligence above, the examination of intelligence in television representations has often had a narrow focus, often being tied to education level or occupational status. Social class has also been examined as a possible reflection of intelligence (see, for example, Butsch, 2003), but largely other variables have been the dominant focus of examination, including gender, age, and race (e.g., Press & Strathman, 1993). Some studies have gone so far as to examine representations of intelligence as they relate to these other demographics.

Television presentations of characters as more or less intelligent have often been correlated with other characteristics, such as age. For example, Carmichael (1976) found older people were presented as less intelligent than younger characters. Gender differences have also been identified, as in Vernon, Williams, Phillips and Wilson's (1990) research that found older male characters were rated higher on desirable traits including intelligence than were older female television characters. Bazzini, McIntosh, Smith, Cook, and Harris (1997) examined the film portrayals from 1940 to 1990 of older men and women by a variety of dimensions, including intelligence, and found older women were associated with lower intelligence than were older men. Bell (1992) found the leading characters of the top five programs of older viewers were more likely to exhibit intelligence, power, and affluence rather than foolishness and stubbornness.

Intelligence may also be reflected in one's educational status, a quality that may be associated directly or implied indirectly with television characters, and may also differ in dramatic versus comedic character representations. Research by Buselle and Crandall (2002) found that college student viewers evaluated African American characters' educational achievement higher in situation comedies than in television dramas. While such distinctions were found between viewers of dramas and situation comedies, it is unclear how viewers of reality television program might compare in this type of analysis.

Representations of social class have also been found to reflect patterns regarding portrayals of intelligence. For example, Butsch (2003) found working class men are often portrayed as buffoons, as "dumb, immature, irresponsible, or lacking in common sense" (p. 576). In contrast, working class wives were portrayed as rational, sensible, and intelligent. When considering middle-class characters in relationships on television, Butsch found both men and women were typically portrayed as wise and rational.

Connections to intelligence, knowledge level, or education level have been examined in studies of television character occupations as well. Signorielli's (2004) research considers occupational status and prestige of occupation when assessing television representations by age, gender, and race. She found television portrays women and minorities as having less prestigious jobs, those typically requiring less formal education, than white male characters. Her approach to occupational status included categories of professional, white collar, and blue collar jobs, which may reflect prestige and potentially educational status as well. Glascock's (2001) analysis of occupations of television characters used a different methodology, relying on Kranz's (1999) rankings of 250 occupations' status and prestige as tied to income. Glascock found male characters typically had more prestigious jobs on television than did female characters and men were more often in supervisory roles than were women. In order to assess representations of recognition and respect given to female television characters from the 1960s through the 1990s, Signorielli and Bacue (1999) examined occupational status instead of intelligence level, skills, or education (all of which may also reflect recognition and respect).

Analyses of status and occupation that do not directly examine intelligence may have connections to various conceptualizations offered above. For example, Vande Berg and Streckfuls (1992) analyzed prime-time television occupational roles of both men and women and overall found men held higher status occupational positions than did women. Regarding organizational actions of occupations (as outlined by Vande Berg & Trujillo, 1989), they found women were more likely to be portrayed in interpersonal functions than were men and men were more likely than women to be portrayed in decisional, political, or operational functions. These interpersonal functions, such as motivating, socializing, and counseling may relate to Eagly et al.'s (1991) notion of social competence, while skills of other functions may reflect factors of academic intelligence.

Concepts of Intelligence in Reality Television

Contemporary analyses of reality television have ignored issues related to intelligence, instead focusing on issues of representations of homosexuality (LeBesco, 2004), race (Kraszewski, 2004), religion (Engstrom & Semic, 2003), and social class (Palmer, 2004). To date, intelligence has not been a focus of research on reality television.

Social competence, which Eagly et al. (1991) contrast to intellectual competence, has some connection in reality television research, primarily in the study of relationships. Biltereyest (2004) acknowledges that contemporary reality television focuses on issues of interpersonal friendships as well as romances. In their study of gender and gossip on *Big Brother*, Thornborrow and Morris (2004) found participants worked to make themselves appear likeable and attractive to their housemates (potential factors of social competence), but they did not discuss issues of participants working to make themselves appear intelligent to others.

While representations of knowledge or intelligence haven't been examined in reality television, factors of knowledge of viewers (or lack thereof) have been the focus of past research. While they acknowledge viewers' concerns for reality television stories being contrived and manipulated by editing, Nabi, Biely, Morgan and Stitt (2003) found that reality television viewers reported learning from the personal dynamics of reality television programs. This may reflect viewers' potential for gaining social competence themselves.

Most critiques of reality television concerning issues of knowledge or intelligence deal not with the content of the programs themselves, but rather concerns about the audiences' lack of knowledge of critical viewing in their willingness to accept altered and edited versions of reality that are portrayed. Harwood (2005) raises concern with the lack of reality of such programs and audiences' willingness to "buy in" to such stories, saying:

> Television reality programs, for example, offer tours through some of the wide border between fact and fiction. Here non-professional actors confront competition and conflict under conditions specified by television professionals. Large audiences seem to recognize, accept, and enjoying the hoaxing (p. 54).

TLC's *Faking It*

Each hour-long episode of *Faking It* reviews a sequential series of events from the participants' month-long experience. Beyond the camera crew recording events of the training, mentoring, and testing, participants also are recorded in pseudo-confessional style, common on many current reality programs, to share their personal reflections of the experience. It should also be noted that participants apply to go through the *Faking It* experience, they are neither unknowingly nominated by friends or family, as with other lifestyle changing reality programs, nor are they confronted randomly on the street to participate in this process.

Participants agree to leave home for four weeks and enter intensive training toward a new skill, profession, or trade. They are paired with coaches and mentors, typically two individuals involved in the profession they will learn. Mentors conduct a "crash course" in their profession and work one-on-one with the participant.

At the conclusion of the training, the participant goes through a test of his or her newly-acquired skills and is judged by a panel of three experts in comparison to two other participants, both of whom are more experienced in that profession. Judges evaluate each of the three participants and only then apparently learn that one of the three participants just learned the skill in the past month; they are then asked to guess which one of the three was the "faker." While there is no prize for "faking out" the judges, the program creates much anticipation and has the mentors/coaches reveal to the participants how successful they were in faking their way through the test.

Analysis

The eighteen episodes of the 2003 and 2004 seasons were analyzed to identify representations of intelligence given the training experiences of participants, as well as their final challenges.

First, the episodes were analyzed for portrayals of social competence as compared to intellectual competence, given Eagly et al.'s (1991) distinction discussed above. Specifically, programs were categorized as reflecting aspects of social competence when they involve training and challenges involved primarily dealing with people as the primary goal and were categorized as reflecting intellectual competence when the training involves primarily intellectual development and obtaining understanding of particular content. Each episode was assessed by the author on a scale of zero (low) to five (high) on both social competence and intellectual competence. Given these measures, particular experiences of *Faking It* participants could be rated on both social and intellectual competencies, rather than propose they are opposite ends of the spectrum. As seen above with Sternberg's (1990) descriptions of academic and everyday intelligence, both involve some degree of social competence, but perhaps on different levels or types of interpersonal relationships.

Second, episodes were further analyzed to identify examples that reflected the multiple intelligences first described by Gardner (1983). Because a wide variety of occupations and behaviors were learned, we can consider how common gaining a particular type (or types) of intelligence was in the program. Given the nature of the four-week training and challenge test, we may find particular types of intelligence easier to be learned and gained in a short period of time that was reflected in more of these experiences than others. For this part of the analysis, each episode's experiences and challenges were evaluated overall on a dichotomous scale for each of Gardner's seven intelligences. Given some of the intersections of intelligence, *Faking It* episodes could demonstrate more than one type of intelligence taking place in the training and testing.

Results and Discussion

Social and Intellectual Competence

Social competence scores ranged from a low of two to a high of five, which indicates all training and challenges involved some degree of social competence and interaction with others as part of the occupation or skill learned. Episodes rated highest on social competence included "Simple Life to Social Life" in which an organic farmer learns to pose as a New York socialite, as well as "Super Shy to Superfly," in which a librarian learns what it takes to become a "barmaid" at Coyote Ugly. Others rated a four out of five on social competence included episodes that focused on training for an NFL cheerleader ("Ivy League to Big League"), a beauty pageant contestant ("Lifesaver to Heartbreaker"), a fitness instructor ("Computer Geek to Fitness Freak"), and a Chippendales dancer ("Lectures to Lap Dances").

Intellectual competence scores ranged from a low of zero (in one-third of the eighteen episodes, including the two rated highest on social competence), and a high of five (assigned to only one episode). The program rated highest on intellectual competence was "Six Pack to Chardonnay" in which the captain of the U.S. Beer Drinking Team trains to be a wine sommelier. Compared to many other experiences in *Faking It* episodes, such as learning to be a car salesman (in "Bibles to Bling Bling") or becoming a bodyguard (in "3 R's to Protecting Stars"), this training comes closest to involving formal education in the content and skills required. Five other episodes were rated a four in intellectual competence as each involve further education, training, and, potentially, an official license for occupations such as interior designer ("Toolbelt to Toile"), hairstylist ("On the range to La-La Land"), real estate agent ("Snow Hills to Beverly Hills"), fashion stylist ("Dry Clean to Style Queen") and chef ("Toppings to Top Chef").

The occupation-related training on *Faking It* reflected more social competence than intellectual competence experiences overall. The average social competence score assigned to *Faking It* episodes was 3.0, while the average intellectual competence score was 1.94. A variety of explanations for this distinction may exist; for example, there may be a limitation on what occupations can be trained for in four weeks or what occupations do not require formal and lengthy education or licensing procedures. Conventions of television production may explain this distinction as well, in terms of what experiences will be visually interesting may be more likely to be those involving social interactions with others rather than reading lengthy training manuals.

While these two competencies were not conceptualized as opposite ends of a continuum of intelligence, the results do suggest a reverse relationship between social competence and intellectual competence may exist in some cases. The six episodes rated a four or five on social competence were the same six rated a zero on intellectual competence. These findings may suggest that there may be a particular group of occupational experiences (or that television constructs a particular group of occupational experiences) that rely primarily on one's relationships with others rather than their understanding of specific content or issues related to those professions, for which little education is necessary. Another finding to note in this analysis is that gender differences existed in social and intellectual competencies when evaluating *Faking It* episodes. Ten episodes involved male participants, while eight episodes featured female participants. The average score for men on social competence was 2.4 (on the zero to five scale) and the average intellectual competence score was also 2.4. For female participants, the social competence average score was 3.75, while the intellectual competence average was only 1.38. These patterns are of concern for a couple of reasons: women's training was focused more on social competence than it was on intellectual competence and women varied dramatically from men on these two competencies. Images of women doing well in social, but not intellectual activities may perpetuate gender stereotypes regarding intellect, ability, and occupational categories that disadvantage them.

Gardner's Multiple Intelligences

In reviewing the descriptions offered by Sternberg (1990) of Gardner's (1983) seven types of intelligence, all eighteen episodes were found to involve intrapersonal intelligence. The process of undergoing four weeks of extensive training, some physical, others more emotional or psychological, resulted in participants having gained a further understanding of themselves. Not only would one learn other potential occupations they may succeed in, or talents they may be able to demonstrate they had not considered previously, *Faking It* participants may also learn what they cannot do, what they are not "cut out" for in skills or professions. That self-learning process certainly enhances one's intrapersonal intelligence through these training and challenge experiences. Given this conclusion, the six other intelligences did have greater variability, which were considered individually as well as summated across experiences (on a scale of zero to six, once intrapersonal intelligence was removed from the calculation).

In considering linguistic intelligence involving speaking, reading, and writing, four of eighteen episodes reflected training that required or improved one's linguistic intelligence. Examples that reflected elements of linguistic intelligence were those that involved either considerable reading or studying by participants ("Six Pack to Chardonnay") or particular skills of listening and speaking ("Drag Racer to Drag Queen," "Simple Life to Social Life"). Three of eighteen episodes involved logistic or mathematical intelligence, demonstrated in various quantitative skills necessary for particular occupations, such as interior design ("Toolbelt to Toile") or real estate ("Snow Hills to Beverly Hills"). Spatial intelligence was present in four of the *Faking It* episodes, typically in the form of developing an artist's vision, whether it was for designing a physical space ("Toolbelt to Toile") or coordinating a fashion wardrobe ("Dry Clean to Style Queen"). Given the types of challenges participants faced, none involved skills related to musical intelligence. Bodily-kinesthetic intelligence appeared in ten episodes, and involved a wide variety of physical experienced related to dancing ("Super Shy to Superfly"), cheerleading ("Ivy League to Big League"), modeling ("Hard Work to Hard Body"), as well as sports ("Computer Geek to Fitness Freak," "Briefcase to Body Slam"). Consistent with the above findings that all occupations were rated to some degree as involving social competence, fourteen of eighteen experiences had elements of interpersonal intelligence. The vast majority of occupations featured in *Faking It* required interactions with others whether it involved selling an idea or a product and interacting with potential clients of customers to provide a particular service or develop a relationship.

In summing the number of intelligences reflected in the *Faking It* experiences (with a potential low of zero and high of six), five occupations exhibited greater complexity in that they involved three different intelligences. These involved a mix of content specific intelligences such as linguistic or mathematical, as well as the more physical and relationally oriented intelligences. The episodes that involved three different intelligence areas featured training for a realtor ("Snow Hills to Beverly Hills"), a socialite ("Simple Life to Social Life"), a drag queen ("Drag Racer to Drag Queen"), a hair stylist ("On the Range to La-La Land"), and a fashion stylist ("Dry Clean to Style Queen").

It is interesting to note that these five episodes were not featured as examples of high intellectual competence given the first part of the analysis, but were more broadly involved in factors of intelligence when considering Gardner's (1983) conceptualization. It confirms the challenge Sternberg (1990) identified that there are multiple approaches to defining intelligence that vary in their approach regarding academic, social, and practical skills.

Conclusions

The analysis of representations of intelligence in the reality program *Faking It* confirms the fractured nature of understanding intelligence in the variety of portrayals discussed. Intellectual competence is clearly not the focus of the development of skills in the training participants undertake, but rather factors of social competence, learning to relate to others, to convince them of your ideas, is much more common. The variety of intelligences reflected in occupations portrayed on the reality program support Gardner's (1983) notion of multiple intelligences rather than a conceptualization of a single intelligence.

Why would viewing this type of transformation of intelligence and skills, and the challenges faced in the transformation be appealing to viewers? The appeal of such experiences and episodes can be explained in part by Reiss' (2001) findings that viewers of reality television watch out of a desire for prestige or status. Reiss and Wiltz (2004) confirmed these findings, saying "millions of people are interested in watching real life experiences of ordinary people. . . . Ordinary people can watch the shows, see people like themselves, and fantasize that they could gain celebrity status by being on television" (p. 374). In witnessing the experiences on *Faking It*, viewers can witness the transformation of participants and consider changes they could pursue in their own lives.

In contrast to Thomas and Callahan's (1982) critique of television shows being morality plays that "have portrayed the alleged virtues of poverty and the corresponding evils of wealth" (p. 184), *Faking It* suggests a very different message to audiences. Thomas and Callahan argue "the publication and dissemination of this myth of the happy poor is central in limiting social mobility so as to preserve the status quo" (p. 184). However, the experiences and training in *Faking It* suggests quite the opposite; the experiences of *Faking It* participants suggests that change can occur, that one can obtain new knowledge and skills.

Rather than reinforcing concepts of social control, as Biltereyst (2004) acknowledges others have argued regarding reality television, *Faking It* suggests the potential of what can be attained or accomplished. As Biltereyst acknowledges, "people can gain knowledge or learn from how some RT [reality television] formats . . . work upon specific issues in terms of identity politics, interpersonal relations, or on issues dealing with the relationship between the individual and society" (p. 21). The occupational experiences featured in *Faking It* suggest to viewers what potential they could reach with initial training and focus and also suggest future development one could pursue.

Regarding "life experiment" reality television such as *Faking It*, Hill (2005) reports, "Sometimes the experiment ends with a life-affirmative message—the participants want to change their lives for the better; more often the experiment ends with a negative message—the participants are judgmental of other people and their different life experiences" (p. 37). While differences may be apparent between one's past experience and current training on *Faking It*, more participants respond with a newfound appreciation for their "old lives" or a healthy respect for an occupation they had not taken seriously in the past.

While *Faking It* does suggest participants' potential for pursuing new directions, the reality television program could go even further in stretching the challenges participants face, especially intellectually. In Engstrom and Semic's (2003) analysis of the portrayal of religion in the reality program *A Wedding Story,* they argue the program could provide even further diversity in the portrayal of religion. So too could *Faking It* provide greater representation of intellectual transitions people (female participants, in particular) could face in their four-week challenges and provide occupational opportunities that both participants as well as viewers might seriously contemplate and possibly pursue.

Note

1. For example, *Faking It* participant Haley Holmes pursued auditions for Coyote Ugly after her experience, and Adam Mahler provided a demonstration of his Chippendales' performance when he returned home to UCLA.

References

Bazzini, D. G., McIntosh, W. D., Smith, S. M., Cook, S., & Harris, C. (1997). The aging woman in popular film: Underrepresented, unattractive, unfriendly, and unintelligent. *Sex Roles, 36,* 531-543.

Bell, J. (1992). In search of a discourse on aging: The elderly on television. *Gerontologist, 32*(3), 305-311.

Biltereyst, D. (2004). Media audiences and the game of controversy: On reality TV, moral panic and controversial media stories. *Journal of Media Practice, 5*(1), 7-24.

Busselle, R., & Crandall, H. (2002). Television viewing and perceptions about race differences in socioeconomic success. *Journal of Broadcasting & Electronic Media, 46* (2), 265-282.

Butsch, R. (2003). Ralph, Fred, Archie, and Homer: Why television keeps re-creating the white male working-class buffoon. In G. Dines and J. M. Humez (Ed.), *Gender, Race, and Class in Media: A Text-Reader*, 2nd Edition. Thousand Oaks, CA: Sage.

Carmichael, C. W. (1976). Communication and gerontology: Interfacing disciplines. *Journal of the Western Communication Association, 40,* 121-129.

Eagly, A. H., Ashmore, R. D., Makhijani, M. G., & Longo, L.C. (1991). What is beautiful is good, but . . . : A meta-analytic review of research on the physical attractiveness stereotype. *Psychological Bulletin, 110*, 109-128.

Engstrom, E., & Semic, B. (2003). Portrayal of religion in reality TV programming: Hegemony and the contemporary American wedding. *Journal of Media and Religion, 2*(3), 145-163.

Gardner, H. (1983). *Frames of mind: The theory of multiple intelligences.* New York: Basic.

Glascock, J. (2001). Gender roles on prime-time network television: Demographics and behaviors. *Journal of Broadcasting and Electronic Media, 45*(4), 656-669.

Harwood, K. (2005). Television hoaxes ahead. *TV Quarterly, 36* (1), 51-54.

Hill, A. (2005). *Reality TV: Audiences and popular factual television.* London: Routledge.

Kranz, L. (1999). *Jobs Rated Almanac.* New York: St. Martin's Press.

Kraszewski, J. (2004), Mediating race, reality, and liberalism MTV's *The Real World.* In S. Murray and L. Ouellette (Eds.), *Reality TV: Remaking Television Culture.* New York: New York University Press.

LeBesco, K (2004). Got to be real: Mediating gayness on Survivor. In S. Murray and L. Ouellette (Eds.), *Reality TV: Remaking Television Culture.* New York: New York University Press.

Nabi, R. L., Biely, E. N., Morgan, S. J., & Stitt, C. R. (2003). Reality-based television programming and the psychology of its appeal. *Media Psychology, 5*, 303-330.

Neisser, U. (1976). General, academic, and artificial intelligence. In L. Resnick (Ed.), *Human intelligence: Perspectives on its theory and measurement.* Norwood, NJ: Ablex.

Palmer, G. (2004). New you: Class and transformation in lifestyle television. In S. Holmes and D. Jermyn (Eds.), *Understanding Reality Television.* New York: Routledge.

Press, A., & Strathman, T. (1993). Work, family and social class in television images of women: Prime-time television and the construction of postfeminism. *Women and Language, 16* (2), 7-15.

Reiss, S. (2001). Why America loves reality TV. *Psychology Today, 34*(5), 52-54.

Reiss, S., & Wiltz, J. (2004). Why people watch reality TV. *Media Psychology, 6*, 363-378.

Signorielli, N. (2004). Aging on television: Messages relating to gender, race, and occupation in prime time. *Journal of Broadcasting and Electronic Media, 48*(2), 279-301.

Signorielli, N. & Bacue, A. (1999). Recognition and respect: A content analysis of prime-time television characters across three decades. *Sex Roles, 40*(7/8), 527-544.

Sternberg, R. J. (1990). *Metaphors of the mind: Conceptions of the nature of intelligence.* Cambridge: Cambridge University Press.

Sternberg, R. J. (1996). *Successful intelligence: How practice and creative intelligence determine success in life.* New York: Simon & Schuster.

Sternberg, R. J., Wagner, R. K., Williams, W. M., & Horvath, J. A. (1995). Testing common sense. *American Psychologist, 50*(11), 919-927.

Teven, J. J. (2004). *Survivor the Amazon*: An examination of the persuasive strategies used to outwit, outplay, and outlast the competition. *Texas Speech Communication Journal, 29*(1), 52-64.

Thomas, S., & Callahan, B. P. (1982). Allocating happiness: TV families and social class. *Journal of Communication, 32*(3), 184-190.

Thornborrow, J., & Morris, D. (2004). Gossip as strategy: The management of talk about others on reality TV show "Big Brother." *Journal of Sociolinguistics, 8*(2), 246-271.

Vande Berg, L. R., & Streckfuss, D. (1992). Prime-time television's portrayal of women and the world of work: A demographic profile. *Journal of Broadcasting, 36,* 195-208.

Vande Berg, L. R., & Trujillo, N. (1989). *Organizational life on television.* Norwood, NJ: Ablex.

Vernon, J. A., Williams, J, A, Jr., Phillips, T., & Wilson, J. (1990). Media stereotyping: A comparison of the way elderly women and men are portrayed on prime-time television. *Journal of Women & Aging, 2*(4), 55-68.

Index

About the Contributors

Cheryl Campanella Bracken (Ph.D, Temple University) is an associate professor in the School of Communication at Cleveland State University. Her research interests include media effects, with an emphasis on the psychological processing of media.

Joan Conners (Ph.D., mass communication, University of Minnesota, Twin Cities) is an assistant professor of speech communication at Randolph-Macon College in Ashland, VA. Her research interests include media representations and diversity, and political communication.

Mary T. Conway (Ph.D., Temple University) is an assistant professor of English at Community College of Philadelphia. She has published articles in *Journal of Advanced Composition Theory*, *Camera Obscura*, *Wide Angle*, *Parallax*, *Discourse*, and *Road Bike Magazine*. Currently she is investigating epistemologies of the subject in museum exhibitions and hybrid learning.

Marilyn Ellzey is assistant professor of broadcast journalism in the School of Mass Communication and Journalism at the University of Southern Mississippi. Her professional experience includes ten years as anchor and reporter for WDAM-TV, the NBC affiliate in Hattiesburg, Mississippi. She holds bachelor's and master's degrees in mass communication from the University of Southern Mississippi and a Ph.D. in mass communication from the University of Alabama in Tuscaloosa. Her specialty is process and effects of mass communication. Her research interests include portrayals of women and minorities in mass media, particularly broadcast news coverage of civil rights issues and activities.

Amy Franzini (Ph.D., Temple University) is an assistant professor of communication studies at Widener University in Chester, Pennsylvania. Her research focuses on television content representing adolescence and/or specifically aimed at adolescent audiences.

Kylo-Patrick Hart (Ph.D., University of Michigan) is chair of the Department of Communication Studies at Plymouth State University, where he teaches courses in film studies, television studies, and popular culture. He is author of the book *The AIDS Movie: Representing a Pandemic in Film and Television*, as well as numerous research essays pertaining to film and television that have appeared in academic journals (including *Journal of Film and Video*, *The Journal of Men's Studies*, and *Popular Culture Review*) and media anthologies (including *Television: Critical Concepts in Media and Cultural Studies* and *Gender, Race, and Class in Media: A Text-Reader*).

Lisa Holderman (Ph.D., Temple University) is associate professor of communications in the Department of English, Communications, and Theater Arts at Arcadia University, in Glenside, PA. She teaches courses in media theory, research methods, popular media, and public speaking. Her research interests include mass media sociology and popular culture and she has published articles in journals such as *Mass Communication and Society, Popular Communication,* and *The New Jersey Journal of Communication* (now the *Atlantic Journal of Communication*) and has recently contributed a chapter to *Stardom and Celebrity: A Reader.*

Susan Kahlenberg (Ph.D., Temple University) is assistant professor of media and communication at Muhlenberg College in Allentown, Pennsylvania. Her research interests focus on media images and effects, with an emphasis on gender issues. She teaches courses in gender, communication and culture, health communication, media theory and methods, media and society, and mass persuasion and propaganda.

Christine Mains is a Ph.D. candidate at University of Calgary, currently writing a dissertation on the representation of the wizard/scientist as teacher and the transmission of knowledge-power in popular culture. Her MA thesis focused on the quest of the female hero in the work of American fantasist Patricia A. McKillip, and she has published several articles on McKillip as well as on the figure of the wizard, the work of Canadian fantasist Charles de Lint, and the television series *Stargate SG-1*. She is the editor of the *SFRA Review* and serves on the executive board of the International Association for the Fantastic in the Arts.

Alison Miller (M.A., B.A., Auburn University) is currently a visiting instructor at East Carolina University, and is a Ph.D. candidate at the University of Southern Mississippi.

Gary R. Pettey is an assistant professor in the School of Communication at Cleveland State University. His research interests include media effects, with an emphasis on the social psychological impact of media.

Holly Randell-Moon is a Ph.D. student in critical and cultural studies at Macquarie University, Sydney. Her thesis explores the intersections between religion and politics under the Howard government. Other research interests include representations of gender and subjectivity in popular culture and television, and she has published on *Desperate Housewives* in *Topic: The Washington and Jefferson College Review* (forthcoming).

Amy H. Sturgis (Ph.D., Vanderbilt University) is assistant professor of interdisciplinary studies at Belmont University. The author of three books and editor of two, Sturgis has published articles in scholarly journals and popular magazines such as *Mythlore, Apex Science Fiction and Fantasy Digest, CSL,* and *Reason Magazine*, among others, and has a regular column ("Attack of the 5'2" Woman") with *Pop Thought*. She has presented research with such organizations as the Media Studies Working Group, the International Conference on Medievalism, and the Mythopoeic Society, and been interviewed on National Public Radio's "The Talk of the Nation" as a scholarly expert on science fiction culture. Her most recent projects include editing Baron de la Motte-Fouqué's *The Magic Ring* (2006) and contributing to *Tolkien on Film: Essays on Peter Jackson's The Lord of the Rings* (2005), *The J.R.R. Encyclopedia: Scholarship and Critical Assessment* (2006), and *Women in Science Fiction and Fantasy: An Encyclopedia* (2007). Visit her online at http://www.amyhsturgis.com.

Sari Thomas (Ph.D., University of Pennsylvania) is the director of the Center for Accuracy in Media Study in Los Angeles. She has published in numerous journals and anthologies including *Journal of Communication, Communication Research, Philosophy of the Social Sciences* and *Social Science Quarterly*. She served as editor-in-chief of *Critical Studies in Mass Communication*. In addition to her scholarly work, she has written screenplays and hosted her own radio show.

Cynthia W. Walker is an assistant professor in the Department of Communication, St. Peter's College. Walker earned her Ph.D. from the School of Communication, Information and Library Studies (SCILS) at Rutgers University. She is active in the media literacy movement in New Jersey, designing curriculum for middle school classrooms. Among her publications are "The Gun as Star and the U.N.C.L.E. Special" in *Bang Bang, Shoot Shoot! Essays on Guns and Popular Culture*, edited by M. Pomerance & J. Sakeris, several entries in the *Encyclopedia of Television*, edited by Horace Newcomb, and several teacher's guides, including *The Teacher's Guide to Star Trek*. In addition to teaching, Walker has been a public relations specialist in the private and public sectors as well as a working journalist for over thirty years. Currently, she covers professional regional theater for *The Home News Tribune*, a daily New Jersey newspaper. Her book, *Work/Text: Investigating The Man From U.N.C.L.E.*, proposes a new dialogic model of mass communication.